CW01272751

Natures in Translation

NATURES in TRANSLATION

Romanticism and Colonial Natural History

ALAN BEWELL

Johns Hopkins University Press
Baltimore

© 2017 Johns Hopkins University Press
All rights reserved. Published 2017
Printed in the United States of America on acid-free paper
2 4 6 8 9 7 5 3 1

Johns Hopkins University Press
2715 North Charles Street
Baltimore, Maryland 21218-4363
www.press.jhu.edu

Library of Congress Cataloging-in-Publication Data

Names: Bewell, Alan, 1951– author.
Title: Natures in translation : romanticism and colonial natural history / Alan Bewell.
Description: Baltimore : Johns Hopkins University Press, 2016. | Includes bibliographical references and index.
Identifiers: LCCN 2016007278 | ISBN 9781421420967 (hardcover : alk. paper) | ISBN 9781421420974 (electronic) | ISBN 1421420961 (hardcover : alk. paper) | ISBN 142142097X (electronic)
Subjects: LCSH: English literature—19th century—History and criticism. | English literature—18th century—History and criticism. | Nature in literature. | Natural history in literature. | Romanticism—English-speaking countries.
Classification: LCC PR468.N3 B49 2016 | DDC 820.9/36—dc23
LC record available at https://lccn.loc.gov/2016007278

A catalog record for this book is available from the British Library.

Special discounts are available for bulk purchases of this book. For more information, please contact Special Sales at 410-516-6936 or specialsales@press.jhu.edu.

Johns Hopkins University Press uses environmentally friendly book materials, including recycled text paper that is composed of at least 30 percent post-consumer waste, whenever possible.

For Carmen and Janet

CONTENTS

List of Illustrations ix
Preface xi

Introduction: Natures in Translation 1

1
Erasmus Darwin's Cosmopolitan Nature 53

2
Traveling Natures 87

3
Translating Early Australian Natural History 123

4
An England of the Mind: Gilbert White
and the Black-Bobs of Selborne 153

5
William Bartram's *Travels* and the Contested Natures
of Southeast America 196

6
"I see around me things which you cannot see": William Wordsworth
and the Historical Ecology of Human Passion 226

7
John Clare and the Ghosts of Natures Past 270

8
Of Weeds and Men: Evolution and the Science of Modern Natures 296

9
Frankenstein and the Origin and Extinction of Species 327

Notes 341
Works Cited 351
Index 383

ILLUSTRATIONS

Frontispiece to Pierre Sonnerat, *Voyage à la Nouvelle Guinée*	45
Emma Crewe, *Flora at Play with Cupid*	61
Henry Fuseli, *Flora Attired by the Elements*	63
Thomas Gosse, *Transplanting of the Bread-Fruit-Trees from Otaheite*	97
Anon., *Plan and Section of Part of the Bounty Armed Transport*	112
Anon., *A Sketch of Part of the Sheer and Middle-deck of His Majesty's Ship Providence*	113
Thomas Gosse, *Founding of the Settlement of Port-Jackson*	129
Sarah Stone, *Great Brown King's Fisher*	145
Sarah Stone, *Port Jackson Thrush*	147
Sarah Stone, *Yellow Eared Fly Catcher*	149
William Westall, *Chasm Island, Native Cave Painting*	151
William Bartram, *Sarracenia Flava or Pitcher, insectivorous plant with snail, snake eating a frog*	211

PREFACE

Among the lengthy taxonomic lists of plants that the botanist Joseph Dalton Hooker included in *Flora Antarctica*, one entry speaks of a certain kind of encounter between a European and the colonial natural world which merits attention. Late in November 1840, on Auckland Island, a cold, bleak, and desolate piece of land in the sub-Antarctic region south of New Zealand, Hooker was surprised to come upon some chickweed (*Stellaria media*), along with annual bluegrass (*Poa annua*), growing on the grave of a sailor. Listed under the order Caryophylleae, Hooker first identifies this common British weed and then designates its habitat:

> 2. Stellaria *media*, With.—*Engl. Bot.* t. 537. *DeC. Prodr.* vol. 1. p. 396. *Alsine*. L. Hab. Lord Auckland's Islands; covering the tomb of a French sailor, and growing along with *Poa annua*, L.: undoubtedly introduced. A straggling, very common European form of the plant, still retaining all its characters. (*Botany of the Antarctic Voyage* 1:8)

A botanist standing at the grave of an unknown sailor is a provocative image, and one might expect it to have produced, in a Wordsworth or Gray, a graveside meditation on time, change, and the susceptibility of all things—even empires—to decay and ruin. In this instance, however, Hooker appears to have been less interested in the sailor (whom he at times refers to as being French and at others as English) than in the two plants he saw growing on the grave. What might have been a narrative about colonial contact—about a traveler's encounter with a new culture or nature—is in this case the story of a meeting, across time and space, between fellow travelers: an English scientist, a European sailor, and two familiar European plants growing in a forsaken spot in the middle of nowhere.

The island that Hooker had landed on was uninhabited but not unvisited by Europeans. The Lord Auckland's Islands had been discovered thirty-three years earlier, in 1807, by the whaler Captain Abraham Bristow, and eight months before Hooker arrived they had been visited by a French scientific expedition commanded by d'Urville on the *Astrolabe*. Hooker did not have to be told that Europeans had visited the island before him—he could tell just by looking at the plants around him. Recalling this encounter thirty years later, Hooker declared that "the first evidence I met with of its having been previously visited by man was the English chickweed; and this I traced to a mound that marked the grave of a British sailor, and that was covered with the plant, doubtless the offspring of seed that had adhered to the spade or mattock with which the grave had been dug" ("Distribution of the North American Flora" 155). Where the mariners thought they were using their spades to bury one of their own, they were actually using them to plant European biota.

What interests me here is that, as a trained botanist, Hooker could see the traces that these Europeans had left in the landscape, not only in man-made artifacts, which are subject to ruin and decay, but also, more importantly, in the physical nature they had brought with them and left behind. Here we have evidence of a certain kind of seeing, historical in ways that we rarely consider, which could read human history as it was registered in the changes that were taking place in the natural world. To Hooker, a landscape was as much a document of colonial history as a garrison, a farm, or a trading post, and close attention to the plants that were "literally" in the landscape could provide access to a history that was unavailable to someone untrained in botany.[1] Current readers of colonial texts do not see nature in this way, so they miss the history that is often in plain sight in colonial texts and the distinctive ways in which that history is being addressed.

Seeing these plants growing on an anonymous grave, Hooker could also see that one nature, long isolated from biological contact with the rest of the world, was being replaced by another. The grave thus marks both an end and a beginning. By linking certain people with certain kinds of plants, it suggests that, at least to the eye of a botanist, the best way to say "A European lies here" is not to erect a granite monument, but instead to leave behind a "European" weed. During the colonial period, plants provided living testimony of the history of contact. Whereas current historical ecologists are able to extend their knowledge of the hidden pasts of natural environments by analyzing sedimented pollen and spores, eighteenth- and nineteenth-century naturalists saw colonial history in the distribution of plants and animals. During this period, natures that had long been associated with specific places and peoples traveled to new places, where, in turn, they often became linked to new peoples. The landscapes composed of these natures thus documented colonial history even as the mobility and mixing of natures and people during this period were changing these associations forever. Thus, the science of botany was ideally suited to providing a special kind of historical seeing; it was a science specially attuned to registering the changes being wrought by colonialism. For writers of this period, it was not enough simply to recognize a plant or animal; one also needed to know where it had come from, why it had traveled, and where it might go.

In the same lecture in which Hooker recalled his visit to the Aucklands, he also remembered an earlier occurrence on the same James Clark Ross voyage. Anchored off the Falkland Islands, he could not wait to begin studying its native vegetation, so he asked an officer who was being sent ashore to communicate with the governor to bring back any plants he could lay his hands on when he returned to the ship. Imagine Hooker's disappointment when the officer returned with an armful of a common English weed—shepherd's purse (*Capsella bursa-pastoris*). Impatient to get his hands on samples of the indigenous nature of the Falklands, Hooker discovered that a familiar nature had already settled there before him. Nature seemed to be acting very differently during the colonial period than it had in the past, and many plants were just as successful at travel and settlement as Europeans, some even more so. A decade earlier, John Clare, having moved from Helpston to the village of Northborough, felt that he was out of place, having lost the nature that was familiar to him. Feeling that the three miles separating Northborough from Helpston might just as well have been the Atlantic, Clare found solace when he came upon the same plant—shepherd's purse—feeling "what I never felt before / This weed an ancient neighbour here" ("The Flitting," *Poems of the Middle Period* 3:479–89, lines 197–98). For Clare, seeing this weed linked him to another place. For him, the weed spoke about continuities and survivals, about things that, perhaps, do not change. Yet the context of this seeing was not a belief in the permanence of nature, but of what it means to stand on moving ground, when "Nature herself seems on the flitting" ("Decay A Ballad," *Poems of the Middle Period* 4:114–18, lines 4–10).

In this volume, I seek to recapture this kind of historical seeing by discussing how writers during the late eighteenth and early nineteenth centuries, from Erasmus Darwin to his grandson Charles Darwin, understood a world in which natures were traveling and resettling the globe like never before. Arguing that there is no Nature—only natures—and that most of these have undergone translation in time or across space, this book traces a major shift in how the natural world was understood, from a belief in a nature that was universally stable, unchanging, and rooted in place to one in which natures were seen as having histories shaped by mobility, conflict, and change. For imperial naturalists, such as Sir Joseph Banks and Erasmus Darwin, natural history held the promise of ushering in a "cosmopolitan" nature in which all valuable species, through trade and exchange, might become true "citizens of the world," bettering the lives of people everywhere (or, at least, those who controlled their movements). Others, such as Gilbert White, William Bartram, William Wordsworth, and John Clare, were anxious about these changes and sought to preserve local or nativist conceptions of nature in the face of the threat posed by natures coming from somewhere else.

Charles Darwin, who had witnessed during his voyage on the *Beagle* the life-and-death struggles of colonial natures, stood out from his contemporaries in recognizing that the global movement and change that characterized these natures were not anomalous, but instead might serve as a model for the historical evolution of all natures since the beginning of time. Thus, with Darwin emerges a theory that links nature to modernity and situates biological form within a nexus of mobility, settlement, and change—the nature that Mary Shelley first described in her groundbreaking account of the coming-into-being of a new species, *Frankenstein*.

I see colonial natures not simply as physical entities but also as products of translation, as complex materialities deeply linked to language, history, and cultures. It is my hope that readers of this book will find a vocabulary and a historical way of seeing these natures which will allow them to recognize what was both gained and lost in the colonial translation of natures. I hope also that it will allow them to recognize the degree to which the science of colonial natural history and the extraordinary literature that it produced constitute an important afterlife of these natures, allowing them to continue to speak to us "that what we are, and have been, may be known" (Wordsworth, "Hart-Leap Well," *Poetical Works* 2:249–54, line 170).

In writing this book, I have incurred lots of intellectual debts that I would like to acknowledge here. I benefited from a John Simon Guggenheim Foundation fellowship that gave me time to write (a rare commodity this past decade) at a crucial stage in the book's history. Further research funding from the Social Sciences and Humanities Research Council of Canada also proved tremendously valuable, as it has for many Canadian academics. The University of Toronto is an extraordinary place to teach and to do research. The library resources are outstanding, but more important to a writer are the energy and excitement that I have daily found in lectures and seminars, in departmental hallways, and among the faculty and students here. Much of what I know about the responsibilities of scholarship and what teaching is all about I learned here, from my colleagues, so my love and admiration for this university run deep. I want particularly to thank Angela Esterhammer, Heather Jackson, Terry Robinson, Cannon Schmitt, Karen Weisman, and Dan White. They set a high bar for collegiality and scholarship in Romanticism and the nineteenth century. At the University of Toronto, we have regular meetings of informal discussion groups, such as the Eighteenth-Century Group and WINCS, or Work in Nineteenth-Century Studies. For a chair in a large academic department, these meetings have proven vital to me. Academic scholarship happens in libraries and studies, but its lifeblood is one's intellectual community. My thanks to all those who support and keep vital the North American Society for the Study of Romanticism (NASSR) and the International Conference on Romanticism (ICR).

I also greatly appreciate the opportunities I have had to present work related to this book at the following universities and institutions: Aberystwyth University, Arizona State University, Columbia University, the Huntington Library, Ludwig Maximilian University of Munich, McGill University, National Taiwan University, Princeton University, Purdue University, Queen's University, Stony Brook University, Tulane University, UCLA, University of Iowa, University of Maryland–College Park, Université de Montréal, University of Notre Dame, University of Southern California, University of Sydney, University of Vermont, University of Washington, Vanderbilt University, Western University, and the William Andrews Clark Memorial Library. On these occasions I have met people and friends who have supported and challenged me, suggested new paths of inquiry, and generously shared ideas and bibliographies. Let me single out some of these people for special thanks: Melissa Bailles, Christoph Bode, Gabriel Cervantes, Will Christie, Deirdre Coleman, Jeff Cox, Adriana Craciun, Neil Freistat, Geraldine Friedman, Tim Fulford, Eric Gidal, Kevin Gilmartin, Kevis Goodman, Arden Hegele, Kevin Hutchings, Peter Kitson, Greg Kucich, Debbie Lee, Mark Lussier, Katherine

Magyarody, Peter Manning, Anne Mellor, Nahoko Miyamoto, Jonathan Mulrooney, Judith Pascoe, Dahlia Porter, Tilottama Rajan, Molly Rothenberg, Margaret Russett, Charles Rzepka, Michael Sinatra, Rasheed Tazudeen, Richard Marggraf Turley, Kathleen Wilson, and Ya-Feng Wu.

Over the years, I have turned to Gary Handwerk repeatedly for much-needed advice on how to be a better chair and a better ecological critic. Many of the ideas that most matter to me first emerged in the classroom, and I thank my students, graduate and undergraduate alike, for helping me to distinguish the good ideas from the not so good ones and the ideas that needed more work from those that needed lots more work. My thanks also to Elizabeth Bohls, who provided superb guidance and criticism of the book manuscript; it could not have had a better reader.

It has been a great pleasure to be able to work again with the editorial staff at the Johns Hopkins University Press. My thanks to Matthew McAdam, Juliana McCarthy, and Catherine Goldstead, and especially to my keen-eyed and constantly vigilant copy editor, Jeremy Horsefield.

Parts of this book have been published in different forms elsewhere. A version of chapter 1 appeared as "Erasmus Darwin's Cosmopolitan Nature," *ELH* 76 (2009): 19–48. Also, part of chapter 2 appeared as "Traveling Natures," *Nineteenth-Century Contexts* 29, nos. 2–3 (2007): 1–22. A version of chapter 7 was published as "John Clare and the Ghosts of Natures Past," *Nineteenth-Century Literature* 65 (2011): 548–78. I wish to thank the editors of these journals for permission to use this material here.

I learned how to write seriously about literature during the early 1980s when working on my dissertation. From the beginning, academic writing was mixed up with parenting, as my best discoveries as a writer often occurred while I was hearing the voices of my two young daughters—Carmen and Janet—as they played together in our small townhouse. Often their voices were interspersed with another's, that of their mother, Sharon. Frequently, I couldn't help but join them. Who could resist taking part in a march, a dance, or a song with the Smurfs on the phonograph? At other times, however, I could not leave my makeshift study, because the writing was going well, so I would continue to work, finding that the pleasure of writing was somehow heightened by their presence nearby. Somehow, through some weird alchemy, my writing ever since has always been strangely mixed up with memories of those experiences. My girls have grown up now, but I still seem to need to imagine them being nearby when I write. As my intellectual commitment to ecology has grown, they seem to be even more in my thoughts and writing these days, and, as an added surprise, during the past four years a couple of new voices, not unlike the voices of old, have joined in the mental chorus, their tiny voices merging with and bringing back to life those in my memory. Along with Sharon, my two daughters have been near me in some of the best moments of my writing—they are near me now—so I am glad to have the opportunity to dedicate this book to them.

Natures in Translation

INTRODUCTION

Natures in Translation

> You cannot rename a whole country overnight
> (Friel, *Translations* 41)

That nature is central to British Romanticism is commonplace. For many critics, Romanticism *is* nature writing, and not without justification, for representations of the natural world—both domestic natures, seen close at hand, and foreign and exotic natures, brought from afar—appear during this period as if seen for the first time, with a freshness, concreteness, depth, and intensity that have rarely been equaled. This turn to nature in all its concrete richness had been gaining momentum over the course of the eighteenth century in natural history, scientific illustration, art, gardening, and textiles, but it was not until the Romantic period that it came to inform most spheres of British culture. At the same time, nature as a generalized, abstract idea also pervades Romantic thought, with notions of naturalness, harmony, and organic form, of vitality, process, and the interdependence of all living things appearing everywhere in this literature as the standard against which artistic, ethical, and political values are measured. Personified as nurse, guide, lawgiver, healer, teacher, economist, or muse, nature is invoked by poets, priests, philosophers, moralists, doctors, and prophets alike as the ultimate source of beauty, truth, health, and happiness. In Meyer H. Abrams's justly famous essay "Structure and Style in the Greater Romantic Lyric," nature is the necessary precondition of poetic self-understanding, for in turning to nature one finds oneself, and for many of these poets, notably Wordsworth in his claim that "Love of Nature Lead[s] to Love of Man" (1805 *Prelude*, title of bk. 8), it serves as the basis for community and social renewal. Human beings have not always accorded nature such a privileged status, and certainly this viewpoint nowadays is in the minority, so Jerome K. McGann is not far from the mark, though for the wrong reasons, in seeing this commitment to nature as a kind of "Romantic ideology." Ideologies rarely go unquestioned, however, and in *Pride and Prejudice* Jane Austen makes it an object of satire when Elizabeth Bennet, entertaining the possibility of a trip to the Lake District while suffering from disappointment in love, exclaims to Mrs. Gardiner, "What are men to rocks and mountains?" (154).

Although in recent years there has been a greater cultural commitment to ecology, it has not necessarily been shaped by an increased commitment to the natural world. Indeed, some critics, such as Timothy Morton, have even argued that it would be best for ecological thought simply to do away with the highly fraught concept of "nature." One of the biggest challenges one faces in writing about or teaching British Romantic poetry to a mainly urban audience is how to explain why most of these poets, with the notable exception of Charles Lamb, spent so much time talking about landscapes and rural scenery, describing the seasons and the weather, and meditating on birds, flowers, mountains, rocks, and trees. As Kenneth R. Johnston has remarked, "there is a tendency for even the finest minds to glaze over whenever Wordsworth starts 'talking nature'" ("The Romantic Idea-Elegy" 35). Theresa M. Kelley remarks that "English reading audiences of the late eighteenth and early nineteenth centuries were attracted to botany and natural history to a degree that most modern readers would find puzzling or boring" (*Clandestine Marriage* 63). Surely it is quite reasonable for a modern reader to ask why a poet, such as Charlotte Smith, would write a high-flying ode "On an Olive Tree," or an "Apostrophe to an Old Tree," or would compose poems on bees, butterflies, chafers, and spider gossamer; on owls, doves, jays, swallows, wheatears, mistle thrushes, and nightingales; on violets, geraniums, mulberries, and snowdrops; and on hedgehogs. To this already mind-glazing list might be added the poem "To the Goddess of Botany" and another to the goddess "Flora," along with others that catalog the flora and fauna that she observed at the seaside, on the heath, in the New Forest, Hampshire, in "A Walk in the Shrubbery," and at Beachy Head. As if writing about external nature were not enough, Smith also devotes three poems to remarking on other peoples' drawing of plants. Like Anna Laetitia Barbauld, who used a trapped mouse in "The Mouse's Petition" to address the issue of slavery, Smith also entered these debates not by writing a poem on slaves, but instead by discussing "The Fire-Fly of Jamaica, seen in a collection." The centrality of nature during the Romantic period is obvious, but surely modern readers have some right to complain that enough has been said and written about it, by both poets and critics. Yet why nature mattered so much to these writers, why it played such an important role in their understanding of themselves and the world, and why it no longer speaks to us in the same way as it once did—these are questions that should nevertheless trouble any contemporary reader. That Percy Bysshe Shelley, wanting to make a major political and philosophical statement, would choose as his interlocutor a Swiss mountain seems to make perfect sense on its own terms, and he clearly believed that "the wilderness" had "a voice . . . to repeal / Large codes of fraud and woe; not understood / By all" (*Poetry and Prose* 96–101, lines 76–77). Nevertheless, a modern reader, not yet inducted into the Romantic culture of nature, may perhaps be forgiven for concluding that "Mont Blanc" is a perfect title for a poem in which nature's voice is as cryptic as the message communicated by Ahab's white whale.

A fundamental difference between our world and that of the Romantics is that nature no longer speaks to us as it did for these writers, nor do political, aesthetic, or philosophical appeals to "nature" have the same traction that they once did. In an age that is as much "post-natural" as it is "post-modern" and "post-colonial," "nature" appears far less frequently in contemporary discourse, and when it does, it is usually treated as a highly suspect term. The stones of Mont Blanc have thus returned to their primordial silence, even if we decide to interpret that silence, following Annie Dillard, as "Nature's . . . one remark" on human beings and their failings (*Teaching a Stone to Talk* 69). Modern transportation, communication, and information technologies, urbanization, and biotechnology have changed what nature is and how we interact with it to such a degree that Bill McKibben is probably right to speak of "the end of nature." Given the enormous role that human beings now play in creating the environments that sustain them, it is often difficult to see the relevance of the Romantic idea of nature as an external presence and power that everywhere environs us and shapes human destinies. We still depend on nature (how else would we exist?), but our relationship to it is so highly mediated that we can easily lose sight of its importance, especially when the natures that we depend on most are located elsewhere. In modern societies, despite the increasing importance of local food movements, most of the food that people eat does not come from where they live, but instead arrives by rail, roads, ships, or planes. Nor is this food linked to time and the seasons, for what usually changes from month to month is not the food we eat, but the places from which it comes. McKibben rightly remarks that "people in Manhattan are as dependent on faraway resources as people on the Mir space station" ("A Special Moment in History" 57), so it is understandable that people nowadays forget or even ignore, as other generations could not, their dependence on those material natures existing at great distances from them. It is also easy to conclude that transportation systems are just as important in the production of food as fields. Furthermore, most of what we call nature has been so modified, genetically or otherwise, to suit our needs that it hardly seems appropriate even to call it by that name. As human beings control more and more of the natural world, nature, when we recognize it as such, appears on the margins of our concerns.

In our new, post-natural world, "nature" is primarily marketed as an "event" or an "experience" for consumers. For the tourist industry, nature is equated with distance and mostly refers to places that one can visit on vacations or weekend outings. "Nature" in this sense is valued because it has become increasingly hard to find—so hard that many are willing to fly or drive great distances to join others in appreciating it. Nature is most frequently preserved and displayed in "wilderness parks" or in brochures for ecotourism, where it has more in common with theme parks and museums than with anything intimately connected with our everyday lives.[1] Ironically, as the pressure increases to accommodate a greater number of people seeking wilderness experiences or encounters with the "natural world," most wilderness areas are becoming zoological drive-throughs. They have become, as David Louter suggests in his study of Washington's national parks, a "windshield wilderness, places where the relationship between automobiles and nature seemed to be mutually beneficial" (*Windshield Wilderness* 165). With the Internet and GPS now leading crowds of bird watchers to places where a rare bird has been spotted, with the burgeoning of ecotourism as an industry, nature appreciation and tourism are beginning to look a lot alike, as they also did during the late eighteenth century. Packaged by the media for stay-at-home consumption, "nature" on the *Discovery, Oasis HD*, or *National Geographic* channels or in David Attenborough's *Life on Earth* appears in all its glossy exoticism as a vivid spectacle. Elsewhere, in advertising, nature has been fused with a healthy lifestyle to promote beauty and health products, adventure, and travel. Often less something to be loved or appreciated than to be challenged, it appears as an "extreme sport"; the big-game hunting of the late Victorian and early modern period has thus been replaced by ATVs, dirt bikes, rock climbing, and television shows like *Survivor*. Unfortunately, in a consumer society, loving something usually means being sold on it, so even our wildernesses depend on marketing for their continuance. As John Clare realized, modern natures present a paradox: the only ones that human beings leave alone are those that they do not value. In such a context, we need to revisit what we mean by a term and a concept like "nature," so that it can be used in a more critically sophisticated manner.

A variety of explanations have been given for why nature achieved such cultural prominence in Britain during the late eighteenth and nineteenth centuries. A number of important studies have linked this turn to nature to the emergence of modern ecological consciousness.[2] Seminal in this regard has been Keith Thomas's *Man and the Natural World*, which argues that a fundamental change took place in human sensibility over the course of the eighteenth century as the deep-seated European anthropocentric belief in man's ascendancy over nature gave way to a more progressive recognition that human beings have much in common with animals and that they have a responsibility to preserve the nonhuman world. Writes Thomas, "The explicit acceptance of the view that the world does not exist for man alone can be fairly regarded as one of the great revolutions in modern Western thought, though it is one to which historians have scarcely done justice" (*Man and the Natural World* 166). What is striking about Thomas's account of how the British came to be a nature-loving people is that prior to the eighteenth and nineteenth centuries, they seem to have displayed an almost unlimited capacity for not being encumbered by such sentiments. It is clear that their sympathy for nature is a historical phenomenon. And attitudes that come into being at a certain time can also pass away, as the declining cultural interest in nature in recent decades would suggest. In Thomas's account, no sooner did this new sensibility appear than it came into conflict with "the material foundations of human society" (303), that is, with the industry and commerce that characterize modern societies. Thomas thus adopts the conventional view that the discovery of nature was intrinsically an antimodern gesture, seeing ecological sentiments and modern social demands as being pitted against each other. Most Romantic ecological critics have adopted a similar line, associating nature with nonhuman origins and with the past, a nature bathed in idealism and nostalgia, recovered from childhood or from the pages of Rousseau, Wordsworth, Thoreau, or Ruskin. They see nature as being tied up with tradition, as something inherently at odds with modernity. Other critics treat "nature" as a suspect term and hope that by using the word "environment" they can remove its connection to tradition and the past. As Anthony Giddens remarks, "we begin to speak about 'the environment' only once nature, like tradition, has become dissolved" (*In Defence of Sociology* 31). Rather than doing away with the idea of "nature," however, we might more properly seek to understand how natures have functioned in the past and in contemporary society. Perhaps by understanding why our natures do not speak to us like those of the past, we can understand more about ourselves and what we have lost.

This book argues that the historical emergence of nature was not so much opposed to modernity as one of its primary expressions. British attitudes toward nature during the eighteenth and nineteenth centuries, I would argue, were shaped by contradiction, as the British came to see nature as something that stood apart from the modern world of "getting and spending" (Wordsworth, "The world is too much with us," *Poetical Works* 2:18–19, line 2)—of mobility, exchange, and transformation—at the same time as they were actively engaged in translating it into the very forms that would allow it to be accessed from a distance, marketed, exchanged, and improved, the very activities that led to its achieving cultural priority in British society. The cultural preoccupation with nature was thus often just as much a product and expression of modernity as a reaction against it.

It is crucial to recognize that nature meant different things to different people. For both Wordsworth and John Clare, nature was powerfully linked to the past. Whereas for Wordsworth it provided a spiritual alternative to the pressures of modern industrial, urban, and commercial life, Clare's was a nature that was irretrievably lost, a ghostly world that spoke of his displacement and exile. In this regard, Saree Makdisi rightly suggests that the "romantic period in Britain marks the earliest sustained (though largely doomed) attempt to articulate a form of opposition to the culture of modernization—including but not limited to imperialism—from its very beginnings" (*Romantic Imperialism* 9). Alongside those writers who saw nature as being local, rooted in place, and separated from commercial concerns, there were others who understood it very differently. For a legion of new middle-class professionals—of amateur and professional naturalists; travelers and tourists; artists and illustrators; urban, suburban, and estate gardeners, planters, and nurserymen; doctors, merchants, agriculturalists, colonial administrators, and settlers—for a multitude of individuals who claimed expertise in local and global natures or who made their living from studying, writing about, exchanging, or transporting natures, the interest in nature was not an escape from the present, but was on the cutting edge of new forms of knowledge, new disciplinary formations, and new economic and aesthetic possibilities. For these people, whose ranks also included nature writers of all stripes and political persuasions, nature existed to be mobilized, represented, studied, exploited, traded, and exchanged. Even Wordsworth, in his contradictory manner, captures the excitement inherent in seeing nature as a new world available to the writer when he declares, "The earth is all before me" (1805 *Prelude* 1:14). In seeing and writing about nature in all its particularity and diversity, writers during the Romantic period were recognizing worlds that preexisted their writing, but the activity of seeing, writing, and publishing on these natures was new and inseparably bound up with the technologies and preoccupations shaping modern societies.

Rather than seeing nature as either a thing or an idea, as a physical entity existing apart from human life or as an expression of a specific ideological position in regard to it, be it conservative, imperialistic, "Romantic," or otherwise, I understand nature during the colonial period as being the site of intense political, social, discursive, and material struggle. What Pierre Bourdieu says about literature, that it is a field of "position-takings [within] . . . a *field of forces*, but it is also a *field of struggles* tending to transform or conserve this field of forces," fully applies to natures (*Field of Cultural Production* 30). Every nature, both materially and intellectually, I would argue, represents a position taken by human beings, as individuals or as groups, in relation to the natural world around them. This was especially the case during the colonial period, when natures everywhere were the primary arenas of social, political, and cultural struggle. Colonial settlers did not travel alone; they brought their own natures with them, and they helped these natures settle alongside them. At a time when different natures were associated with different peoples, we need to ask, "Whose nature are we talking about?" Nature was where history was being made at this time, and within this context the turn to nature expressed positions being taken in relation to this struggle. In its focus on nature, the literature of this period documents this struggle, a long history of material and ideational positions taken and lost by the British and the people whom they sought to colonize. It is thus not a coincidence that the history of European colonialism and the turn to nature occur simultaneously because these histories were deeply intertwined. Far from being separate from the dominant concerns of British imperial culture, nature was integrally bound up with the business of empire.

In light of work in the history of science over the past two decades, it is hardly controversial to claim that science during the late eighteenth and early nineteenth centuries was shaped by a commitment to empire.[3] This scholarship has demonstrated the degree to which the global perspective of modern science—the extraordinary circulation and pooling of knowledge and the expansive information networks within which this knowledge was produced—proceeded in tandem with the expansion of Western colonial interests. Nowhere was this mutual interdependence clearer than in the case of natural history, which, over the course of the eighteenth century, underwent a revolution to become the premiere colonial science, the most important intellectual arm of colonial thought and practice, and, succeeding geography, the primary scientific branch of the literature of contact. It was, indeed, as Mary Louise Pratt observes, "a European knowledge-building enterprise of unprecedented scale and appeal" (*Imperial Eyes* 25). Deepak Kumar reminds us that "colonisation primarily meant exploration and exploitation. Natural resources were the star attraction and this brought the practitioners of natural history into the limelight" ("Evolution of Colonial Science in India" 51). The seventeenth and eighteenth centuries witnessed the emergence of "natural history" as a modern science, and with it, as David Elliston Allen has shown, the rise in the popular cultural interest in natural history, collecting, and gardening. In both global and domestic spheres, the naturalist thus came to define the terms by which the British engaged with the natural world and colonial societies. Indeed, E. C. Spary remarks that French naturalists at the end of the eighteenth century saw themselves as "experts in defining the natural, while simultaneously renegotiating their definitions before a succession of different audiences" (*Utopia's Garden* 9).

In focusing on the literary representation of English domestic natures without reference to this broader colonial context, one can easily forget that over the course of the eighteenth and nineteenth centuries the natures that materially mattered most to the British were those that existed at a distance from England, as offshore, colonial concerns. At the forefront of British commercial activity and of Enlightenment knowledge building, the material, commercial, and intellectual interest in these natures provided the material and discursive contexts within which the British cultural interest in their own domestic or national nature eventually developed. The natures brought into being by colonial activity, both the new physical natures that were transferred and resettled across the planet and those that appeared as representations on the pages of philosophical, literary, and scientific texts, constitute one of the most important social, intellectual, and material legacies of the period. Natures of all kinds existed before European colonialism, and they still exist today, often precariously and in much diminished and degraded forms. Yet if we are to understand why "nature" came to occupy center stage in British culture during the eighteenth and nineteenth centuries, we need to relinquish an islanded perspective, by setting its scientific and cultural appearance in the larger imperial context that gave it meaning and importance. Kipling was right when he asked, "And what should they know of England who only England know?" ("The English Flag," *Poems* 1:291, line 2). To the extent that we read Romantic nature poetry as the product of isolated geniuses, who either documented the powerful feelings arising from their solitary encounters with a highly subjective nature or sought to escape from the historical present into a nostalgic conception of nature, linked to a traditional rural past, we isolate this writing from the social and cultural contexts within which it circulated and from which it derived much of its contemporary relevance.[4] Furthermore, as Elizabeth DeLoughrey and George B. Handley make eminently clear, to understand the environmental problems that we currently face and to begin to address them, we need to think in global terms about the environmental legacies of colonialism: "the mutually constitutive relationship between nature and empire over the past few centuries has produced a rich legacy of interpretive lenses for postcolonial ecology, and a long global conversation about how to define and attain environmental sovereignty at local, regional, national, and planetary scales" (*Postcolonial Ecologies* 20).

Nature or Natures?

Although the turn toward a more historical understanding of Romantic literature has introduced a notable range of new concerns over the past thirty years, not enough attention has been given to the ways in which writers responded to the extraordinary changes that were taking place at this time in the natural world across the globe. Much of this neglect derives from the assumption, common among literary historians, critics, and the general public, that nature neither changes nor initiates change. Whereas history is ongoing, as human beings make and remake their worlds every day, nature is usually treated as if it were a singular creation that does not change in any truly essential way. Understood in general terms, it is largely viewed as if it were little more than the static backdrop against which human actions and values are registered or the basic raw material with which human beings manufacture their unique worlds.[5] As one of the great conceptual triplets born from the Enlightenment, "Nature" lives in the shadow of her two upwardly mobile siblings, "History" and "Culture." The grand narrative of Western civilization repeats in innumerable ways the story of how our species has used its ingenuity to change both our own nature and the natural world around us, and will and freedom are defined in similar terms. "Every where man is what he was capable of rendering himself, what he had the will and the power to become," writes Johann Gottfried von Herder in the first major articulation in 1783 of the idea of "culture." For Herder, nature needs human beings to fulfill its destiny. "Whatever a nation, or a whole race of men, wills for its own good with firm conviction, and pursues with energy, Nature . . . will assuredly grant" (*Outlines of a Philosophy of the History of Man* 441). Karl Marx presents an even more subject-orientated model of the human capacity to transform nature. "By thus acting on the external world and changing it," he writes, man "at the same time changes his own nature. He develops his slumbering powers and compels them to act in obedience to his sway" (*Capital* 177–78). "Where man is not nature is barren," agrees William Blake ("The Marriage of Heaven and Hell," plate 10).

In Romantic literary studies, this story is told in many versions. In earlier criticism, although there is much talk about uniting nature and mind, much of this criticism reaffirms the Enlightenment and humanist belief that nature is inherently separate from culture, even as nature continues to be valued as an alternative to modernity.[6] Among the newer, more historically oriented critics, extreme positions are the norm. Nature is commonly understood as a huge cover-up—as a symptom of essentialist thinking, a false category, a screen that hides or displaces history, or a conservative strategy that allows ideological constructions and socially produced differences to masquerade as things that exist apart from human meanings and values. It is assumed that to claim that something is "natural" is to dehistoricize it, for when social differences are understood as existing in the "nature of things," they resist change and alternatives. Since nature is seen as an obstacle to both the history that human beings make and the histories that they write, and since it places limits on human freedom, the task of most historicist criticism of Romantic literature has been to penetrate or dissolve nature so that the human agency that stands behind it can be recognized. Inevitably, historicist criticism finds man and his works hiding everywhere behind what a less sophisticated criticism had seen as only "rocks, and stones, and trees" (Wordsworth, "A slumber did my spirit seal," *Poetical Works* 2:216, line 8). Thus, in his groundbreaking study of "The Ruined Cottage," Alan Liu interprets Wordsworth's turn to nature—his "vegetable tropics" of revision—as a gesture by which "vegetation, in sum, substitutes for the human being and the human place" (*Sense of History* 317–18). Nature, we learn, is the favorite hiding place of the political, the social, and the ideological, and where an earlier criticism had seen through nature to discover the workings of the Romantic imagination, historicist criticism now pierces through the Romantic ideology of nature to discover history, society, or politics. The task of Romantic criticism thus becomes intrinsically one of enfranchising poems and their readers from their confused entanglement with nature. Unfortunately, both early and recent forms of criticism view nature in the same way, as something that is inherently ahistorical and antithetical to human culture and the mind.

The social, economic, and cultural centrality of nature during the Romantic period is manifestly an important historical phenomenon, which our recent denatured and denaturing histories have not been able to grasp, not because they have been too historical, but because they have not been historical enough. In this regard, this study adopts the perspective of Bruno Latour, who has argued that the intellectual separation of nature and politics fundamentally limits our understanding of both. He writes that "politics does not fall neatly on one side of a divide and nature on the other. From the time the term 'politics' was invented, every type of politics has been defined by its relation to nature." Therefore, he argues against "*distinguishing* between questions of nature and questions of politics," suggesting that we treat "these two sets of questions as a single issue that arises for all *collectives*" (*Politics of Nature* 1). By dividing nature from politics, by seeing nature as something that exists in its separation from human life, as a wilderness or an externalized object available to human exploitation and manipulation, human beings since the Enlightenment have abstracted themselves from the natural world and thus lost the capacity to understand their real historical circumstances. Nature, in turn, has been treated as an abstraction, a fiction, or an object of control. This study adopts the alternative view that the nature of any place or time is the historical product of that place and time, sharing in the twofold view of the relationship between history and environments articulated by Emily W. B. Russell, who remarks "on the role that history, specifically human history, has played in shaping communities, ecosystems, and landscapes and conversely, the role that changing environments have played in human history" (*People and the Land through Time* xvi).

Arthur O. Lovejoy once discriminated sixty-six different uses of the word "nature" without even taking into account the role of the word in aesthetics. What is extraordinary is that none of these uses involved a plural ("Some Meanings of 'Nature'" 447–56). Raymond Williams reminds us, in *Keywords*, that "nature" is arguably the "most complex word in the language," and yet, despite a vast number of meanings, the word "nature" remains steadfastly singular (219). Unlike "culture," which is regularly employed in both singular and plural forms, "nature" in common usage lacks a plural form. I would argue that the impact of this understanding of "nature" has been disastrous. Consider, for instance, that when one is not using the word "culture" to refer to "the works and practices of intellectual and especially artistic activity," it is commonly used in three distinct ways: referring to values and behaviors that human beings share in common, to "a general process of intellectual, spiritual and aesthetic development," or to "a particular way of life, whether of a people, a period or a group" (90). In speaking, for instance, of German or French culture, in distinguishing between "high" and "popular" culture, in noting the differences between working-class culture and court culture, or in talking about scientific, indigenous, teen, or youth cultures, one is assuming that the more general meaning of "culture" encompasses a *plurality of cultures*, each with its own integrity, history, and conditions of existence. The value in recognizing those aspects of human culture that are universal, such as our use of language, tools, or social structures, cannot be gainsaid, yet what has most engaged the critical attention of contemporary anthropologists, sociologists, historians, and cultural theorists has not been determining what cultures share in common, but instead their differences and the changes that they have undergone in history.

Natures are every bit as complex, and yet we display no corresponding critical recognition of this fact. We talk as if there were only *one* nature. This use of "nature" in its generalized singular sense, though useful in lots of ways, prevents us from speaking meaningfully about the historical, geographical, and material diversity of natures across time and space and about the different kinds of nature that might exist in a given place or time. Just as any reasonable account of the history of a cultural tradition would need to differentiate it from other traditions while registering how it changed under new circumstances, so too the historical study of the natural world requires us to stop thinking of nature solely in generalized terms, as if it were a singular entity separately existing from us across all times and across all places, to recognize a *plurality of natures*. Only by so doing can we begin to see that the natural world in its general meaning is composed of many natures that are *materially, historically,* and *culturally* distinct from one another, that these natures can succeed or evolve in relation to each other across time and space, and that a given environment can be composed, like a society, of many different and often competing natures, reflecting different social relationships and values. Politics, in other words, are embedded in natures, and natures embedded in politics, behaviors, and ideas. Nature is not more present in the countryside than in the city, nor is a wilderness less a product of human choices than a woodlot. "Everything we know about environmental history," writes William Cronon, "suggests that people have been manipulating the natural world on various scales for as long as we have a record of their passing" ("Trouble with Wilderness" 83). Natures have changed in response to the positions that human beings have taken in relation to them, and human beings and their histories have, in turn, been changed by those natures.

In *Contested Natures*, Phil Macnaghton and John Urry argue for a pluralistic conception of nature, stressing that "there is no singular 'nature' as such, only a diversity of contested natures; and that each such nature is constituted through a variety of socio-cultural processes from which such natures cannot be plausibly separated" (1). Their insistence on the importance of social structures in the formation of natures is salutary; however, their social constructivist position leads them to adopt the same historicist perspective that I have been criticizing, which dissolves the materiality of the natural world into social forces and relationships. "Nature is in some senses as cultural as the content of television" (249), they write. They thus fail to recognize that, as Donald Worster puts it, "there are different forces at work in the world and not all of them emanate from humans" ("Doing Environmental History" 292). This study is more sympathetic to the kind of ethnographic ecological work done by Jake Kosek in *Understories*, a study of the political struggles that center on the forests of New Mexico. There Kosek seeks "to write natural histories and engage in forest politics without recourse to essentialist ideas of nature, while at the same time acknowledging the consequential materiality of the forest in political struggles" (285).[7] There are no natures that exist without any connection to human life. All are embedded within often competing forms of human agency. Yet these natures are never wholly within human control. This study follows Lawrence

Buell's definition of an "environmentally oriented work" in seeking to demonstrate that "the nonhuman environment is present not merely as a framing device but as a presence that begins to suggest that human history is implicated in natural history" (*Environmental Imagination* 7–8).

Natures can be said to occupy a continuum extending from those whose continuance is inextricably bound up with human beings to those that have been pushed to the margins or have disappeared because they cannot live with us or we with them. Those natures over whose reproduction we maintain control—notably domestic animals, food and commercial crops (genetically modified or not), and ornamental plants—are still natures, even if, like zoo animals or garden plants, they would have great difficulty surviving in their current forms outside of the new environments in which they are maintained and confined. A garden, a park, or a cultivated field is just as much a form of nature as a meadow or a woodland, even though the social, cultural, and biological conditions of each are quite different. Consequently, it can be said that where one nature stands—be it a garden, a pasture, a field of crops, a desert, or a wilderness—another nature has been displaced. Also, what is meant by "nature" varies from one place to another. A wild or indigenous plant in one place may be an economic crop, a garden plant, or a weed somewhere else. In fact, Chris Bight has observed that one of the "curious and lasting effects" of botanical commerce has been that "economic production usually occurs where the plants are not native. South American rubber is grown mostly in the Malaysian archipelago, South American cocoa in West Africa, African coffee in the New World tropics, Southeast Asian bananas in Central America, and so on" (*Life Out of Bounds* 143). North Americans often equate nature with "wilderness," and they have devoted significant resources to establishing parks that preserve, through extensive management, this kind of nature. In Britain, on the other hand, there is a commitment to preserving earlier forms of nature that were more clearly the product of human agency, such as National Trust parklands and coppiced woods. What is thus being preserved is not solely a nonhuman landscape, but a cultural landscape. Japanese gardens are even more focused on sustaining the horticultural and aesthetic practices that have created a certain kind of nature for more than a thousand years. Neither wilderness advocates nor those seeking to preserve historically produced natures are closer to "nature as it really is." One group cannot claim to have a better idea of what nature truly is than the other, but both recognize that the natures that they seek to preserve would not continue without human assistance. Every nature is thus an expression of the politics of the people who depend on it. Recognizing the plurality of natures allows us to consider the complex ways in which societies and the natural worlds that sustain them are mutually bound up with each other. It allows us to apprehend their unique histories, to grasp their changing geographies, and to assess the mutual impact that each has had on the other.

Nowhere is the need for a conceptual framework that recognizes the plurality of natures more necessary than in British colonial studies, for at no other time in history did the natural world undergo such rapid change, nor has the struggle between competing natures and peoples been fiercer. As a growing body of work in ecology and environmental history has demonstrated, European colonialism introduced changes within global ecologies on a scale never seen before.[8] A static conception of nature in the abstract singular, however, has prevented us from appreciating these changes or grasping their historical significance. That natures are rarely homogeneous was nowhere more evident than during the colonial period, when the contact between indigenous and foreign biota was as significant as the contact between peoples. A useful way of thinking about the colonial contestation of natures is provided by Paul Carter's idea of "spatial history," by which he means "the spatial forms and fantasies through which a culture declares its presence. It is spatiality as a form of non-linear writing; as a form of history" (*Road to Botany Bay* xxii). Carter is interested in how physical spaces do not preexist but instead are brought into being through spatial acts of writing and naming: "choosing directions, applying names, imagining goals, inhabiting the country" (xxi). For example, a Dutch captain, Dirck Hartog, blown off course in 1616 onto an island off the northwest coast of Australia, names the island after himself and claims possession of it by attaching an inscribed pewter plate to a tree. About eighty years later, another Dutchman lands on the same spot and, after transcribing Hartog's inscription onto a new plate, adds his own notation. Two years later, William Dampier visits the place and renames it Shark Bay. A century later, in 1801, Captain Emmanuel Hamelin discovers the pewter plate and renames the place Cape Inscription (xxiv). Here we see a history that is being written on maps and on the ground, as places are built up through a succession of inscriptions (in letters, journals, plates, markers, and maps). What makes Carter's work of particular interest is that he recognizes that spatial inscriptions are polyvocal, potentially containing the traces of many voices and cultures. Every inscription is also a form of erasure, every possession an act of dispossession, every settlement a form of unsettlement, as one name or one material nature supplants and displaces another.

Carter is primarily interested in cultural inscriptive practices, notably the manner in which Europeans colonized Australia by changing the land and the names of the plants and animals that existed on it. A more complex understanding of the polyvocal character of these spaces emerges, however, when one recognizes that settlers were not simply writing themselves into the land, or renaming what was already in existence, but also seeking to change the very plants and animals that composed these natures from the ground up, bringing new natures into being that would replace those that had existed there since time immemorial. Settlement was thus as much a biological event, a matter of settling European plants and animals in new places, as it was a human event. We need to speak of colonialism in terms of the biological resettlement of the globe. Natures were not only being renamed but also being remade, transformed, or translated into something else. Every time a newly introduced plant rooted itself in a new place, whether that plant was a cultivated plant or a weed, it replaced a nature that had preceded its appearance. Entire ecologies could be changed by the appearance of a single plant. To register what was happening on the ground, we need to be able to recognize the biological changes registered by such settlements, knowing that where one nature stands, another nature has been displaced. Every nature exists by virtue of some form of erasure, even in those rare cases where the agency of human beings is not at work. When Anna Laetitia Barbauld thought of empire, she understood it primarily as *planting* people in new parts of the globe:

> Wide spreads thy race from Ganges to the pole,
> O'er half the western world thy accents roll:
> Nations beyond the Apalachian hills
> Thy hand has planted and thy spirit fills. ("Eighteenth Hundred and
>   Eleven," *Poems* 152–61, lines 81–84)

Here there is great pride displayed for the British capacity for migrating to and settling in new places. We should also consider colonization as an activity of transplanting and resettling natures. "Colony" comes from the Latin word *colere*, meaning "to cultivate," which is a fitting word because British colonialism was as much about planting natures as people. Instead of understanding British colonialism as a grand narrative of the advance of civilization through the control of Nature in the abstract, therefore, we should recognize that it was more materially the story of the plurality of ways in which a people living on a small island in the North Sea became a nation of planters and transplanters and the impact that these activities had on other peoples and places. It is also the story of the many ways in which local and indigenous natures, which had been largely stable and isolated from one another, underwent a process of globalization, as they were integrated into and circulated within larger networks. Natures that had served the needs of indigenous societies were thus often displaced by or subsumed within new natures that largely served European needs, practices, and values. That is why the British did not so much possess the indigenous natures of the regions that they colonized as translate them into something else.

To a great extent, the British Empire possessed other natures *in* and *through* translation, and that is why the natures they possessed were different from those that indigenous peoples lost. In *The Gay Science*, Nietzsche was critical of the extent to which the Roman Empire used translation as "a form of conquest." "The degree of the historical sense of any age may be inferred," he writes, "from the manner in which this age makes *translations* and tries to absorb former ages and books." The Romans, he argues, looked on the past as something that was dead and ugly: "They did not know the delights of the historical sense; what was past and alien was an embarrassment for them; and being Romans, they saw it as an incentive for a Roman conquest. Indeed, translation was a form of conquest" (*Gay Science* 136–37). The idea of a single "Nature" is susceptible to the same critique, but our tendency to understand the natural world in a singular, ahistorical manner has prevented us from adequately recognizing the degree to which European colonialism leagued with science produced a nature that largely absorbed other natures into itself. To have any hope of understanding the biological history of contact, we need to read the natures that the British came to know during this period as translations. The continuity that we see between the natures of today and those that existed prior to European contact is, from this perspective, pretty much a product of historical amnesia. Continuities exist, yet they do so in a context of radical change. Nature during the colonial period was, indeed, the site of conflict and struggle between Europeans and indigenes, not only over how to understand, name, and interact with a diversity of physical natures, but also over the very terms of their existence. Recognizing the plurality of natures makes it possible to begin to recognize the polyvocal dividedness inherent in these natures themselves and to recognize this heterogeneity as an expression of the life-and-death struggles played out between traditional natures and the new "settler natures" that were supplanting them.

One of postcolonial criticism's more pressing tasks is thus not to get rid of "nature," but to recover the many natures that have been dissolved in the colonial and imperial formation of the contemporary idea of nature. Here the goal is not simply to see natures in local terms, as essentialized entities destroyed by the coming of Europeans, but to understand how the local was being transformed by broader global forces, to grasp the changes that took place in the very nature of places when the plants and animals that had originally composed them were replaced by others, or when indigenous plants were given new scientific names and used for new purposes, and to grasp the political struggles between competing positions with regard to the natural world which were embedded within these changing and often contradictory natures.

In describing the history of the mind, De Quincey conceptualized it in a manner very similar to that of Paul Carter, seeing it as a palimpsest in which each successive layer of writing is made possible by effacing the traces of the writing that preceded it: for "a new succession of thoughts," the inscribed vellum or parchment had to be cleaned. At this point in *Suspiria de Profundis*, he shifted from the palimpsest to a horticultural metaphor, suggesting that "the soil, if cleansed from what once had been hothouse plants, but now were held to be weeds, would be ready to receive a fresh and more appropriate crop" (*Confessions* 146–47). De Quincey believed that the imagination might serve as a kind of new chemistry that would reverse this process, calling back all the words that had been successively erased from the vellum over the centuries. Historical ecology can function in a similar manner, by seeking to recover the many natures—"hothouse plants" become "weeds"—that lie buried in natures that through time have undergone enormous changes, both materially and linguistically. By recognizing colonial natures as palimpsests, in which one nature has been overwritten by another, erasing and translating its predecessor in that process, we can begin to recognize the history of the natural world as a series of positions taken and lost: why nature came to matter so much both to those colonizing nations who prided themselves in their capacity to remake and resettle other peoples' natures and to those indigenous nations that watched as a familiar nature that they had always known disappeared before their eyes, knowing that not even the names that they used to speak about this nature were likely to remain. In recognizing colonial landscapes as having been built on the contestation of competing natures, we can begin to recognize the struggles—ideational as well as material—that underlay their appearance and disappearance. Modernity is not simply the story of the destruction of nature, but of the colonization, replacement, and resettlement of traditional natures by a globalized nature profoundly linked to science, cities, and trade.

This struggle, over what nature both was and should be, made nature the focus of intense political, cultural, and literary concern, and the turn to nature, which coincides with the rise of European colonialism, was an expression of this struggle. The literature of this period, particularly in the area of natural history, documents this struggle. Take, for instance, John Josselyn's *New-England's Rarities* (1672), one of the best sources available for understanding the changing ecological landscape of seventeenth-century New England. After devoting two chapters to listing and commenting on (1) the wild plants that New England and England shared in common and (2) those that were found only in New England, Josselyn added a third chapter, entitled "Such Plants as have sprung up since the English Planted and kept Cattle in New-England," enumerating the introduced plants that now grew wild in New England. Among the twenty-two plants listed there, his comment on the broad-leaved plantain (*Plantago major*) powerfully suggests the manner in which settlers and First Nations people saw colonial history registered in the changes taking place in the nature that surrounded them. Josselyn reports that "the *Indians* call [it] *English-Mans* Foot, as though produced by their treading" (*New-England's Rarities* 137). Almost a century later, in his *Travels into North America* (1748), the Swedish-Finnish botanist Peter Kalm confirmed this anecdote, declaring that First Nations people, "who always had an extensive knowledge of the plants of the country," claim "that this plant never grew here before the arrival of the *Europeans*. They therefore gave it a name which signifies the *Englishman's foot*; for they say, that, where a *European* had walked, there this plant grew in his footsteps" (1:92–93). One would be hard-pressed to find a better example of how a detailed knowledge of the natural world could provide an understanding of the material dimensions of European and indigenous contact. In an almost mythic rendering of colonialism as an ecological event, every advancing step of the Europeans leaves (quite literally) a trace, a biological footprint, on the land. For Native Americans, the British and their weeds traveled in tandem. In *The Song of Hiawatha*, a poem based on the legends of the Ojibwe of northern Michigan, Wisconsin, and Minnesota, Longfellow drew on this image in the penultimate canto, entitled "The White Man's Foot."

Where Robinson Crusoe is terrified by the discovery of a single footprint in the sand, for it provides him with an advance warning that he is not alone on the island and his survival is threatened, Hiawatha welcomes the advance notice that nature provides—through the plantain and the European honeybee (the "Ahmo")—of the imminent arrival of the men in black robes:

> Wheresoe'er they move, before them
> Swarms the stinging fly, the Ahmo,
> Swarms the bee, the honey-maker;
> Wheresoe'er they tread, beneath them
> Springs a flower unknown among us,
> Springs the White-man's Foot in blossom. (*Poetical Works* 162)

Although *The Song of Hiawatha* promotes the idea that the history of the Ojibwe finds its appropriate end in the coming of the white man, its use of an insect and a flower as the primary signs of colonial change tells us much about the manner in which nature became powerfully linked with history and change during this period. Furthermore, it suggests the manner in which the coming of a people could be registered in the nature that they brought with them even as the corresponding disappearance of others might be marked in the nature that they left behind.

Traveling Natures

J. G. A. Pocock has argued that British society at the end of the seventeenth century underwent a fundamental change as the British developed a new sense of their history and of their political identity based on commerce, "upon the exchange of forms of mobile property and upon modes of consciousness suited to a world of moving objects" ("Mobility of Property" 108–9). If we are to understand what nature became during this period, we need to recognize the degree to which colonial natures had the impact they did in England and elsewhere because they had become an integral part of this world of moving things. Marx recognized that mobility had played a central role in forging the modern world, but his language basically excludes nature from the equation: "Nature builds no machines, no locomotives, railways, electric telegraphs, self-acting mules etc. These are products of human industry; natural material transformed into organs of the human will over nature, or of human participation in nature" (*Grundrisse* 706). The less strident phrasing of a "human participation in nature" nevertheless captures why mobility is one of the most striking characteristics of modern natures. Whether mobilized by human agency or traveling on their own, natures during this period spread across the globe on a scale not seen before, traveling in many forms, as living biota, scientific specimens, commodities, images, or printed texts.

Take, for instance, the ship that brought three-times-transported swindler and compiler of the first dictionary of "flash language" James Hardy Vaux back to England from the penal colony at Sydney in 1806. It was "literally crowded, so as to resemble Noah's Ark. There were kangaroos, black swans, a noble emu, and cockatoos, parrots, and smaller birds without number." A Maori was also on board, "punctured or tattooed in a most fanciful and extraordinary manner from head to foot," and it was hoped that he, along with the microcosm of Australian nature that accompanied him, would prove a profitable "object of curiosity in England" (*Memoirs* 116). Having been the focus of public attention for almost two decades, Australia's indigenes, its animals and its people, were on their way to becoming attractions in the "shows of London."[9] The Maori never made it, for he died of smallpox before the ship reached the Brazils. Neither did most of the antipodean nature that had joined him, mostly because of the "severity of the weather" off Cape Horn. A cockatoo survived, along with a half dozen black swans that ended up in "the royal Menagerie in Kew Gardens" as a gift to George III. Although decidedly not a successful journey, the image of a modern Ark crammed with strange animals setting sail for a distant part of the globe vividly conveys an aspect of nature during the colonial period which has not been adequately recognized. This was a time when plants and animals from all quarters of the globe began to travel in earnest, and yet representations of nature during this period seem almost inevitably to emphasize nature as being statically rooted in place, as if traveling and migration were a distinctly human privilege. Nature appears in this literature as a place to be visited and revisited, something to be brought home as a memory, a poem, or an image, not something that itself travels, coming across the seas and fundamentally changing lives, for better or worse.

How did the British encounter the many natures that were appearing at this time? With the rise of modern transportation networks—canals, roads, ships, and rail—they were able, in a far greater degree, to travel physically to the countryside, or to Devon, Wales, or the Lake District. The rise of recreational travel during this period and the literature pertaining to it reflects this fact. However, as the example from Vaux's *Memoirs* suggests, the British did not have to travel to the countryside to experience nature; it often came to them, through the ready availability of new means of transport and through media. Frequently, foreign natures arrived by post. A letter from the royal physician James Lind to Sir Joseph Banks dated 5 April 1796 is typical in this regard: "I enclosed send you some Beans, that I received of a Gentleman from Madras. They are esteemed the best sort of Kidney-Beans in the Carnatic, and were introduced into that Country by Mr Duffin (a Surgeon) and generally called after him the Duffin-bean. There are some of them sown at Frogmore, but your taking the charge of a few of them will insure their succeeding in this part of the World" (*Scientific Correspondence* 4:420). Banks was probably as excited about receiving these beans as the titular hero Jack in the English fairy tale published five years earlier, and for both the fairy-tale entrepreneur and the imperial naturalist, they held great promise of future wealth. By the time that these beans arrived at his doorstep, this variety of *Phaseolus lunatus*—the lima bean, as it came to be known in 1806—was already a seasoned traveler. Having been cultivated for centuries in Central and South America, the beans had made their way to the botanic garden at Mauritius sometime after 1735, from whence Dr. Duffin, who was stationed at Vellore in the 1780s, introduced them to India. From there, they traveled to Cochin, China (1790), to South Africa, to Queen Charlotte's country retreat at Frogmore, and from thence to Soho Square (Sturtevant, "Kitchen Garden Esculents" 452–54). As one of the primary leaders in a correspondence network focused on the global transfer and exchange of plants, Banks was the ideal person to keep this botanical chain letter going and these beans moving.

Biologists since Charles Lyell and Charles Darwin have noted that natures have been traveling for a long time. On winds, rivers, and tides, or in the stomachs, fur, or feathers of birds and beasts, plants have found ingenious ways to spread themselves across the earth. As William Bartram noted, the Aristotelian idea that plants differ from animals because they lack locomotion, the ability "to shift themselves from the places where nature has planted them," does not adequately recognize the degree to which "vegetables have the power of moving and exercising their members, and have the means of transplanting or colonizing their tribes almost over the surface of the whole earth" (*Travels* lv). Human beings have been instrumental in this process. Wherever they have traveled, they have brought with them their favorite plants and animals. The image of the refugee Aeneas departing from the burning towers of Troy carrying his father Anchises and the family's hearth gods on his shoulders is thus only part of the picture of human migration and the founding of new nations. Like Noah, facing catastrophe, he would have also brought with him whatever he could salvage of the physical environment—the seeds, roots, cuttings, and domesticated animals—that the Trojans depended on. Human beings differ from animals, therefore, not only in their capacity to manipulate and exploit nature but also because they are one of the few animals that carry their nature with them. "Man carries whole floras about the globe with him," writes Edgar Anderson. "He now lives surrounded by transported landscapes." Most of "our everyday plants," he further suggests, should be seen as artifacts, for "that is what many of our weeds and crops really are. Though man did not wittingly produce all of them, some are as much dependent upon him as much a result of his cultures, as a temple or a vase or an automobile" (*Plants, Man and Life* 9). Writers during the Romantic period were not unaware of the extent to which human beings were transforming the natural world by engaging in plant transference. "The migration of plants is evident," wrote Alexander von Humboldt and Aimé Bonpland. "Some plants, which were made the objects of gardening and agriculture from the most ancient times, have accompanied man from one end of the globe to another" (*Essai sur la Géographie des plantes* 25). "Man himself has done more for the dispersion of plants than winds, or seas, or rivers, or animals," writes D. C. Willdenow. "The wars which nations wage with one another; the migrations of different people; the pilgrimages to Palestine; the travels of merchants, and trade itself, have brought to us great numbers of new plants, and have carried our plants to many distant regions" (*Principles of Botany* 419). As the midwife of modern natures, colonialism dramatically accelerated this mobility; the migration of natures, both by intention and by accident, was as much a characteristic of this period as the migration of people.

In his groundbreaking work on the historical ecology of colonialism, Alfred W. Crosby has coined the term "Columbian exchange" to refer to the enormous globalized transference of biota that took place at this time. European settlers, he argues, did not travel alone, but brought with them their domestic animals, crops, weeds, pests, parasites, and diseases, in what he calls "a grunting, lowing, neighing, crowing, chirping, snarling, buzzing, self-replicating and world-altering avalanche" (*Ecological Imperialism* 194). Crosby's innovation lies in his argument that the success of Europeans in settling the globe is not to be explained solely in terms of their technological domination over others, but also in terms of the biological success of the natures that they brought with them. The ground of expansion was laid not only by guns, warships, and information networks but also, he suggests, by European germs, domesticated animals, and weeds: "On the pampa, Iberian horses and cattle have driven back the guanaco and rhea; in North America, speakers of Indo-European languages have overwhelmed speakers of Algonkin and Muskhogean and other Amerindean languages; in the antipodes, the dandelions and house cats of the Old World have marched forward, and kangaroo grass and kiwis have retreated" (7). The sun may have set on the British dream of a global empire, but it still "never sets on the empire of the dandelion" (7). Long after the decline of the British Empire, those who remain must still live with the new environments that were created by the expansionary diffusion of European nature across the globe.

For those natures that were portable, colonialism offered enormous opportunities for expansion. Crosby's term, "portmanteau biota," designates a huge number of plants and animals that accompanied the British on their journeys abroad, whose power lay in their portability. Unlike carry-on baggage, however, these natures were as much settlers as their human counterparts and did as much to change entire landscapes and environments. "Environments ... are not necessarily fixed entities to which outsiders have either to accommodate themselves or perish," writes David Arnold. "They are not, as it were, rooted to one spot: they, too, can cross the seas, march with armies, and conquer entire continents" (*Problem of Nature* 99). If natures prior to this time were to a great extent restricted in their movement by climatic or other physical barriers, colonial natures were redefined when human beings began in earnest to assist biota in moving across the globe. Many of these plants and animals would even become as cosmopolitan as human beings. It was also in colonial environments that their impact was displayed most clearly and dramatically.[10] Alongside our assumption of nature's stability and permanence, therefore, another story needs to be told: of natures mobilized; of natures uprooted, deterritorialized, and transplanted to new parts of the globe; of immigrant, creole, and transnational natures; of newly emergent natures composed of the entanglement of indigenous species with exotic, foreign, or introduced ones; and of the appearance on the beaches and littorals of the world of biological newcomers who settled alongside others that had occupied these places since time immemorial. For some species, the colonial period was a time of golden opportunity. For those that could travel—for crops, such as wheat, rice, potatoes, corn, tobacco, cotton, and beans; for domesticated animals, such as pigs, goats, horses, sheep, cattle, chickens, ducks, cats, and dogs; for Old World diseases, for

European weeds, and for any number of animals, from English sparrows, starlings, and pigeons to rats, rabbits, and lampreys—colonialism offered enormous opportunities to increase their numbers through the immigration and naturalization services provided by Europeans. In the new environments shaped by colonial expansion, it was not the meek but the well-traveled that inherited the earth.

One of the primary distinctions that the British would make between themselves and other nations was that they were a seafaring people who could go anywhere. During the heyday of colonial expansion, they sought to extend this capacity to nature too. However, this expanded movement of natures did not in the long run lead to an increase in ecological diversity; as Charles Lyell was among the first to observe, it led to a greater global presence of fewer plants and animals. For the vast number of other species, those that had not been given free passage to new worlds, for those species that were not given the opportunity, in Peter Townshend's phrase, of "going mobile," from dodos and greater Auks to New Zealand kiore and *rango pangos* and Australian marsupials, the changes wrought by traveling natures were largely tragic. The journey that many of these creatures took was a journey into oblivion. In thinking about the new natures ushered into being by colonialism and the impact that they had on human societies, therefore, we should attend not only to the natures that emerged as the winners in the struggle for territory but also to those that were destroyed or marginalized by this process, the wild, useless, or waste natures that confronted the altogether new possibility of being evicted from their ancestral lands—their nature—by foreign landlords who brought with them their own preferred biotic tenants.

Natural History and the Globalization of Natures

When Sir Joseph Banks sent George Caley to Australia as a botanical collector, he remarked that "there are many objects, both in the vegetable and mineral kingdoms, hitherto undiscovered that will, when brought forward, become objects of national importance, and lay the foundation of a trade profitable to the mother country with that hitherto unproductive colony" (Caley, *Reflections on the Colony of New South Wales* 21). By disembedding plants and animals from their traditional physical and cultural locales and reembedding them in new locations, natures across the globe, which had long been seen as being deficient in various ways and in need of improvement, could be remade from the ground up. Eighteenth- and nineteenth-century Britons increasingly embraced the idea that they were a cosmopolitan people, whose strength lay in their power to move from one place to another, exploring, trading, and collecting information about other places and people. Liberty, as William Blackstone defined it, "consists in the power of loco-motion, of changing situation, or removing one's person to whatsoever place one's own inclination may direct; without imprisonment or restraint, unless by due course of law" (cited in Blomley, *Law, Space and the Geographies of Power* 180). What is distinctive about this period is that the British sought, on a global scale, to extend these rights to nature, envisioning a world in which no useful plant or animal, by reason of its natural origin or habits, could not also become a full "citizen of the world." Natural history would emerge as the cutting-edge science of empire because it held the promise of restructuring global natures to meet European economic needs. Its descriptive and documentary activities functioned in tandem with its overall goal of collecting, redistributing, and managing anything that was "useful" in nature. Marie Louise Pratt has noted the importance of natural history in the emergence of "Europe's 'planetary consciousness' ... the construction of global-scale meaning through the descriptive apparatuses of natural history" (*Imperial Eyes* 15). But natural history was not simply a form of consciousness; it embodied technologies and practices aimed at transforming the planet itself. Embodying a fundamental transformation in Europeans' relationship to nature, its revolutionary character lay in the project of *producing natures that could travel*. The multitudinous new species that were coming under the view of naturalists were no longer seen as being necessarily rooted in place. Instead, they were named and classified so that they could be exchanged and transferred within the broader project of biologically refashioning the globe. A second, new and improved creation was under way, based on the mobilization of natures, one that would banish useless natures—weeds, wildernesses, and wastes—for natures that could better serve humankind.

The great promise of Enlightenment natural history was that it would improve or complete locally insufficient natures through commerce and knowledge. The global distribution of plants and animals would not be governed by God's original design, but by human needs. As Thomas Jefferson remarked, "the greatest service which can be rendered to any country, is to add a useful plant to its culture" ("Memorandum (Services to My Country)," *Writings* 703). For Henry Hamilton, who served as lieutenant governor of Canada, the introduction of a new plant to a country was a "patriotic undertaking" (Hamilton to Banks, 12 May 1795, Banks, *Scientific Correspondence* 4:363). Although local floras and faunas were one of the most popular forms of archiving natures during the period, their primary purpose was not to celebrate local or traditional forms of knowledge, but to allow for comparisons to be drawn across localities. Local plants and animals were also renamed and reclassified within a new system of knowledge which allowed them to be integrated into a commercial world system.[11] Plants that had once been understood as being part of an established, unchanging order of nature were now subject to being transplanted to new places. Eighteenth-century natural history was thus intellectually structured around the enormous project of modernizing natures. By removing the traditional geographic boundaries that kept natures in place, naturalists were able to bring them into new relationships with each other, routing and rerouting them across vast distances. To adopt Anthony Giddens's phrasing, local natures came to be "penetrated by and shaped in terms of social influences quite distant from them" (*Consequences of Modernity* 19). In this way, precolonial natures that had been insular were brought into contact with each other, and one can see the beginning of the process by which local ecologies were integrated into a world system. During the eighteenth century, the ground on which natures had stood since time immemorial began to move. The effects of this mobilization on certain natures at the expense of others are central to understanding the history of colonialism and what natures are today.

Given natural history's cultural primacy in mobilizing local natures, it should not be surprising that in addition to collecting, cataloging, and exchanging the earth's "natural productions," naturalists were profoundly interested in studying the factors governing the global distribution or geography of plants and animals. Producing inventories of where certain plants and animals were located, learning about their individual climate and soil requirements, and figuring out how to adapt them to new ecological conditions were all fundamental to understanding how they might be moved. By the 1840s the most innovative thinking about nature had adopted the idea that plants and animals had a tremendous capacity for travel. In the work of Charles Darwin, a very modern conception of nature appeared, one that, instead of taking for granted the idea that species were created to be rooted in place, now saw them as being, by their very natures, predisposed toward mobility. By the 1850s, the many acclimatization societies that were springing up across the globe popularized ideas about the transformation of local natures which had been gaining momentum since the previous century. At the inaugural meeting of the "Acclimatisation Society of France," Isidore Geoffroy Saint-Hilaire declared, "The prospect was nothing less than to people our fields, our forests, and our rivers with new guests; to increase and vary our food resources, and to create other economical or additional products" (cited in Low, *Feral Future* 31). The optimism expressed in this vision of a total reordering of the nature of France may have been misguided, but it fully reflected ideas that achieved dominance within nineteenth-century natural history circles.

Although we have long recognized that colonial spaces exhibited a dramatic intermingling of peoples and cultures, we have not adequately addressed the degree to which their physical environments were just as diverse. These were natures shaped as much by the immigration of plants and animals as by the immigration of people. What emerged were *hybrid* natures, brought into being by the mixing of indigenous plants and animals with newcomers. These new natures, like the societies that produced them, were the products of conflict, displacement, and exchange, because Europeans brought with them not only their customs, languages, and values but also, both by calculation and by accident, their own portable natures. With contact, colonial natures thus became divided environments, susceptible to rapid change. Ecological balances that had been built up over vast periods of time were often destroyed in a matter of decades by the introduction of a single new species of plant or animal. William Stork comments, for instance, that "the new introduction of but a single grain or plant, as the rice in Carolina, or the turnip in Norfolk, will sometimes totally change the face and condition of a country" (*Description of East-Florida* 2:2). Colonial environments should not be seen, then, as simply a fusion of natures (in coexistence or productive adaptation), but instead as sites of intense struggle. Composed of overlapping natures with different histories and geographical origins, configurations of *old* and *new* species, of traditional and modern natures, of relicts and newcomers putting down roots, these natures embodied in diverse and unique ways the fundamental dynamics and conflicts of modernity.

The success of European plants and animals in traveling and displacing colonial flora and fauna and, likewise, the apparent failure of colonial natures to return the favor in this age of transplants did not go unnoticed by nineteenth-century biologists who studied plant and animal distribution. The Victorian botanist Joseph Hooker, whose interest in the migration of plants was stimulated by his correspondence with Darwin, was puzzled by the "total want of reciprocity in migration" (*Botany of the Antarctic Voyage* cv) between metropolitan and colonial biota. "Whatever countries beyond the seas we may visit, in the temperate regions of the globe," he writes elsewhere, "we find that their vegetation has been invaded, and in many cases profoundly modified by immigrant plants from other countries, and these are in almost all cases natives of North-western Europe" ("Distribution of the North American Flora" 155). Darwin spent a year studying British weed populations in the abandoned fields around his home in Down and concluded that European plants were biologically superior to their colonial counterparts not so much because they were inherently so but because they had adapted themselves over generations to living in circumstances where land was under cultivation. With the worldwide spread of European methods of farming, indigenous plants were thus at a considerable disadvantage to their foreign competitors in the struggle for cultivated land. The plow was thus a form of biological warfare, and the same could be said for horses, cattle, and sheep.

For the British, the global expansion of European biota was an extraordinary success story, productive of both astonishment and pride. Reading through Susan Fenimore Cooper's *Rural Hours* (1855), Darwin came upon a passage in which she recounted the incredible number of European weeds that were showing up everywhere in the United States:

> A very large proportion of the most common weeds in our fields and gardens, and about our buildings, are strangers to the soil. It will be easy to name a number of these:—such, for instances, as the dock and the burdock, found about every barn and out-building; the common plantains and mallows—regular path-weeds; the groundsel, purslane, pigweed, goose-foot, shepherd's-purse, and lamb's quarters, so troublesome in gardens; the chickweed growing everywhere; the pimpernel, celandine, and knawel; the lady's thumb and May-weed; the common nettles and tazel; wild flax, stickseed, burweed, doorweed; all the mulleins; the most pestilent thistles, both the common sort and that which is erroneously called the Canada thistle; the sow thistles; the chess, corncockle, tares, bugloss, or blue-weed, and the pigeon-weed of the grain-fields; the darnel, yarrow, wild parsnip, ox-eye daisy, the wild garlick, the acrid buttercup, and the acrid St. John's wort of the meadows; the nightshades, Jerusalem artichoke, wild radish, wild mustard, or charlock, the poison hemlock, the henbane,—ay, even the very dandelion, a plant which we tread under foot at every turn. Others still might be added to the list, which were entirely unknown to the red man, having been introduced by the

European race, and are now choking up all our way-sides, forming the vast throng of foreign weeds. (*Rural Hours* 64–65)

When Cooper looked around her, she saw a New England landscape that would have been "entirely unknown to the red man," with roads everywhere choked with a "vast throng of foreign weeds," a new version of the contradictory myth of America as a land of opportunity for immigrants and as a place threatened by successive waves of immigrants. "Does it not hurt your Yankee pride that we thrash you so confoundedly?" Darwin asked his American botanist friend Asa Gray, who had listed 260 naturalized plants in his *Manual of the Flora of the Northern United States* (1848): "I am sure Mrs. Gray will stick up for your own weeds. Ask her whether they are not more honest, downright good sort of weeds." For her part, Mrs. Gray was quite willing to stand her ground in this nationalistic debate over the interpretation of biogeographical change, countering that American weeds were "modest, woodland, retiring things; and no match for the intrusive, pretentious, self-asserting foreigners" (6 Nov. 1862, *Correspondence of Charles Darwin* 10:505).

A century earlier, John Bartram, who was probably personally responsible for about 25 percent of the American plants that were introduced into England during the eighteenth century, was less diplomatic in talking about these plant aliens. In a short essay that he sent to the British merchant gardener Peter Collinson in 1759 entitled "Introduced Plants Troublesome in Pennsylvania Pastures and Fields," Bartram asserts that most of the plants that have proven "most troublesom in our pastures & fields in Pensilvania" have been "brought from europe." Included in his list of unwelcome newcomers are the yellow thistle (*Cirsium horridulum*), brought to America by a Scotch minister in a mattress filled with thistledown; St John's wort (*Hypericum perforatum*), "a very pernicious weed" that "spreads over whole fields & spoils their pasturage"; the oxeye daisy (*Leucanthemum vulgare*), "a very destructive weed, in medow & pasture ground choaking ye grass & takeing full posession of ye ground"; and *Saponaria* (soapwort), dockweed, wild garlic, and dandelion. Worst of all, however, was the yellow toadflax (*Linaria vulgaris*), "ye stinking yellow linarya," "ye most hurtfall plant in our pastures that can grow in our Northern climate." Originally introduced by a Welsh Quaker named Ranstead, as a medicinal herb used to treat insect bites, the plant quickly became a common weed across the New England states. Neither "the spade plow nor hoe can destroy (eradicate) it," Bartram writes, and it "is now spread over great part of ye inhabited parts of pensilvania" (1759, *Correspondence of John Bartram* 451). A decade earlier, in his notes to Thomas Short's *Medicina Britannica*, Bartram labeled it as "a troublesome, stinking Weed" that "is no Native, but we can never I believe eradicate it." Whereas the British made the plant seem like a tasty item on a breakfast menu, calling it "Butter and Eggs," its popular American name—"Dog Piss Weed"—speaks volumes about how the Americans viewed this plant intruder (287).

Underlying these debates also lay darker and crueler conclusions. For Alfred Russell Wallace, the march of European natures in step with European people confirmed their superiority. Civilized man increases at the expense of the savage, he writes, "just as the more favourable increase at the expense of the less favourable varieties in the animal and vegetable kingdoms, just as the weeds of Europe overrun North America and Australia, extinguishing native products by the inherent vigour of their organisation, and by their greater capacity for existence and multiplication" ("Origin of Human Races" 48). European settlers were quick to draw similar conclusions. German-born Julius von Haast would write to Darwin from Christchurch, New Zealand, noting "the wonderful capabilities of European products, to settle themselves on antipodean ground & to destroy or drive away the indigenous inhabitants.... Yes, it is really wonderful to behold the botanical and zoological changes, which have taken place since first Captn Cook visited the shores of New Zealand" (21 July 1863, *Correspondence of Charles Darwin* 11:551). British colonists promoted the idea that the settlement of the globe by European people and plants was ordained not only by reason, economy, and faith but also by a mysterious biological destiny. Speaking of the wild chicory, oxeye daisy, and mayweed that he saw everywhere in New England, Joseph Hooker commented, "These, and more than two hundred and fifty other Old England plants, which are now peopling New England, were for the most part fellow-emigrants and fellow-colonists with the Anglo-Saxon, having (as seeds) accompanied him across the Atlantic, and having, like him, asserted their supremacy over and displaced a certain number of natives of the soil" ("Distribution of the North American Flora" 155).

How these changes in the natural world were understood by writers of the eighteenth and nineteenth centuries needs more consideration. Where some, such as Haast, viewed them as being "really wonderful," others understood these events as tragedies being played out across the globe. When, in 1856, Charles Darwin learned that the naturalist W. B. D. Manellin was returning to New Zealand, he was quick to remind him that "the aboriginal rat (& the Frog) are great desiderata in Natural History" (5 June 1856–59, *Correspondence of Charles Darwin* 6:130). The first of these, the kiore, was no more native to New Zealand than the Maori, who brought it with them as a food source when they arrived on these isolated shores a thousand years earlier. By the middle of the nineteenth century, however, both the rat and the people who depended on it were threatened. Ironically, the closer these natures came to oblivion, the more valuable they became to the same culture that was destroying them. Even the New Zealand blue bottle housefly, the *Rango pango* of the Maori, was facing extermination. Wherever the housefly appears, writes Haast, "it expels the blue bottle fly, which seems to shun its company . . . the Settlers knowing the useful quality of their old home acquaintance have carried it in bottles and boxes to their new inland stations." For Haast, as for Darwin, a strange form of biological manifest destiny was being enacted in the manifest ecological changes that were taking place in the colonial world. To see English plant and animal communities—a biotic diaspora—successfully settling in new regions of the globe and displacing the indigenous natures was to discern a greater force at work in the world, shaping the history not only of humankind but also of nature itself. The Maori, for their part, were not oblivious to the significance of the biological changes taking place around them, for Haast comments on a common saying among them: "So as the white man's rat has driven away or killed our Kiore (native rat) the European housefly drives away our own (Bluebottle) the clover kills our fern (*Pteris esculenta*) so will the Maori disappear before the white man" (21 July 1863, *Correspondence of Charles Darwin* 11:551).[12]

Here Haast was referring to a familiar Maori proverb or *whakatauki*, which Charles Wentworth Dilke would later use as the epigraph to his chapter "The Two Flies" in his book on the British Empire, *Greater Britain* (1868):

As the Pakéha fly has driven out the Maori fly;
As the Pakéha grass has killed the Maori grass;
As the Pakéha rat has slain the Maori rat;
As the Pakéha clover has starved the Maori fern,
So will the Pakéha destroy the Maori. (*Greater Britain* 1:390)

This "as/so" logic encapsulates another dimension of colonialism: that it was everywhere the story of the struggle of indigenous peoples to hold on to their natures in the face of forms of interpretation that read an inscrutable destiny in the changes that were taking place in the land. Although attention has been given to the overt struggle between settlers and indigenes, we also should consider the battles that were fought over competing physical natures and ideas about those natures. To lose this battle was to lose a world, to see one's nature made obsolescent, or reduced to an isolated patch reserved on the margins of the new natures that were coming into being. It was also to be forced to inhabit another nature, increasingly composed of "alien" plants and animals, which had never before been part of one's culture or world, but which now tenaciously claimed their ground. Surely H. G. Wells was right when, upon learning the history of the eviction of the Tasmanian Aborigines from their native land, he rewrote this event as a "war of the worlds." Should we not begin to think of colonialism as the history of the global mobilization of natures as much as of peoples, a history in which natures were transplanted, displaced, and ultimately set in conflict with each other? If so, we need also to recognize, as Tim Cresswell has observed, that "mobility becomes meaningful within systems of domination and resistance, inclusion and exclusion, and is embedded ... [in] systematically asymmetrical power relations" ("Mobilities" 9). The parallel between the displacement and eradication of peoples and of biota was not the expression of an inscrutable fate, but the expression of the power that lay in the colonial management of the movement of people and natures—the power that lay in translation.

Natural History as Translation

In Frances Brooke's Canadian novel *The History of Emily Montague*, army colonel Edward Rivers remarks in a letter that "an old Indian told me, they had also songs of friendship, but I could never procure a translation," the reason being, he goes on to say, that "on my pressing this Indian to translate one into French for me, he told me with a haughty air, the Indians were not us'd to make translations, and that if I chose to understand their songs I must learn their language" (1:20–21). Hurons, it appears, do not do translations; if Rivers is to experience their songs of friendship, he will need to learn Anishinaabemowin himself. Although this anecdote might be interpreted both as a classic statement that songs can only be understood in their original languages and as a conventional assertion of the nobility of the American Indian, the word "haughty" suggests that the "height" to which Rivers is referring is only in the Indian's self-estimation and that his bias against translation is being construed as an expression of pride or arrogance. Whatever the merits of this view of Indian culture, it tells us a good deal about how the

British understood themselves by the middle of the eighteenth century, for it indicates that they had begun to distinguish themselves from their indigenous counterparts as a people committed to translation. As many historians and critics have recognized, translation and empire went hand in hand.[13] In a world that increasingly was accessed through translations, knowledge, power, pleasure, and wealth lay in the capacity to move ideas, people, and things across distances and boundaries. Britain in the late eighteenth century and early nineteenth century was, in all aspects, a *translational culture*, one that augmented its power and enriched itself poetically, scientifically, culturally, and materially by translating things from one place, state, form, or condition to another. In this regard, by seeking to carry something valuable back with him, a translation of an Indian song, Rivers was doing exactly what any British colonialist would have done in similar circumstances. In turn, the Huron's response may be less an expression of native pride than of reticence in knowing how such documents of friendship might ultimately be used.

During the colonial period, natural history was the science most attuned to the globalizing possibilities opened up by translation. It was not a new science, for its Western origins go back to Aristotle's *Historia animalium*, but European colonialism helped to form its character as a modern translational science. Throughout the sixteenth and seventeenth centuries, colonial natural histories appeared sporadically, beginning with the publication in 1526 of the first natural history of the New World, Gonzalo Fernández de Oviedo's *Natural History of the West Indies*. These books were the direct outgrowth of colonial settlement, in the Caribbean, Spanish America, the American colonies, and the Dutch-controlled East Indies, India, and Surinam. Important as these studies were, they had a limited reach, because they were written in different languages and there was no universally accepted system of nomenclature, classification, and description which would allow naturalists to compare and interpret this information across different parts of the world. These early books thus stand in marked contrast to the practice of natural history during the eighteenth century, a field that stands out for its global reach and ambition. As William Clark, Jan Golinski, and Simon Schaffer suggest, "Natural history in all its branches ... assumed a central place in topographies of enlightened science as a global enterprise" (*Sciences in Enlightened Europe* 27). It was not simply that European naturalists were by this time searching every nook and cranny of the earth for natural resources and for new plants and animals to bring home, to describe, and to name, but just as importantly, their innovative methods of organizing and disseminating information provided the model for other forms of imperial knowledge, administration, and control. Naturalists were among the vanguard in colonial administrations because the business of empire was largely organized around the production and exchange of natural knowledge and commodities. Naturalists were seen as "colonial experts," the go-to people, in Mary Louise Pratt's view, for "territorial surveillance, appropriation of resources, and administrative control" (*Imperial Eyes* 39), the people best suited for surveying the human and natural resources of a locality and providing advice on how best to utilize them.

One of the most important innovations of natural history lay in its success at disembedding global natures from the material entanglements and the local forms of knowledge that had previously governed their understanding so that they could be integrated into a world system of information exchange. Natural history constituted a project of material and cultural translation on a global scale. European print technology played a central role in this process as it allowed natures that would otherwise have been little known beyond their traditional locales to appear in easily reproducible forms communicated across wide distances to expert and amateur naturalists alike. As Anthony Giddens remarks, "printed materials cross space as easily as time because they can be distributed to many readers more or less simultaneously.... Today the printed word remains at the core of modernity and its global networks" (*Modernity and Self-Identity* 24). Alongside the enormous material changes introduced by the globalized movement and exchange of plants, animals, parasites, and even pathogens, therefore, it is important to recognize that by far the most common way in which colonial natures appeared in Britain during the eighteenth and nineteenth centuries was in printed form. Natural history was decidedly a textual activity whose natures were largely produced on paper, and that is how they moved. New plants, new animals, new natures inundated Britain as print culture—in travel narratives, natural histories, calendars, floras, faunas, engravings, poems, and novels. By the last third of the eighteenth century, nature had become a mainstay of the book trade. Each year would bring a new batch of expensive natural history books, accompanied by reams of gardening books and magazines, nursery catalogs, deluxe folio engravings, and a plethora of dictionaries and introductory manuals that sought to introduce nature, gardening, and horticulture to an avid public. By the 1830s, the *Magazine of Natural History* could thus declare that "of such works now there is no lack: we have Floras and Faunas, Magazines, Miscellanies, Registers, Cabinets,

Monographs, and Enumerations, in abundance, together with Illustrations, zoological, entomological and ornithological, besides a formidable phalanx ... of Transactions" (Anon., "Remarks" 297). Just as importantly, in addition to supplying accounts of exotic cultures, travel narratives were expected to include substantial information on the natural history of foreign localities. For instance, the publication of the official account of the convict settlement at Botany Bay was delayed by a year as the government awaited the preparation of natural history materials. By turning nature into something that was primarily experienced as print, European natural history produced a revolution in natural knowledge and its reception.

In addition to being mobilized and exchanged in printed words and illustrations, colonial natures were also circulated and studied as material culture, in a wealth of specimens, wanted alive (for orangeries, botanic gardens, nurseries, menageries, and zoos) or dead (as skinned, stuffed, pressed, mounted, or polished specimens in herbariums, cabinets, and hobby collections, or for the many public natural history museums that were springing up everywhere). From the decorative arts to clothing fashions, from the rarest objects collected by private connoisseurs to the popular main-stage attractions of the "shows of London," exotic and domestic natures were collected, reproduced, and consumed across an entire nation. Nature and natural history were all the fashion. In this regard, it is fitting, perhaps, that the most famous early eighteenth-century English nature collector, Sir Hans Sloane, a man whose collection of specimens and artifacts from Jamaica formed the core of the British Museum when it opened in 1759, also introduced milk chocolate into England. The interest in colonial natures may have been an acquired taste, but by the end of the century, they were flooding into Britain from all parts of the globe—from Australia, Tahiti, the Americas, Africa, and Asia—on a scale never before seen. Where reading about and collecting rare and exotic natures had once been a prerogative of the wealthy, these activities were now increasingly associated with the middle and working classes, who had come to consider natural history as a valuable recreation and mode of self-improvement (Durling, *Georgic Tradition* 146–47). Also, as is clear from Charlotte Smith's *Conversations Introducing Poetry: Chiefly on Subjects of Natural History for the Use of Children and Young Persons* (1804) and *A Natural History of Birds, intended chiefly for Young Persons* (1807), natural history now occupied an important place in the moral education of children. A nation of shopkeepers was fast becoming a nation of amateur naturalists, avid collectors, and dedicated gardeners. Thus, in manifold ways, the natural world for the British during the eighteenth century was much larger, richer, and far more diverse and complex than it had been in previous centuries. And amid this enormous influx and exchange of natures, the British began to notice that they too had a nature that was worth collecting and writing about. In this regard, it is worth recognizing that the Romantic turn to nature, the preoccupation of these writers with transforming British nature into a reading experience, was not so much an activity unique to literature, but instead was an expression of the production of textualized natures which was very much a defining feature of the period.

Bruno Latour, upon whose arguments on the translational aspects of science I have been drawing, claims that the global dominance of European colonial science over other forms of knowledge arose from its capacity to produce and transport textualized information across extensive networks. Enlightenment science, he suggests, developed technologies that were aimed at the mobilization of things. Through these technologies, Europeans were able "to act at a distance on unfamiliar events, places and people" by "*somehow* bringing home these events, places and people" by rendering them in forms that were both permanent and mobile ("immutable mobiles"). "The history of science," he argues, "is in large part the history of the mobilisation of anything that can be made to move and shipped back home," from specimens of rocks, plants, and animals to art, cultural artifacts, and people (*Science in Action* 223–25).[14] Inscriptions played a key role in this process, for anything that could not be moved physically was mobilized in the form of maps, surveys, drawings, written texts, samples, or specimens. For Latour, whose model of colonial science was powerfully influenced by Elizabeth Eisenstein's *The Printing Press as an Agent of Change* (1979), the capacity of colonial science to accumulate local knowledges derived from its being fundamentally a textual activity.[15] The goal of travel was not just *to see* things well, but *to record* them, to produce inscriptions, texts, maps, and records that could then be brought back to European metropolitan centers, where they could be compared, amended, and preserved so that later travelers, charged with doing the same thing, would not so much visit these places as revisit them. Constructed out of the texts, specimens, and inscriptions gathered from colonial expeditions and encounters, colonial natures were thus imported into European metropolitan centers, where they stood in for the natures that they were in a large degree replacing. Natural history was thus far more than simply about collecting and classifying specimens: it was a translational world-making activity, which mediated colonial natures through their portable forms. Like the local natures from which they were drawn, yet also fundamentally different, these modern textualized natures allowed Europeans to know and act on the world from a distance. They thus transformed their own position, as a point among many, into a center acting on others. "Without travelling more than a few hundred metres and opening more than a few dozen drawers," writes Latour, the zoologists in a natural history museum could thus "travel through all the continents, climates and periods" (*Science in Action* 225). Thus, as Jim Endersby comments, "in his herbarium, [Joseph] Hooker could see the plant world spread out at his feet" (*Imperial Nature* 155).

This activity of transforming nature into specimens, images, and texts—a "nature trade" in the fullest sense—changed how the British saw themselves and their relationship to the natural world. For the British upper and middle classes, colonial natures no longer existed at a great distance, but instead were increasingly ready to hand in a variety of forms. Surrounded by a diversity of mobilized natures, available to them in their drawing rooms, studies, libraries, kitchens, hothouses, gardens, storefronts, museums, and other public institutions, and experiencing these translated natures on a daily basis, the British could easily imagine that they existed at the center of a global nature. To be British and part of an empire throughout the late eighteenth and nineteenth centuries was to have other peoples' natures and the commodities derived from them at your fingertips: in a book, as a print illustration, on the table, or in the garden. It was to live, in the fullest sense, in a world reconstituted by translation. "If I have not gone to foreign countries, young man, foreign countries have come to me," remarks the auctioneer Mr. Sapsea in Dickens's *The Mystery of Edwin Drood*. "Put it that I take an inventory, or make a catalogue. I see a French clock. I never saw him before, in my life, but I instantly lay my finger on him and say 'Paris!' I see some cups and saucers of Chinese make, equally strangers to me personally: I put my finger on them, then and there, and I say 'Pekin, Nankin, and Canton.' It is the same with Japan, with Egypt, and with bamboo and sandalwood from the East Indies; I put my finger on them all" (37). The same could be said about the plants during this period. The cultural emergences both of a global nature and of what came to be seen as England's traditional rural nature were the products of the mobilizing powers of natural history, which produced its own "inventories" and "catalogues." Benedict Anderson's argument that nations are "imagined communities" made possible by print capitalism certainly applies to late eighteenth-century British ideas of the natural world, for a burgeoning culture of natural history allowed the British living in cities or rural villages or spread across the far corners of the earth to see themselves as being part of a shared nature, centered in the country but expanding outward to encompass the globe.

The manner in which a global conception of nature could provide a sense of identity and communal belonging stretching across the earth can be seen in Mungo Park's *Travels in the Interior Districts of Africa*. At the very nadir of his journey, robbed of all that he possessed and "in the midst of a vast wilderness . . . naked and alone; surrounded by savage animals, and men still more savage," Park writes that he was about to give up, to "lie down and perish," when, as he says, "the extraordinary beauty of a small moss, in fructification, irresistibly caught my eye." Although the plant "was not larger than the top of one of my fingers," he writes, "I could not contemplate the delicate conformation of its roots, leaves, and capsula, without admiration" (226–27). Park is here adopting the conventional language of religious conversion to announce his discovery of an all-protecting God even in the African wilderness, but what the episode really expresses is his dedication to botany, as intellectual resource and psychological mainstay, for without it he would not have noticed this miniscule plant. Science is a "seeing" that saves. Here Park commits himself to natural history and to the idea of building a global nature that it sustained. Indeed, it was his botanical knowledge that first brought him to the attention of Banks and led to his being chosen to explore the source of the Niger River.[16] By doing his job, using the cultural capital that could not be taken away from him even if his material capital could and was, Park is able to see himself as part of an "imagined community" stretching across the globe and infiltrating every corner of the earth. And Park's readers, part of this same community, would have had no difficulty understanding the point of this story. They did not have to go to Africa to see the flower that he described: to be with him in this nature, to be translated there with him, they needed only to open the pages of his book.

Intellectual innovation in natural history was primarily driven by the sheer amount of new information that was being produced every year about the natural world in print. By the eighteenth century, the "book of nature" had become a vast "library of natures" composed of many volumes from many places in many languages using many different modes of taxonomy and nomenclature. That is why the appearance of the Linnaean taxonomic system in the 1730s, culminating in the introduction of binomial nomenclature in the *Species Plantarum* in 1753, was so important, for it allowed naturalists to share and manage the massive amount of information that was now available. Linnaeus's system, which can be seen as the natural history equivalent of the Dewey Decimal Classification, provided a simple and effective method of using names to organize, archive, and retrieve knowledge about the natural world. For Linnaeus, names functioned like call numbers, according each species a unique place in the system. "If you do not know the names of things," he wrote in the *Philosophica Botanica* (1751), "the knowledge of them is lost too" (169). Linnaeus is famous for having thought of himself as a second Adam, completing the project of naming the world first begun in Eden. Such a view, which placed the Swedish systematist at the center of Creation, was profoundly misleading in many ways, not least of all because the plants and animals that came under his scrutiny were not so much being named for the first time as scientifically renamed and culturally recoded, that is, translated in the fullest sense of the word. Many of the organic beings that Linnaeus named already had names, and often many names, as the same species in a different culture would have a different name and could be understood differently. Given the Babel of natures that existed prior to the advent of the Linnaean system, there can be little question of the value of such an activity. Through Linnaean nomenclature and the taxonomies that it enabled, any plant or animal could be disembedded from its locality and culture and reassigned a unique spot in an archive, where, bound and marked with its equivalent binomial call number and catalog entry description, it could easily be found by anyone with even the most rudimentary understanding of the system. Linnaeus's system thus constituted a revolution in the management of global information achieved by reducing the many languages that had been used to refer to the natural world to a single universal language. Its value did not lie in the relationships that it established among organic beings, but in the simplicity with which it allowed its practitioners to organize, transfer, store, and retrieve knowledge about the natural world.[17] Local natures with all of their accompanying complexities were thus transformed into translocal natures, accessible and comprehensible by anyone anywhere. Linnaean classification also required its practitioners to focus on specific aspects of a plant or animal as being significant, to the exclusion of other aspects. It thus dictated what features of a species were relevant to its proper scientific representation and thus where it was placed in this vast biological archive.

To summarize what I have been suggesting, the Linnaean system can be understood as inherently a project of translation. It allowed local natures, rooted in different cultures, to be remade so that they could cross the physical and cultural boundaries that had previously separated them. Although Linnaeus, like any translator, insisted on the importance of accuracy in the representation of the natural world, the radical character of colonial natural history lay less in the accuracy with which it represented these natures than in the ways in which it made these natures portable, that is, capable of crossing worlds. Much of the energy and excitement of this period derives from Europeans' enthusiasm in experiencing the new natures that were thus made available in translation and in knowing them in new ways. In the best of these, the reader is made aware that he is reading these natures in translation, that there are other ways of knowing and talking about them; in the worst of these, the translation replaces the original without signaling the transformative and mobilizing processes that brought it into being.

Of additional importance is the fact that this was not simply a situation of a text replacing a physical nature, but instead a much more complex relationship in which global physical natures were themselves being changed by this activity. As I have already suggested, Enlightenment natural history was not satisfied with simply translating local natures into a universal language. It also sought to use translation to physically remake these natures themselves. Naturalists were at the forefront of this new conception of the world, because they were fundamentally engaged in the dual activity of both transplanting physical natures across the globe and developing ways of reading these natures in translation. In the same manner as linguistic translation can be understood as a twofold process that makes possible the "migration-through-transformation of discursive elements (signs)" and their reinterpretation or recontextualization "according to different norms, codes, and models," colonial naturalists sought to manage the migration of plants and animals across the natural and cultural barriers that had previously limited their movement and devoted much of their attention to reembedding them in new ecological and cultural contexts (Lambert and Robyns, "Translation" 3:3604).

Although today we commonly understand translation primarily in linguistic terms, as the communication of meaning from one language to another, the word originally had a much broader meaning having to do with the change that a person or a thing undergoes when it moves from one place, condition, or state to another. Part of a family of words that address boundary crossing, referring to change through movement or movement through change, such as "transfer," "transit," "transport," "transmit," "transplant," or "transcend," "to translate" means etymologically "to carry across," and it originally meant "to bear, convey, or remove from one person, place or condition to another" (*OED*). It could refer to the material conveyance, transport, or transplantation of persons or things, as in the translation of the bones of a saint from one place to another, a bishop from one see to another, or a tree from one place to another. "Plante and translate the crabbe trée, where, and whensoeuer it please you," remarks John Lyly, "and it will neuer beare sweete apple" (*Euphues* 6). "Grain out of the hotter countries translated into the colder," remarks Francis Bacon, "will be more forward than the ordinary grain of the cold countrey" (*Works* 3:114). For Sir William Blackstone, one of the primary requirements of property is that it be capable of being "alienated" or "translated" from one person to another (*Commentaries on the Laws of England* 2:294), and it is in this sense that John Adams speaks of "the rapid translation of property from hand to hand" (*Works* 9:470). During the medieval period, universal history was understood as a *translatio imperii et studii*, as a transfer of rule and learning from one people or place to another. Translation provides a useful way of considering the colonial mobilization of people, words, and things, because, unlike other forms of exchange, such as "transfer" and "transit," which speak of things moving without being changed, translation intrinsically recognizes *change to be constitutive of movement* and, vice versa, *movement as being constitutive of change*. Nothing that moves stays the same, and things must change if they are to move. Translation and transformation go hand in hand. Translation is thus closely related to another group of boundary-crossing words that refer to a change of nature or condition without necessarily considering movement, words such as "transform," "transmute," or "transmogrify." Hence, Enoch and Elijah were said to have been translated to heaven without death. As both the lovers and the mechanicals of Shakespeare's *A Midsummer Night's Dream* discover, a change of place can lead to a change in one's nature. A journey into a forest can be transformative. Quince, coming upon Bottom the weaver metamorphosed into a monster with the head of an ass, memorably exclaims, "Bless thee, Bottom, bless thee! Thou art translated" (3.1.122). The fundamental task of eighteenth-century colonialists—"planters," as they liked to be called—was physically to translate the world. "Planting," as Peter Collinson commented, "is another means of painting with Living pencils" (cited in Laird, *Flowering of the Landscape Garden* 64).

Colonial natural history was the disciplinary expression of this commitment to managing natures in translation and the primary discursive embodiment of this *translatio naturi*. Through its translational practices, natures that had previously been restricted to a single locality and understood in local terms were planted in new contexts where they were understood in new ways and put to new uses. Plants that had grown in one part of the world since time immemorial—sugarcane, indigo, breadfruit, okra, pineapple, and eucalyptus—were transplanted to new locations, where they put down cultural as well as natural roots. They became something else. Having been translated from South America to Ireland, potatoes would come to define the Irish during the nineteenth century. A tall grass with a sugary pulp would make its way from Polynesia to China and India, then to the Middle East, and from thence across the Atlantic to the Caribbean, where it would become the source of misery for millions of black slaves. What it was, how it was understood, and what it could do changed as it moved. An animal known in Marathi as *mangus*, which in its native habitat fed on eggs and small vertebrates, traveling under the new name *Herpestes auropunctatus*, would become a showpiece in European metropolitan museums and zoos, traveling thence to the Caribbean and Hawaii, which surely must have seemed to this creature to be mongoose heaven, where, after feeding on rats and cane frogs, it caused the wholesale extinction of many species of animal on these islands. In Kipling's "Rikki-tikki-tavi," the same animal that had proved itself to be one of the most rapacious and destructive animals on earth appears as a loveable and heroic defender of British colonial domestic life (its adoptive family) from the threatening malevolence of Indian nature. "This is the story of the great war that Rikki-tikki-tavi fought single-handed, through the bathrooms of the big bungalow in Segowlee Cantonment," writes Kipling in the opening line of the story (*Jungle Books* 117). From childhood onward, Victorians learned about their place in the imperial natural world through their encounter with translated animals, in zoos, museums, natural histories, and storybooks. Yet translation is intrinsically a selective activity. Certain aspects of what these creatures had once been in their native environments were carried over in these translations, while others were lost, ignored, or transformed. By the same token, natures that did not travel were not exempted from translation, for as Europeans expanded across the globe, they also brought with them their own ideas about nature and its uses; their own science, art, and languages to describe the new environments they encountered; and their own (often domesticated) natures. Natures were thus translated at home and abroad.

All translation is intrinsically a cross-cultural activity, and colonial natural history was no different. Although Linnaeus, at home in his study, could minimize how much the natural knowledge that he prized was a product of colonial contact and cross-cultural translation, naturalists in the field were less likely to adopt such a stance, because their success in accumulating global natural knowledge depended on their ability to lift natures out of the local languages and cultures in which they traditionally had been understood. The frontispiece to Pierre Sonnerat's *Voyage à la Nouvelle Guinée* (1776) is exemplary in this regard, for there the naturalist is portrayed as someone positioned at the intersection of two worlds. The frontispiece is structured by colonial power relations, as it depicts a hierarchical division of labor that sharply differentiates European scientific culture from the indigenous localities that it translates. Nevertheless, it still presents Sonnerat and the Papuans as being engaged in a shared activity: in exchange for the local nature that they have collected and brought to him (figured in the enormous pile of dead animals that litter the foreground), Sonnerat provides the European scientific knowledge and the pen that will picture and describe their natural history, putting their knowledge and their nature into his book. As much a translator as an author, Sonnerat thus occupies a "middle ground" between two cultures and two natures. He provides the means by which this people and their nature will enter the mobilizing textualized space of the book whose title page lies in the lower right-hand corner of the frontispiece. Patrick Browne, the author of *The Civil and Natural History of Jamaica*, argues, for instance, that "a circumstantial account of the species commonly found in every country, their properties and mannerisms," is the "first step" in discovering which of their properties might prove to be "the greatest use to mankind." This task cannot be achieved, he insists, without "the observations of the vulgar, who by a long experience frequently learn both their genius and qualities" (ccclxxix).

Even though the goal of eighteenth- and nineteenth-century natural history was to establish a universal monolingual natural world, meeting the needs of European science and commerce, a primary prerequisite for achieving this goal was to know the indigenous languages and local knowledges by which local natures had previously been understood. It was thus a

Frontispiece to Pierre Sonnerat, *Voyage à la Nouvelle Guinée*, National Library of Australia, ANL RARE—RB MISC 2554

common practice for early natural histories to include within them native vocabularies (particularly of indigenous plants and animals), and naturalists were often as much collectors of languages as of colonial natures. In his directions to the Hudson's Bay Company naturalist Andrew Graham, Thomas Pennant requested, for instance, that, wherever possible, he include the native names in his descriptions of the skins that he was supplying to him. In *Fauna boreali-americana*, which describes the animals encountered on Franklin's northern overland voyages, the author John Richardson (assisted by William Swainson for the birds) provides not only the scientific but also the First Nations' names for these arctic animals. Thus, for the American black bear (*Ursus americanus*), we learn not only that it was called a "black bear" by Pennant, David Bailie Warden, and John Godman and that it received its scientific name from Peter Simon Pallas, but also the names that it was called by the Chippewa, Cree, and Algonquin:

> Black Bear. PENNANT. *Arct. Zool.*, vol. i. p. 57, and INTRODUCTION, p. cxx. *Hist. Quad.*, vol. ii. p.11.
> WARDEN. *United States*, vol. i. p. 195.
> GODMAN. *Nat. Hist.*, vol. i. p. 194.
> URSUS AMERICANUS. PALLAS. *Spicel. Zool.*, vol. xiv. p. 6–26.
> HARLAN. *Fauna*, p. 51.
> Sass. CHEPEWYAN INDIANS.
> Musquaw (pl. musquawuck). CREE INDIANS, or, when reference is made to the black colour of the fur, it is termed cuskeeteh musquaw. The cinnamon-coloured variety is named oosaw-wusquaw, the first letter of the proper name being altered *euphoniae causá*.
> Mucquaw. ALGONQUINS. Macoush (a young bear.) (14)

Early colonial naturalists used the indigenous names of plants and animals in order to communicate with native people, who often functioned as field experts, collectors, and informants. Their descriptions in the book of the uses of plants and of animal behavior thus incorporate a substantial amount of information about First Nations peoples' attitudes toward and relationships with the natural world. As these texts went through subsequent editions or were incorporated into newer natural histories, however, the acknowledgment of the indigenous knowledge and the understanding of nature on which these studies were initially built disappear. The status of these texts as translations thus becomes invisible, and the indigenous natures and peoples that originally made them possible undergo erasure. We continue to read in translation, but now without knowing it.

Edwin Gentzler has usefully observed that translation is "less something that happens between separate and distinct cultures and more something that is *constitutive* of those cultures" (*Translation and Identity in the Americas* 5). The point is particularly relevant in regard to colonial nature translation, for the natures of both Britain and its colonies were reconstituted through this activity. To the degree that these translations were inscribed with the values and expectations of their target audience, to the extent that their activities were seen as a form of cultural and material enrichment, colonial naturalists played a critical role in the formation of ideas of the distinctiveness of Britain as an imperial nation whose primacy lay in the leadership role that it was playing in modernizing global natures through science and commerce. One of the primary tasks of early naturalists was to translate foreign natures into familiar terms, and numerous studies have shown how naturalists sought to accommodate their descriptions of the strange environments that they encountered to the knowledge and aesthetic frameworks of their audiences back home.[18] George Steiner, in *After Babel*, argues that translation, as an "act of importation, can potentially dislocate or relocate the whole of the native structure" (315). Colonial natural history played a similar role in introducing new plant and animal species into Britain, which substantially changed its nature; on many islands across the globe the changes have been even more dramatic.

Although most seventeenth- and early eighteenth-century natural histories draw attention to, and acknowledge their status as, translations, as they seek to carry across local natures and indigenous understandings of them to a scientifically and commercially oriented European audience, the history of natural history is the story of the progressive aspiration toward a *monolingual* understanding of world natures, an orientation manifest in Linnaeus's emphasis on the need for a universal language for the science. Naturalists were thus not unlike other translators in aspiring to make themselves and the collaborative, cross-cultural activities that were essential to the production of their natural histories invisible. Translators are self-effacing, Lawrence Venuti suggests, for they gain their authority by not being seen, by pretending to speak in the voice of the authors that they translate and hoping that their translations will be read as if they were the originals, yet their success as translators depends on their ability to adapt their translations to the cultural assumptions and stylistic conventions of their target audience. "Translation never communicates in an untroubled fashion," he writes, "because the translator negotiates the linguistic and cultural differences of the foreign text by reducing them and supplying another set of differences, basically domestic, drawn from the receiving language and culture to enable the foreign to be received there" ("Translation, Community, Utopia" 468). In this view, a Constance Garnett translation of *Anna Karenina* is very much a British cultural text, which conveys as much information about how the

British understood "Russia" and "Tolstoy" as it does about Russian culture. Colonial natural histories are similar in that they translated foreign natures to conform to the demands of European science and aesthetics. By so doing, the local or "foreign" dimensions of these natures are often lost. Colonial natural history translation was all about claiming the power to speak for nature. British naturalists and travel writers did not usually seek to represent natures through indigenous eyes, but saw themselves as the standard bearers of European Enlightenment, framing their written accounts of colonial natures to meet the expectations of metropolitan scientists, of an imperial administration increasingly concerned with controlling and exploiting global natural resources, and of a reading public that saw foreign natures as places of adventure, pleasure, luxury, or gain. Even though Enlightenment nature representation frequently emphasizes accuracy and factuality, its primary objective was to integrate indigenous local and vernacular knowledges into scientific and commercial modes of understanding. Although few translations claim to be better than the texts that they translate, colonial natural history set out to do just that, improving on the natures that it acted on through the very act of translating them in word and deed. No doubt, Enlightenment natural history, in its effort to create a "universal language of nature," made it possible to understand nature in global terms, across time and space. It took natures that were embedded in local cultures and rewrote them so that anyone anywhere could experience them. However, it is easy to forget that one is reading these natures in translation, and we do so at our peril.

The primary task of any critical reading of the translational practice of colonial natural history should be to recover, as much as possible, the middle ground out of which these translated natures emerged. By breaking through the invisibility of this translational practice and the monolingualism and monoculturalism it implies, we can begin to see these texts not as straightforward factual representations of a given nature—or of "Nature," for that matter—but as cross-cultural activities that customarily sought to possess these natures in and through the act of translating them. To seek to occupy this middle ground is to enter what Emily Apter has usefully termed "translation zones," "sites that are 'in-translation,' that is to say, belonging to no single, discrete language or single medium of communication" (*Translation/Transnation* 6). Hence, even the most prosaic natural histories and travel narratives can be seen as the site of cultural interaction and contestation between different conceptions of a nature; products of the intersection of worlds, they contain within them the traces of the cross-cultural co-operations, co-optations, and struggles that brought these new natures into being and allowed them to travel at the expense of the local natures that preceded them. Polylingualism is a condition of these texts, even when it has undergone erasure. The task of criticism, therefore, should be to understand what was gained and lost by this activity by recovering in these texts their translational qualities, recognizing the natures that they stand in for even as one discerns the manner in which they adapt colonial natures to European value systems. Through its capacity to transform and thus mobilize other natures, colonial natural history was the crucible within which Europeans created the natures on which they came to stand. By recognizing colonial natures as products of translation, in physical and textual terms, we can begin to appreciate the extraordinary achievement of European natural history in bringing a global nature into being, even as we can assess the alternative natures on which this science was built.

Just as colonial natural histories can tell us much about how the British came to see themselves in relation to the rest of the world, they can also be read to gain a better understanding of the indigenous natures that they translated. This goal is not without inherent challenges, however, for whereas normally one can compare a translation with its original, most of the local and indigenous natures of the past have long since passed into oblivion and do not exist apart from the material translations that made them available to a European audience. They are thus often worlds lost in translation. If one focuses on those aspects of colonial natures which translated well across cultures and environments, one cannot but be impressed by the extraordinary success of early naturalists in bringing out the complexity and diversity of global natures and making them available to a wide audience. Their success, in this regard, makes the colonial period a momentous period in the history of mankind's understanding and appreciation of global natures. Yet this success was achieved by changing these natures, by simplifying them and seeking commonalities, because local natures, grounded in local practices and vernaculars, were not easily adapted to European colonial needs and concerns, aesthetic, economic, scientific, or otherwise. At the same time, even more profoundly, these translations were one of the primary means by which the British came to know and to possess the natures that they discovered. Translation was thus a way of doing and getting things, as much as it was a form of representation. To the extent that we conflate these natures, reading about the colonial world as if we were not reading in translation, we miss a fundamental dimension of the politics of the period, a politics centered on what a given nature was or could be. Colonialism represents a struggle not just over land, territory, or space, but also over the natures and peoples that occupied them. Scientific and cultural translations proceeded in tandem with the more material, physical changes that were at the time taking place in the natures that they represented. Bound up with the broader objectives of European colonialism, these nature translations thus intrinsically speak of possessions and dispossessions, of natures gained and lost.

Brian Friel powerfully captures this aspect of British colonialism in his play *Translations*, which deals with the political and cultural impact of the Ordnance Survey of Ireland conducted by the British Army Engineering Corps in the 1830s. The immediate goal of the survey was to map Ireland so that it could be controlled. Captain Lancey, who is in charge, comments that through this work "the military authorities will be equipped with up-to-date and accurate information on every corner of this part of the Empire" (34). In addition to seeking to control Irish space, however, the long-term objective of the survey was to erase the Irish and their culture from the landscape by replacing the Irish names for places, which were preserved orally, with new English names printed on a map. It was, as Friel notes, "to take each of the Gaelic names—every hill, stream, rock, even every patch of ground which possessed its own distinctive Irish name—and Anglicise it, either by changing it into its approximate English sound or by translating it into English words" (38). The Irish claimed the land around them through the names that they had given to these places, so by renaming them, through translation as renaming, the British were able to make their own claim on these places. Colonial translation thus broke the connections that had been traditionally built between languages, nations, territories, and peoples and forged them anew. At one point in the play, the British lieutenant George Yolland, whose role as orthographer is to rename the places being mapped, when asked what he thinks is going on, replies, "It's an eviction of sorts" (52). At the apex of its powers, colonial natural history, which was explicitly involved in this process of renaming nature and, thus, inherently of breaking the claim that indigenous peoples had on the natures around them, also speaks of the dispossession and displacement of the natures that it studied and disseminated. Naturalists did not necessarily post eviction notices on the doorways of indigenous people and indigenous natures; they did not see themselves as being part of a military campaign. Nevertheless, their books often achieved the same result.

Knowing that much of the world documented in colonial natural histories and travel narratives has disappeared or been changed makes reading these books a troubling experience, like walking through a ghost town. At every point one sees natures that exist no more or that were in the process of becoming traces. Only fragments of these once vital and coherent natures remain. In *Views of Nature*, Humboldt tells the story of an old parrot that proved to be the last living creature on earth to speak the language of a South American tribe known as the Atures. He writes, "In Maypures (a curious fact), there lives an old parrot of which the natives maintain that no one can understand him because he is speaking the language of the Atures people" (129).[19] Humboldt's parrot stands as an emblem of the fundamental connection between indigenous natures and societies during the colonial period and of the fate reserved for those natures that proved untranslatable or were not thought worthy of translation. In this sense, in very important ways, colonial natural history can be said, in Walter Benjamin's terms, to constitute the "afterlife" of these natures. A translation, he writes, proceeds from the original, "not so much from its life, as from its afterlife" ("Task of the Translator" 254). Natural histories are filled with ghosts, translated beings who live a strangely posthumous existence in glass cases, engravings, and texts, speaking of peoples and natures that no longer exist, yet which haunt the edges of the new, modern natures that natural history brought into being. Reading in translation, since that is often all that is left us, we should seek to recover these lost natures, not to satisfy what Lévi-Strauss calls "the nostalgic cannibalism of history with the shadows of those that history has already destroyed" (*Tristes Tropiques* 41), but in order to understand as much as possible what has been lost, not only the biota, but also alternative ways of naming, knowing, and relating to the nonhuman world. Indeed, I would argue that an important if less emphasized aspect of the aesthetic and intellectual fascination of colonial nature translation for eighteenth- and nineteenth-century readers lay in its capacity to speak of what was being lost through modernization, through the evocation of partial knowledge, of the mysterious and unknown, of natures that did not translate well or that remained untranslatable.

Much was gained and much lost through this activity of translating natures. In thinking about the complexity of natural history as a translational practice, my goal is not to insist on a single position for how one should read or interpret natural histories during the period. The natures produced through this activity of translation are not all the same. Instead, I want to suggest that there is not one nature, but many natures. In encountering colonial natures, in books, art, museums, or gardens, we should try to occupy a middle ground, recognizing that what we are experiencing are natures that have been made to be accessible to us and that are the products of complex forms of mediation and negotiation. Some things translate very well, while other things do not translate at all, so our task as readers should also be to hear voices that are not our own. In his essay "The Task of the Translator," Walter Benjamin cites with approval Rudolf Pannwitz when he writes, "Our translations, even the best ones, proceed from a mistaken premise. They want to turn Hindi, Greek, English into German instead of turning German into Hindi, Greek, English" (261–62). In "Imaginary Homelands," Salman Rushdie writes about a different way of looking at translation as he speaks of a new kind of writing produced by people who exist in between cultures, a tradition by exiles, expatriates, and emigrants whose experience of the world is not of belonging to a single place but instead of moving between places and cultures. "Having been borne across the world," he writes, "we are translated men. It is normally supposed that something always gets lost in translation; I cling, obstinately, to the notion that something can also be gained" (17). In thinking about what natures have been and what natures have become, we need to incorporate this perspective too, for colonialism ushered in a world in which natures too were being "borne across the world" as a result of unprecedented forms of translation. For many people, the nature that they knew from childhood was fast becoming a lot like Rushdie's "imaginary homelands." Yet new natures were also being born out of this ferment, and the very ferment made it possible to know natures that we might never have known without it. As we increasingly recognize the degree to which we also stand on moving ground, we too can hope that amid these changes and transformations, amid the ruins of past natures, there are still some things that may be gained.

CHAPTER ONE

Erasmus Darwin's Cosmopolitan Nature

> From China's spicy shore this stranger comes
> To animate Britannia's distant sky.
>
> (H. Jones, *Kew Garden* 10)

"I absolutely nauseate Darwin's poem," Samuel Taylor Coleridge declared in 1796 to John Thelwall (*Collected Letters* 1:216). Although nausea may be a little excessive, responses to Erasmus Darwin's *The Botanic Garden* have rarely been neutral. Horace Walpole could not get enough of it. Describing it as "the most delicious poem on earth," he claimed that "Dr. Darwin has destroyed my admiration for any poetry but his own" (*Correspondence* 11:10).[1] Byron, on the other hand, diagnosed the Lichfield doctor as suffering from a tin ear and complained of the monotonous, "pompous chime" of his overwrought pentameter couplets. In a parody of his proclivity for hyperbole, overuse of adjectives, and alliteration, Byron treated the good doctor to a little of his own medicine, depicting him as "that mighty master of unmeaning rhyme" ("English Bards and Scotch Reviewers," *Poetical Works* 1:227–64, lines 893–94). Darwin was never a poet to use a common everyday word when a rare, erudite, technical, or luxurious one could be found or invented. The fame of *The Botanic Garden* may have been short-lived, but it was nevertheless the most popular and the most controversial nature poem of the 1790s. Between 1789 and 1796, part 2 of the poem, *The Loves of the Plants*, which was published first, went through four English editions and two Dublin printings. The complete poem, with part 1, *The Economy of Vegetation*, now added, appeared in 1791 and saw four English editions along with separate Irish and American printings by 1799. It was also translated into French, Portuguese, Italian, and German. Blake, Wordsworth, and Coleridge all fell for a time under the sway of what Wordsworth would later call the "mischievous influence" of "Darwin's dazzling manner" (*Poetical Works* 3:442), even though they soon held quite different views on poetry and nature. In the absence of Darwin's verse, however, Coleridge might not have cherished the ideal of a "philosophical poem," nor is it likely that Southey's encyclopedic epics, with their heavy load of prose notes, would ever have taken their lumbering flight.[2]

With Darwin's *The Loves of the Plants*, a new nature swam into Romanticism's ken, yet I expect that few people would include it in their list of the most important nature poems of the period. This situation is partly explained by that all too rare occurrence in literary history when there has been almost unanimous agreement about the aesthetic merits of a literary work. With the exception, perhaps, of Desmond King-Hele, Darwin has been universally judged to be a remarkably bad poet. His poetry "is a colossal, often a ludicrous, failure," remarks Dwight L. Durling. "His temple of nature is a palace of ice, its shimmer merely counterfeiting warmth. His botanic garden is finally a monstrosity in waxes, the sapless foliage of mere ingenuity" (*Georgic Tradition* 87). J. V. Logan goes much further, denying Darwin's poetry even the possibility of a future resurrection: he insists that it is "dead eternally for the general reader" (*Poetry and Aesthetics of Erasmus Darwin* 94). Whereas Wordsworth revolutionized poetic language in order to preserve a traditional nature that he believed was being threatened by the modern world—"enshrin[ing] the spirit of the past / For future restoration" (1805 *Prelude* 11:342–43)—Darwin adopted a dated poetic style in order to present a thoroughly modern conception of a natural world that was undergoing ceaseless change and transformation and was inescapably bound up with global commerce, industry, and consumption. No poet during the Romantic period conceived of nature in more modern—even hypermodern—terms, yet, unfortunately, Darwin did so in an outmoded style and in the highly artificial, gilded diction that Wordsworth would condemn in the preface to *Lyrical Ballads*.

Darwin's poetry represents the full flowering of the cosmopolitan and commercial ideals of the eighteenth-century British manufacturing elite. He was a founding member of the Lunar Society of Birmingham, a circle that included such luminaries of industrialism and commerce as Matthew Boulton, Josiah Wedgwood, James Keir, and James Watt. For these men, wealth and power rewarded those who used science to invent new technologies to better exploit the physical world. Producing new goods and ingenious inventions was not in itself sufficient, however, for success also depended on developing ready markets, and in this regard the members of the Lunar Society displayed remarkable commercial acumen. They actively promoted a public taste for innovation, fashion, and change, for the new, the up-to-date, and the modern, by cultivating the patronage of the leaders of fashion and through showrooms, exhibitions, catalogs, and advertising. They were among the first to realize that "the future belonged to societies which were trifling enough, but also rich and inventive enough to bother about changing colors, materials and styles in costume, and also the division of the social classes and the map of the world" (Braudel, *Capitalism and Material Life* 235–36). Boulton's Soho works and Wedgwood's Etruria, as the primary sites of the transformation of nature into the world of goods, were so successful, in fact, that they became tourist attractions in and of themselves. Members of the Lunar Society were thus leaders not only of the Industrial Revolution but also in what has been called "the birth of a consumer society."[3] For them, commerce functioned best on a global scale, across national boundaries. As Boulton declared to James Watt, "It would not be worth my while to make for three countries only; but I find it well worth my while to make for all the world" (quoted in McKendrick, "Commercialization of Fashion" 77).[4] A global vision was thus as much a reflection of their business interests as it was of their enlightened philosophical allegiance to cosmopolitanism. Philosophy, science, and commerce came together in a commitment to global markets and the mobility of goods. Although the Lunar Society has been seen as preeminently focused on industrial manufacturing, it reflected much broader interests. Francis D. Klingender comments that during the late eighteenth century, "the most progressive currents of thought were no longer emerging in the metropolis, but in countless provincial areas, where mining, industry and farming were being remodeled on scientific lines" (*Art and the Industrial Revolution* 21).

Darwin's poetry indicates how much this remodeling also extended to natural history and nature writing. Nowhere in Romantic literature is the impact of consumerism on the representation of nature clearer. For Darwin, nature was inseparably bound up with novelty, fashion, movement, and change. Whereas most of Darwin's contemporaries viewed "novelty" as a rudimentary and fleeting aspect of the appreciation of art in comparison with the classicist emphasis on "truth" and "endurance," Darwin made it central to his aesthetics. "The further the artist recedes from nature, the greater the novelty he is likely to produce," Darwin writes. "If he rises above nature, he produces the sublime; and beauty is probably a selection and new combination of her most agreeable parts" (*Botanic Garden* 49). *The Botanic Garden* stands out as the primary cultural document of the new consumerist commercial vision of nature which would underpin Britain's emergence as an imperial nation. Can nature ever be modern? Darwin thought that this was the only nature that truly mattered. His vision of nature aligns with a nation whose strength increasingly lay in its control and management of global natures and in its power to transform them into something new. In considering the limits of Darwin's idea of nature, then, we are in many ways addressing the limits of eighteenth-century cosmopolitan culture and its legacy in the present. The debate occasioned by *The Botanic Garden* responded, no doubt, to his radical political views and free-thinking attitude toward human and nonhuman sexuality, yet I hope to suggest that it was also very much a response to his modern conception of nature and to the consumerist ethos that brought this remarkably metropolitan nature poetry to glittering life.

Novel Natures

Darwin's poetry is inseparably bound up with the new cosmopolitan or globalized natures that were increasingly appearing at the end of the eighteenth century. These mobilized natures took many forms, particularly in colonial contexts, but their most obvious material appearance in Britain was linked to the rise of gardening as a major interest and leisure activity among the middle class during the last quarter of the eighteenth century. This interest had been steadily increasing from early in the seventeenth century, particularly on rural estates. Nursery businesses came into being to meet estate needs, and with them developed an expanding international trade in ornamental or "curious" plants, particularly in flowers from Turkey and Holland. The trade in plants reached an unprecedented scale when the landed gentry took up landscape gardening after the Restoration. Land contours were reshaped to include prospects and dells, lakes and canals were dug, and, most importantly, huge masses of forest trees and shrubs were planted to produce a mix of parkland and wilderness.

Much of this activity was a response to the fact that, by the middle of the seventeenth century, large tracts of English countryside had been deforested. Aesthetics, economics, and patriotism were behind this change. John Evelyn, in *Sylva* (1664), urged that the only answer to this "epidemical" destruction of forests, threatening both the defense and economic future of the nation, was a massive project of afforestation, "nothing less than an universal Plantation of all ... sorts of Trees," but most of all, oaks (2). For the first time, the English gentry, at home and abroad, began to see themselves as planters. The first large-scale examples of modern forestry, remarks Oliver Rackham, occurred between 1660 and 1700, when "some 11,000 acres of Dean and 1,400 acres of the New Forest were enclosed and planted" (*Trees and Woodland* 156). In parkland and plantation settings, millions of trees were planted, and nurseries expanded to meet this demand, with additional trees being supplied by Holland. In 1714, for instance, Sir Edward Blackett alone transported from Holland four thousand beech and elm trees, along with twenty-seven fruit trees, for the gardens of Newby Hall near Ripon.[5] By the 1720s, country estates were seeking more variety in their wilderness plantings and looking to other parts of the globe; by the 1830s and 1840s the importation of non-European trees, especially from North America, was in full swing. Mark Laird comments that "before the 1730s, most North American shrubs and trees had been coveted but rare curiosities, whereas by the middle of the eighteenth century their availability generated almost manic activity, rather as tulips had done one hundred years before" (*Flowering of the Landscape Garden* 70). Creating "groves," "wildernesses," and "shrubberies" was all the rage. For instance, Robert James Petre, the eighth Baron Petre, planted over sixty thousand trees at Thorndon Hall between 1740 and 1742. When the Quaker wool merchant and gardener Peter Collinson visited his estate and walked through its transplanted "North American thickets," he felt that he had been transported to the New World. In 1740 alone, he notes, Lord Petre "planted out about Tenn thousand Americans ... mixed wth about Twenty Thousand Europeans, & some Asians." He writes that Petre's "Collection of the West & East India plants is beyond ... imagination" (J. Bartram, *Correspondence* 167), including among these guavas, papaws, ginger, and limes better than any he had tasted from Barbados. Collinson ate so many pineapples that he was sick of them when he left. He estimated that at the time of Petre's death in 1742, the Baron had accumulated 219,925 plants in his nurseries, most of them exotics (J. E. Smith, *Correspondence of Linnaeus* 1:10). Over the course of the eighteenth century, nearly five hundred species of hardy trees and shrubs would be introduced into Europe, three hundred of these coming from North America, mostly under the auspices of John Bartram and Collinson (Ewan, *William Bartram* 14). In 1757 alone, fifty thousand of these were planted at the Royal Gardens at Kew. Between 1701 and 1750,

Douglas Chambers notes that sixty-one new trees and ninety-one shrubs were introduced into England, and between 1731 and 1768 the number of cultivated plants doubled (*Planters* 81).

This resulted in a massive transformation not only in the landscapes of the English countryside but also in the actual plants that composed them—a translation in the fullest sense. The species that constituted English nature were fundamentally changed, as many of them now came from somewhere else and thus symbolized a new idea of what English nature was. An entirely novel nature, composed of plants that had been transported from different geographical regions, was in the process of being brought into existence. "Whole counties changed their appearance," Chambers remarks, "and what had at one time been foreign now seemed natural (like conifers in Surrey)" (*Planters* 188). And yet this major historical event rarely receives comment. One of the popular truisms of the eighteenth century, popularized by William Cowper in book 1 of *The Task*, was that "*God* made the *country*, and *man* made the town" (*Poetical Works*, bk. 1, line 749). Such a view needs to be substantially revised, for it does not adequately recognize the degree to which the landed gentry of England during the eighteenth century was earnestly engaged in creating a natural landscape in England that had never existed before. In 1756, having learned that a cargo of over eighty boxes of seeds had been sent to England by Thomas Penn's gardener James Alexander, Collinson remarked that "after this Rate England must be turned up side down & America transplanted Heither" (J. Bartram, *Correspondence* 392).[6] Over the coming decades the flood of new introductions only increased, not just from America but from all parts of the globe. Writing to Linnaeus after visiting a number of gardens on the south coast of England, Daniel Solander would write, "I have now seen all the new plantations and gardens in England worth looking at . . . the new English taste in planting . . . is most excellent. No one who has not seen them can imagine the varieties of foreign trees and shrubs which now adorn the gardens of the nobility and gentry" (cited in Rauschenberg, "John Ellis" 155). Landscapes are expressions of power, and this was particularly the case among the landed gentry of eighteenth-century Britain. With a king affectionately known as Farmer George and with the legendary gardens of the East in mind, it is perhaps not surprising that Coleridge would associate Kubla Khan with landscape gardening or that in May 1802, shortly after the Peace of Amiens, he would see the "Poet Bonaparte" as a "Layer out of a World-Garden" (*Notebooks* 1:1166).

By the 1760s, the interest in landscape gardening and the picturesque was being augmented by gardening as a fashionable leisure activity among all social ranks, with the result that the demand for new and fashionable plants reached new heights. Every year brought new introductions from faraway places (camellias, fuchsias, geraniums), and many of these plants, such as *Rhododendron ponticum*, introduced in 1763, quickly established themselves as naturalized species in the wild. In the 1820s, J. C. Loudon estimated that prior to the eighteenth century "the number of exotics in the country probably did not exceed 1000 species: during this century above 5000 new species were introduced from foreign countries, besides the discovery of a number of new native plants" (*Encyclopaedia of Gardening* 85). Never before had the importation and exchange of plants occurred on such a scale. In addition to the yearly increase of exotic plants, the public was just as taken by, and willing to pay appropriately inflated prices for, the many new varieties of familiar flowers, the novel auriculas, polyanthus, ranunculi, carnations, hyacinths, tulips, and crocuses that were now being bred commercially in Holland, France, and Belgium. The industrious Richard Weston, who set out (among other things) in his *Universal Botanist and Nurseryman* (1777) to provide the eighteenth-century version of a Google search, a "United Catalogue" of every plant grown in all the nurseries in England, listed over one thousand varieties of ranunculus alone (4:51–86).[7] Whereas common ranunculi sold for approximately 5 s. for one hundred plants, named varieties went for 1 s. or 1 s. 6 d. each (the fiery deep red *Allegret* cost 1 s. 10 d., while the price tag for *Otho the Great*, not the tragedy but a fine orange-yellow variety, was 1 s. 8 d.). Prices went up from there, depending on how new or rare the variety was. *Wonder of Our Days*, a yellow aurora ranunculus, cost 5 s., while *Ladies' Favourite*, fiery red with yellow stripes, sold for 8 s., and *Honour of the Florists*, also an aurora, commanded 10 s. 6 d. These varieties paled, however, beside the olive-green *Count Mornay*, worth 15 s., and the "dark purple (foncé)" *Black Eagle*, at 16 s. Then as now, the excitement surrounding new varieties of ornamental plants, fruits, and vegetables was inseparably bound up with their names.

Hyacinths were even more fashionable and expensive. Among the more than 575 varieties listed by Weston, *Golden Sun* sold for £1 12 s., while *La Moderne* and the *Duke of Burgundy* cost, respectively, three and four times that amount. In a separate list "of the newest and most valuable Sorts" of hyacinths from Holland, it is not clear whether *Gold Mine*, a double red, purple, and scarlet variety, was named after its color or its seller's expectations, with a price tag of £15. The rose and red *Princess Gallitzin*, one of many floral aristocrats that populated these lists, commanded £15 15 s., yet even she ranked below *Black Flora*, the most expensive of the hyacinth varieties, which sold for £21 (i.e., approximately £1400 today) (*Universal Botanist and Nurseryman* 4:126–28). This was not the first time that plants had undergone massive commodification (the demand for tulips in Holland during the 1630s was even more extreme); however, the global breadth of the commodification of plants during the late eighteenth century was unequalled. Enthusiasm for them informs all aspects of late eighteenth-century culture, from art and literature to interior design and costume. Against those, such as Wordsworth, who celebrated nature as an alternative to urban life, there were many others who looked to nursery catalogs in their quest for, in Alexander Pope's phrasing, a "*Nature* to Advantage drest," or a nature that "oft was *Thought*, but ne'er so well *Exprest*" ("Essay on Criticism," Poems 1:239–326, lines 297–98). The vogue for gardening was so great that it even influenced the extravagant hairstyles of the 1770s and 1780s. In 1777 Hannah More protested against the elaborate coiffures of eleven young women who had visited her: "I hardly do them justice when I pronounce that they had amongst them, on their heads, an acre and a half of shrubbery, besides slopes, garden plots, tulip beds, clumps of peonies, kitchen gardens, and greenhouses" (cited in Corson, *Fashions in Hair* 348).

Neither these young women's nor Darwin's ideas about nature may have conformed to those of Wordsworth, Coleridge, or Gilbert White, but they do reflect what nature was increasingly coming to mean within consumer culture. In seeking a frontispiece for *The Loves of the Plants*, Darwin looked to Emma Crewe, the daughter of the Fox Whig John Crewe and Frances Anne (Greville), who also produced drawings for Wedgwood's pottery. She was one of those leaders of fashion, "Ladies of superior spirit who set the ton," whose patronage Wedgwood courted as he sought to distinguish his wares from those of his competitors (quoted in McKendrick, "Josiah Wedgwood" 112). As Darwin wrote, "With her waving pencil CREWE commands/The realms of Taste" (*Botanic Garden* 2:291–92). Crewe's frontispiece, entitled *Flora at Play with Cupid*, would later draw the wrath of moralists and conservatives, such as Richard Polwhele, for its explicit sexual subtext, one of its most memorable conceits being its image of Cupid, loaded with garden tools, but wearing the tilted hat of a roué. Nature in the frontispiece is primarily decorative and ornamental, in the elaborately excessive hairstyle of Flora and in the floral festoon that decorates what appears to be a picture

Emma Crewe, *Flora at Play with Cupid*, courtesy of the Thomas Fisher Rare Book Room, University of Toronto

sitting on a mantel frame. Crewe's frontispiece situates Darwin's nature in a private, domestic space and emphasizes its appeal to women. The poet was undoubtedly pleased with the frontispiece, for in his proem he likened his plant descriptions to "diverse little pictures suspended over the chimney of a Lady's dressing-room, *connected only by a slight festoon of ribbons*" (vi).

As he prepared the complete poem for publication in 1791, Darwin was seeking another frontispiece that could serve as "an allegory of the whole work," and the task fell to Henry Fuseli, who also drew a symbolic dressing room picture.[8] In an explicit acknowledgment of Darwin's debt to Pope, Fuseli provided an updated version of Belinda at her toilet, the eighteenth-century epitome of woman as a global consumer. *Flora Attired by the Elements* portrays the goddess in what amounts to an improvised outdoor dressing room as she prepares for a wedding or a ball. The elemental forces of nature—Fire, Water, Earth, and Air—make their appearance, not as the Rosicrucian sylphs of Darwin and Pope, but now as domestic handmaids. Like Crewe, Fuseli emphasizes the link between nature, luxury, and high fashion in the poem, perhaps nowhere more clearly than in the stunningly bizarre butterfly hat sported by Air, a woman who seems more like a courtesan than a servant. Fittingly, the nymph of Fire provides Flora with a mirror (shaped like one of the dressing room pictures) so that she can see herself being transformed. The poem is thus intended to provide a mirror to its female readers, particularly rich ones, within which they can see their own nature delineated.

The Botanic Garden should be seen as part of a flood of books on gardening and botany which first began to appear in the 1770s. "We are very fond of all branches of Natural History," Collinson wrote to Linnaeus; "they sell the best of any books in England" (J. E. Smith, *Correspondence of Linnaeus* 1:18). At this time, nurserymen, seedsmen, and florists first began to produce printed catalogs of their plants and their prices.[9] Linnaean classification also gave rise to a steady stream of local and regional floras, most dealing with the plant life of foreign regions (the Caribbean, Russia, China, South Africa, and North and South America). Books aimed at introducing botany to a general audience, the most famous being Jean-Jacques Rousseau's *Lettres Elementaires sur la Botanique* (1771–73), also began to appear in increasing numbers. William Withering, another member of the Lunar Society, published one of the most successful of these, *A Botanical Arrangement of All the Vegetables Naturally Growing in Great Britain*, which appeared in 1776 and went through numerous editions. About the same time, William Curtis, an apothecary turned gardener, set out to capitalize on the gardening vogue by publishing in installments his monumental *Flora Londinensis*, aimed at providing high-quality folio hand-colored plates with descriptions of all the plants growing within ten miles of London. Despite the quality of the plates, many drawn by James Sowerby and William Kilburn, the publication was a tremendous flop and nearly bankrupted Curtis. The gardening public was clearly not interested in native or local plants, but instead wanted to learn more about the new, exotic ornamentals that were becoming available.

Henry Fuseli, *Flora Attired by the Elements*, courtesy of the Thomas Fisher Rare Book Room, University of Toronto

Like any good businessman, Curtis changed his tactics and recouped his losses by publishing in a smaller octavo format colored plates of foreign botanicals. The first issue of what was to become the first and most successful British flower magazine, *The Botanical Magazine; or, Flower-Garden Displayed*, included engravings of the Persian iris, the purple rudbeckia (from the Carolinas and Virginia), and the winter aconite (from Italy and

Austria). Priced at a shilling, it sold over three thousand copies. Its paper wrapper indicates Curtis's new focus of providing accurate representations "in their natural Colours" of "the most Ornamental FOREIGN PLANTS, cultivated in the Open Ground, the Green-House, and the Stove." Accompanying text gave information on the scientific name and characteristics of each plant, along with horticultural advice. Curtis had a clear sense of his market—"Such LADIES, GENTLEMEN, and GARDENERS, as wish to become scientifically acquainted with the Plants they cultivate"—and since these readers also wanted some assurance that these plants could actually grow in their gardens, Curtis explicitly noted that every plant exhibited in the magazine was growing in "his BOTANIC-GARDEN, at *Lambeth-Marsh*." Soon Curtis had competition, such as the *Botanical Review* (1789–90), providing "figures of the scarcest and most beautiful foreign plants" (Desmond, *Great Natural History Books* 105–6), and Henry C. Andrews's *The Botanist's Repository for New and Rare Plants* (1797–1815). Gardening changed the way that many people viewed the natural world during the Romantic period. Nature had become fashionable, especially anything that was new, rare, exotic, or luxurious. Nowhere was nature more a commodity than in English gardens. For seedsmen, nurserymen, and florists, and for the many suburban and city gardeners who scoured the catalogs and nurseries for the latest introduction or variety, the primary justification for having an empire in the first place would seem to have been its capacity to access new plants to grace English gardens. To them, colonial possessions were little more than a gigantic storehouse for a seemingly limitless supply of choice new ornamentals. Getting one's hands on them was the problem. Having noted "the number of beautiful plants and shrubs ... imported from China since the first British embassy in 1793" and many others illustrated "in the rich collections of drawings in the Banksian and other libraries," J. C. Loudon, in his *Encyclopaedia of Gardening* (1822), envisioned a time when all the plants of India and China would be available to British botanists and gardeners, a golden age that would be ushered in when its emperor Kea-King, "with his army of two millions of men," falls "prostrate before a handful of European troops." Upon reading this grandiose vision of an empire serving the desires of English gardeners, Coleridge wondered whether Loudon could "make a turnip grow as quick, as this Asiatico-European Turn-up has sprung up, flowered & seeded in the hot-bed of his fine imagination" (*Notebooks* 5.2:6569).

Darwin's poem caught the first big wave of this revolution in the consumption of global natures. During the 1780s, at the same time as he was writing *The Loves of the Plants*, he was translating Linnaeus's *Systema Vegetabilium* (1783) and the *Genera Plantarum* (1787). In preparing these volumes, Darwin consulted with Banks and sent out sample sheets to more than forty botanists. Even Samuel Johnson was enlisted to help in the formation of botanical terms. By 1784, *The Loves of the Plants* was completed, and Darwin began negotiating with Joseph Johnson for its publication. Unfortunately, it was not until 1789 that the two men reached an agreement. By this time, Curtis's *Botanical Magazine* had demonstrated that there would be a strong market for a poem devoted entirely to describing flowers. Although, in a letter to Watt, Darwin declared that he wrote poetry "for pay, not for fame" (quoted in Schofield, *Lunar Society* 206), even he must have been surprised by its success.[10] He is said to have received "ten shillings a line" from Johnson for *The Loves of the Plants*, and for part 1, *The Economy of Vegetation*, a sum somewhere between £800 and a thousand guineas (Schimmelpennick, *Life* 207).[11] It was an "immense price," commented Anna Seward, who attributed the publisher's willingness to venture such a sum to the current popularity of botany: "Botany was, at that time, and still continues a very fashionable study. Not only philosophers, but fine ladies and gentlemen, sought to explore it's arcana" (*Memoirs* 167). She also suggested that the poem was formally structured to maximize sales by appealing to at least two audiences. With its copious "philosophical notes," the poem had all the markings of an important scientific contribution to natural history, of interest to botanists, many of whom were male. At the same time, Darwin was attempting, as he famously declared in his advertisement to *The Loves of Plants*, "to inlist Imagination under the banner of Science," using poetry "to induce the ingenious to cultivate the knowledge of BOTANY." In this light, the poem is essentially a very expensive gardening book, perfectly suited to appeal to a well-heeled middle-class audience interested in developing a knowledge and taste for plants. As a versified plant catalog, consisting of lavish verse illustrations of the plants that were ostensibly growing in the eight-acre botanic garden that Darwin had established in 1778 "about a mile from Lichfield" (*Loves of the Plants* 4n11), the poem was ideal for a public wanting both to gain scientific conversancy with plants and to know which new or exotic flowers they should have in their gardens. Confessing that "I am only a flower-painter" (48), Darwin produced the literary equivalent of a horticultural slide show, composed of a succession of verbal flower pictures.

This is not to say that this combination of a scientific treatise and glossy coffee-table poem did not have some very radical things to say about nature and the life of plants. Darwin was almost alone in his belief that "vegetables are in reality an inferior sort of animals" (*Phytologia* 1), sensible creatures who had the capacity to feel, think, dream, and desire, though at a rudimentary level. In *The Loves of the Plants*, he seeks to reverse the direction of Ovidean mythology: "Whereas P. OVIDIUS NASO . . . did by art poetic transmute men, women, and even gods and goddesses into trees and flowers; I have undertaken by similar art to restore some of them to their original animality, after having remained prisoners so long in their respective vegetable mansions" (*Loves of the Plants* vi). Very much an expression of a revolutionary age, the poem seeks to free plants from their vegetable bondage, releasing them from the captivity narratives—geographical, philosophical, and technological—that have so far kept them in place. Much of the technical virtuosity of the poem arises from Darwin's effort to create verbal images that would powerfully convey their free movement and, I would add, their open exchange. Here his commitment to the new and novel is clear. In a period that witnessed the advent of new technologies for producing moving images—the magic lantern, the Eidophusikon, the phantasmagoria, and the diorama—Darwin adopts in the proem the language of a London showman hawking a "camera obscura" exhibit, encouraging those "at leasure for such trivial amusement, [to] walk in, and view the wonders of my INCHANTED GARDEN." Nothing is fixed in Darwin's nature, but instead, through the medium of poetry, nature becomes a moving picture in which the life of plants—their animation—is embodied in their medium of representation. There, each plant is shown, he says, "dancing on a whited canvas, and magnified into apparent life," each specimen playing a starring role if only for a moment. Darwin assures his readers that even if they are not "acquainted with the originals," they will find pleasure in "the beauty of their persons, their graceful attitudes, or the brilliancy of their dress" (*Loves of the Plants* vi–vii). To speak of his poem as a "trivial amusement" reminds us of Darwin's inability, at least in his public declarations, to take poetry seriously, and yet, as Oscar Wilde knew so well, only a certain class of people can afford to take their trivialities seriously. Believing that poetry amuses while prose instructs, Darwin continually undercuts the importance of his "Poetic Exhibition" (174). In Interlude 2, the Bookseller compares the poem to a popular children's educational toy: "in the manner you have chained them together in your exhibition, they succeed each other amusingly enough, like prints of the London Cries, wraped upon rollers, with a glass before them" (83). No longer a high-tech, virtual-reality exhibition for a sophisticated audience, the poem is here likened to a series of crude engravings of London street peddlers and hawkers, crying their trades, mounted as a scroll on spindles within a glass-faced box.[12] Shaped by Darwin's contradictory attitude toward his audience and the medium within which his ideas were given animated form, Darwin's magic-lantern show is never static, but alive and constantly changing. Despite the poem's adoption of an older form of poetic diction dressed up for the occasion, the cinematic dimensions of the poem were aimed at producing a nature that had never been seen before, a translated nature that lived, breathed, and moved in transfigured form. Here the medium of this nature was truly its message.

Nature and the Consumer Revolution

When most people speak of Darwin's *The Botanic Garden*, they are referring to *The Loves of the Plants*. This is not surprising, for his witty exploration of the sexuality of plants, how "Beaux and Beauties croud the gaudy groves, / And woo and win their vegetable Loves" (1.9–10), is certainly a tour de force. This was the poem that made Darwin famous, and it continued to appear under separate covers long after the two-volume *Botanic Garden* had become available. The tendency to read *The Loves of the Plants* in isolation, however, can easily lead one to miss the radical philosophical and commercial vision that informs the poem. Darwin originally planned for *The Economy of Vegetation* to be about four hundred lines in length, that is, the equivalent of an introductory canto to *The Loves of the Plants*, but with "3 or 4 times the quantity of notes" (Thornton, "Letter from Erasmus Darwin" 449). By the time it was complete, it was almost six times that length, and the "supplementary notes" had expanded to more than thirty-nine entries and 126 pages. Never one to avoid paradox, Darwin even included two pages denominated as "Notes Omitted"! The two poems are quite different in style: whereas *The Loves of the Plants* is witty and urbane in its theatrical mock-heroic and sentimentalist treatment of a "trivial" subject—the love lives of flowers—*The Economy of Vegetation* is a Southeyan philosophical barge heavily laden with weighty encyclopedic erudition and grand narratives.

Surprisingly, flowers do not really make an important appearance in *The Economy of Vegetation* until its conclusion. Instead, it consists of four immensely learned lessons on "the theory of the earth," with lectures on cosmology, geology, mineralogy, and chemistry delivered by the Goddess of Botany to the Rosicrucian guardian nymphs of Fire, Earth, Water, and Air, whose job it is to keep the economy of nature running smoothly. The canto devoted to the nymphs of Fire, for instance, dispenses a history of the earth, of the formation of the stars, the solar system, and the planet, followed by discussions of fire, heat, and light in volcanoes, phosphoric lights, *ignis fatuus*, luminous flowers, glow worms, fire flies, luminous sea insects, and electric eels. The fourteen supplementary notes provide information on meteors, primary colors, colored clouds, comets, the sun's rays, central fires, Memnon's lyre, and fairy rings, to cite just a few of the topics. Scientific erudition is everywhere on display. A crucial turn in this natural history comes with mankind's discovery of fire, "the first Art" (*Economy of Vegetation* 1.213). Darwin argues that this knowledge, imparted to human beings by the nymphs of Fire, allowed them to take control of nature and themselves: "Nymphs! your soft smiles uncultur'd man subdued,/ And charm'd the Savage from his native wood" (1.209–10). To be truly human is to manage and control nature. The remainder of the canto recounts further discoveries: we learn about the invention of gunpowder, the steam engine, and the electrical battery, along with advances in the understanding of phosphorus and electricity. The role of heat in the germination and growth of plants is mentioned in passing, but this topic is short-lived, as Darwin instead brings the canto to a spectacular firework finale:

> Red rockets rise, loud cracks are heard on high,
> And showers of stars rush headlong from the sky,
> Burst, as in silver lines they hiss along,
> And the quick flash unfolds the gazing throng. (1.595–98)

Such crowd-pleasing poetic spectacle, itself the human equivalent of the opening lines on the creation of the cosmos, indicates that *The Economy of Vegetation* is less a tribute to nature than a paean to human industry and invention. Scientists, chemists, and industrialists, people like Roger Bacon, Benjamin Franklin, James Watt, and Josiah Wedgwood, hold center stage in this poem, not flowers. And the sylphs are far less mythical embodiments of elemental natural forces than a rhetorical device for conveying up-to-date scientific information to the reader.

When Darwin announced to Watt his intention to let "the world" know about his friend's "improvements of the steam-engine," Watt could not hold back his astonishment that he was being included in a natural history: "I know not how steam-engines come among the plants; I cannot find them in the Systema Naturae, by which I should conclude that they are neither plants, animals, nor fossils, otherwise they could not have escaped the notice of Linnaeus. However, if they belong to *your* system, no matter about the Swede" (Muirhead, *Origin and Progress* 2:232). Watt was right. In Darwin's "economy of vegetation," the distinction between nature and *techné* is not of great importance. Instead, value resides in the knowledge, ingenuity, and inventiveness with which one exploits the laws of nature in order to make something that is new or improved. Darwin enfranchises nature in order to increase its value and exchangeability. He is all about rhetorical fireworks, as he continually draws his reader's attention to his inventiveness as a poet, his talent for coining new words or using them in novel ways. Although many of these were scientific terms, Darwin shows a proclivity for unusual, rare, and showy words, such as "pillowy," "pansied," "aurelian," "wafting," "susurrant," "rimple," "sopha'd," "cinctured," "scintillating," "gauzy," "unturbaned," "halo'd," "tesselated," "tintless," "noduled," and "iridescent."[13] Coleridge rightly complained that "Dr Darwin laboured to make his style fine and gaudy, by accumulating and applying all the sonorous and handsome-looking words in our language" (*Shakespearean Criticism* 2:66). Poetry books were expensive during the Romantic period, and Darwin made sure that the public got their own glittering words worth. "No man . . . had a more imperial command of words," it was remarked in his obituary (Aiken, "Biographical Memoirs" 462). The nature that appears in Darwin's verse has undergone a similar transmutation. These are not common plants, but choice, expensive, curious ones worth owning. When they do finally make their appearance at the end of part 1, they are as much products of human invention and industry—as much a novelty—as anything else in the poem. In this sense, in its highly self-conscious manner, *The Botanic Garden* can be said to reflect on the freedoms that have arisen with the Industrial Revolution. The pleasures that it celebrates are those connected with the metamorphosis of nature into commodities.

Imperial Natures

Given the title of the poem, a contemporary reader of *The Economy of Vegetation* would not have been surprised to discover that the first major appearance of flowers in the poem occurs when the Goddess of Botany gives her concluding encomium to the Botanic Garden at Kew: "so sits enthron'd in vegetable pride / Imperial Kew by Thames's glittering side" (4.561–62). For Darwin, Kew is "imperial," not only because it owes its existence to British colonial expansion and to its concomitant control of global resources, but also because Kew represents an "empire of plants," an imperial nature—like Rome, Paris, or London—composed of plants that have come from across the globe. *The Economy of Vegetation* ends with the appearance of an Imperial Spring, of new plants coming into bloom at Kew, partly because of the season, but also because they have made the long journey across the seas to Richmond: "Obedient sails from realms unfurrow'd bring / For her the unnam'd progeny of spring" (4.563–64). Unknown plants, the "unnam'd progeny" of places whose natures have not yet been visited by the plow, are brought by "obedient sails" to Kew, where they are given a local habitation and a name—at least a scientific one.

Darwin describes Kew as a foster parent and foster home for these foreign natures or "stranger flowers." Those that need special accommodation are provided with glass houses that allow them to be fanned "with milder gales." He muses on the strange heterogeneity of this place, made possible by the combination of the global commercial aspirations of Britain, technology, and the botanical science—"the sons of science"—that supported it:

> Delighted Thames through tropic umbrage glides,
> And flowers antarctic, bending o'er his tides;
> Drinks the new tints, the sweets unknown inhales,
> And calls the sons of science to his vales.
> In one bright point admiring Nature eyes
> The fruits and foliage of discordant skies. (*Economy of Vegetation* 4.571–76)

Here Darwin is echoing earlier georgic poems that celebrate the Thames as a river whose "blessings," in the words of John Denham's "Cooper's Hill," are not "to his banks confin'd" (*Poetical Works* 63–89, line 179) but instead flow outward to all the "grateful shores" (line 182) that it reaches:

> [It] Finds wealth where 'tis, bestows it where it wants
> Cities in deserts, woods in Cities plants.
> So that to us no thing, no place is strange,
> While his fair bosom is the worlds exchange. (lines 185–88)

Denham's Thames "makes both *Indies* ours" (line 184); "to us . . . no place is strange" (line 187).[14] What is striking about Darwin's formulation is that this new nature now lies along its banks. A physical translation has literally taken place. Kew embodies this new diasporic nature (even an "admiring Nature" acknowledges this), and its landscape, which brings together "the fruits and foliage of discordant skies," embodies a *concordia discors*, or "happy discord." Now, however, this popular metaphor extends well beyond the nation, its conventional locus, to include the entire globe and all its biota, now centered in Britain.

Botanic gardens were not new to the eighteenth century, having originally been established during the early sixteenth century, first in Rome (1514) and then in Pisa (1544), Padua (1545), and Bologna (1567), for the study and teaching of *materia medica*. From there, they spread to other medical faculties in Leiden, Amsterdam, Montpellier, Oxford, Edinburgh, and London, notably the Chelsea Physic Garden (founded in 1673). In 1728, John Bartram would establish his famous garden near Philadelphia, which became a clearinghouse for the exchange of American native plants. By the late eighteenth century, with the establishment of a global network of botanic gardens in Cape Town (1652), Mauritius (1735), Calcutta (1787), Madras (1789), St. Helena (1788), Penang (in Malaysia) (1796), St. Vincent (1766), and Jamaica (1779), botanic gardens now played a key role in global plant transfer. These gardens directly served colonial needs, for they were the frontline sites of botanical exchange, the places where native plants were collected for exportation and where exotic plants were introduced and naturalized into colonial regions. Botanic gardens can thus be seen as the rest stops in the globalized routing of plants, the translation sites where native, indigenous, and endemic plants were accorded cosmopolitan natures or where newly introduced cosmopolitan plants were naturalized to new climates and conditions. Here modern natures, characterized by their mobility, were born. At the center of this mobilization of plants stood Kew, "a great botanical exchange house for the empire," as Richard Drayton describes it (*Nature's Government* 108).[15] Under the directorship of Banks, Kew brought botany, imperialism, and monarchy together in the shared task of making a new kind of nature, one whose geography was determined not by a divine plan, but by British imperial needs and desires:

> One tranquil hour the ROYAL PAIR steal;
> Through glades exotic pass with step sublime,
> Or mark the growths of Britain's happier clime;
> With beauty blossom'd, and with virtue blaz'd,
> Mark the fair Scions, that themselves have rais'd;
> Sweet blooms the Rose, the towering Oak expands,
> The Grace and Guard of Britain's golden lands. (*Economy of Vegetation* 4.580–86)

In touring their imperial garden, the "Royal Pair" tend the globe. Like the flowing Thames, they pass through "glades exotic" or visit the nature that they have raised. This is kingship grounded in gardening.

The Economy of Vegetation ends with the Goddess of Botany exhorting her nymphs to comb the globe and to bring the earth's choicest plants back to England to create this novel nature:

> Sylphs! who from realms of equatorial day
> To climes, that shudder in the polar ray,
> From zone to zone pursue on shifting wing,
> The bright perennial journey of the spring;
> Bring my rich Balms from Mecca's hallow'd glades,
> Sweet flowers, that glitter in Arabia's shades;
> Fruits, whose fair forms in bright succession glow
> Gilding the banks of Arno, or of Po;
> Each leaf, whose fragrant steam with ruby lip
> Gay China's nymphs from pictur'd vases sip;
> Each spicy rind, which sultry India boasts,
> Scenting the night-air round her breezy coasts;
> Roots, whose bold stems in bleak Siberia blow,
> And gem with many a tint the eternal snow;
> Barks, whose broad umbrage high in ether waves
> O'er Ande's steeps, and hides his golden caves. (4.591–606)

This is an extraordinary image of ecological imperialism. All that is valuable in the natures of Arabia, China, Siberia, Peru, and India, all spices and perfumes, all medical, ornamental, and food plants, are to be gathered and brought back to England (the new Florence or Padua) to usher in a new imperial nature, truly as much the "bright perennial journey" of plants as it is that of the sun. Darwin's vision includes plants as a central aspect of a world of mobile goods or commodities, and he extols the wealth that is to be gained from exotic natures. Always most in his element when he is promoting commerce, Darwin follows Keats's famous advice to Shelley to "load every rift" of his "subject with ore" (*Letters* 2:323): nature is transformed in this passage by its close association with wealth; it is decked out, "gem[med]," "rich," and "golden," suffused with "glitter" and "gilding." No wonder Henry Crabb Robinson complained of Darwin's "tinsel gawdy lines" (*Correspondence* 1:44). Also like Keats, Darwin can rarely talk about luxury and plants for long without also thinking about women. From the ceramics produced by "China" and his friend Josiah Wedgwood, he turns to the Chinese women who sip tea from them.[16] The association of china, luxury beverages, and women takes the reader back to Pope's *Rape of the Lock* and to Belinda, who also "bends her Head" over the "fragrant Steams" (in this case, of coffee) just before the "rape" (Poems 3:134). Also like Keats, Darwin focuses, almost fetishistically, on the erotic qualities of consumption—on the "ruby lips" of these women, which have become jewels through the Midas touch of luxury.

In representing Kew as a uniquely modern nature, constructed to serve human knowledge and for aesthetic appreciation, Darwin was influenced, no doubt, by Henry Jones's *Kew Garden* (1763), a poem based on Jones's visit to Kew in the early years of the decade in which it was first being recognized as a world-class garden. A specialist in patriotic panegyrics, Jones was an Irish bricklayer whose verses on Lord Chesterfield's appointment as Lord Lieutenant of Ireland brought him patronage and public attention.[17] *Kew Garden* is a loco-descriptive poem that uses a walk through the garden in order to inaugurate the new age, symbolized by the garden and brought into being by the accession of George III to the throne in 1760 and by the Treaty of Paris, which brought the Seven Years' War to an end in 1763. Great Britain had emerged from this global conflict as the world's leading colonial power, with new possessions in North America and control of India. "What new creation rises to my view?" (*Kew Garden* 5): the visit to Kew allows Jones to speak of the building of a new geopolitical reality grounded in a new nature. World, king, and garden are thus brought together in this "spot, where Britain's laurel springs/With stem renew'd" (5). Jones repeatedly stresses that the garden, like the society that it represents, is "coeval" (5) with the king himself. That Kew would come to stand for a new conception of what British nature might become through empire is somewhat ironic, for the garden was built in the most unlikely of places: on about 152 acres of marginal land adjacent to the Thames in Richmond. William Chambers, who designed many of its architectural features, observes that "originally the land was one continued dead flat: the soil was in general barren, and without either wood or water. With so many disadvantages it was not easy to produce anything even tolerable in gardening: but princely munificence, guided by a director, equally skilled in cultivating the earth, and in the politer arts, overcame all difficulties. What was once a Desart is now an

Eden" (*Plans* 2). Jones also speaks of Kew as a place "where niggard nature every boon denied" (*Kew Garden* 5), and he too emphasizes the changes introduced by the king. George III is a second Adam, who "cultivates the spot that gave him birth," reforming "the rude enormous sketch / To order, beauty, harmony and ease" (6).

What is striking about this georgic recovery of Eden in the flatlands of Surrey, this eighteenth-century Disney World with its faux gothic castle, Chinese pagoda, and other garden fakery, is that its nature was almost completely composed of costly exotics. The landscape was redesigned to include a lake and prospects, and, in an effort to reproduce an American wilderness, thousands of North American trees were planted, many of them transferred from the estate of the man Horace Walpole called "the treemonger of Whitton," the 3rd Duke of Argyll, who planted thousands of exotic trees on his estate in the 1740s. Despite the interest produced by this translation of America to England, it was the nine-acre "Exotic Garden," begun in 1759, that produced the most excitement. Lord Bute, who was instrumental in developing it, bragged to the governor of Georgia that "the Exotic Garden at Kew is by far the richest in Europe ... getting plants and seeds from every corner of the habitable world" (quoted in Desmond, *Kew* 42). In one of his letters to the American nurseryman John Bartram, Collinson called Kew "the Paradise of our world where all plants are found that money or interest can procure. [W]hen I am there, I am transported with the novelty & Variety & dont know which to admire first or most" (J. Bartram, *Correspondence* 674). "Every clime its richest growth sends here" (*Kew Garden* 10), writes Jones; Kew is a "treasure-house / Of all that East, and West, and South can yield" (19).

Like Darwin, Jones celebrates Kew's "multi-naturalism," for this is a place "where every stranger finds his native home, / And blooms as if beneath parental skies." As an Irishman who had become established in England through patronage, Jones was sensitive to the issue of cultural difference. His Kew is a nature primarily composed of strangers, of immigrants, of foreign plants that have traveled from far across the globe. In an extraordinary use of the fictional "as if," Jones claims that these deracinated plants, clearly seen as "strangers," nevertheless continue to thrive "as if" they were "beneath parental skies" (*Kew Garden* 19). Social relations within an emergent empire are thus reinterpreted using the language of horticulture. Kew Gardens stands in as a "fost'ring nurse," both a source of nurture and an "adoptive" parent, for these vegetable "orphans":

From Nature's genuine source awhile withdraw,
To visit Art in her laborious cells;
That *fost'ring nurse* that rears those *orphans* up,
From regions far remote beyond the burning line,
From Indian gardens, and from Eden's groves,
To Britain's cold *adopting* climate brought;
Nor there shall die, nor disappoint his hope,
Whose patriot heart and powerful hand are stretch'd
From pole to pole for happy Britain's good
Who brings these denizens of nature, health
And pleasure, home, and makes them flourish here. (12)

These plants "to Britain's fostering arms transfer'd" are symbols of the "patriot heart" and "naturalizing" or patriating power of "Farmer George," whose reach extends "from pole to pole" "for happy Britain's good" (15). It is worth noting that although the term "naturalization" was commonly used in botany and horticulture during the eighteenth century to refer to plants or animals that had succeeded in establishing themselves in new environments, this usage originated in sixteenth- and seventeenth-century civil law, where it referred, as it does today, to the process by which the rights or privileges of citizenship were accorded to foreign persons or immigrants, usually through an act of Parliament. The plants in Kew are truly immigrants that have undergone naturalization, having established their place in this new diasporic nature, at the disposition of the king. It is significant that Jones is here projecting his own hopes as a *patriotic* Irishman to establish his own *patria* and *patron* in a new British nation that includes outsiders. Thus, he extends the same welcome to these plants that he is seeking for himself: "Hail fragrant guests! each privilege enjoy,/That royal hospitality can give/Disclose your virtues, and your worth reveal" (13). The idea of an empire destabilizes simple concepts of nativity and nation. *Kew Garden* uses plants to articulate a new model of what it means to belong to a nation by making global plant transfer into an "as if" homecoming, a nature naturalized by the immigration and naturalization services provided by natural history in the service of a gardener king.[18] Just as the late eighteenth century had witnessed the emergence of a new nation composed of people of different ethnic origins and located in diverse parts of the globe (Britons, as these people would be called), George III—himself part of a line of kings that stretched back to Hanover—was also the sovereign force behind the appearance of a new imperial nature.

Both Jones and Darwin can be said to articulate a cosmopolitan ideal, yet the transnationals to which they refer are not people, but plants. Both construct a cosmopolitan nature that mirrors their understanding of Britain's relationship to the colonial world. The botanic garden embodies a new conception of a global community of plants no longer restricted to their places of origin by natural boundaries, a transplanted nature whose conditions of existence derive not from their being rooted in place but from their capacity to travel. Produced in an age of global migration, at a time when populations of plants, animals, and people were moving across the globe like never before, the botanic garden embodies a new idea of what nature was or what it might be, a new relationship between human beings and the natural world, and a new vision of the role that science and empire might play in the global reordering of nature. Both poets are fully cognizant that this novel nature could not exist without scientific knowledge and technology, the "learned systems" (*Kew Garden* 13) and "fost'ring art" (14) that made plant transfers possible. Jones speaks of the innovation of Smeatman's water pump (which raised water from the Thames for irrigation) and of the technological marvel of the orangery, which used stoves to infuse tender tropical plants with "rich prolific warmth ... gentle steams, that ooze, / Emitted through ten thousand million pores ... The principles of vegetating life" (13). Even the English sun is "envious" as it "reviews / A richer harvest than his beams can give" (14). The result is a strangely hybrid and mixed nature, in which "both Indies in their varied pride, / With Europe's paler progenies contend" (14). This new plant commonwealth presents "a rich variety, where order reigns" (7). In viewing the Chinese and Tartar pheasants and other exotic birds kept in the Chinese Aviary, Jones presents Kew as a nature that combines "a thousand seasons" into an "exotic harmony":

> And now from sight the soul has had her fill,
> With colour, motion, shape and life replete;
> A thousand seasons in them sounds I hear,
> Nature's whole concert pouring on my sense,
> Exotic harmony, Hesperian bands,
> With both the Indies mix'd, where all agree,
> Beneath an artificial hemisphere,
> By Taste's own hand extended far and wide,
> By royal hospitality uprear'd,

They bask in plenty, nor regret their own,
And thankful chaunt their kingly patron's praise. (23)

Translating to England the paradisiacal imagery once reserved for the Indies, Jones conceives of Kew as being geographically in neither the East nor the West, but instead "an artificial hemisphere" of "both the Indies mix'd," a place that has no one season, because it has brought together the biota of all climates: "A thousand seasons in them sounds I hear." These birds do not "regret" their cosmopolitan status, but "chaunt," as does the plebeian poet, "their kingly patron's praise." In Jones's poem, georgic goes global. Just when Britain had assumed the identity of a global imperial power, it already had its eulogist and an idea of a nature that suited its newfound status. Later in 1763, in Jones's *The Royal Vision: In an Ode to Peace*, the Lord of Hosts looks down on a vanquished Spain and France and declares to Britannia's guardian angel that her "glorious" people "on yonder orb below / That globe shall rule, my laurel there shall grow" (1).

Jones's cosmopolitan nature is one in which the world's plants and animals willingly become orphans so that they can find a suitable foster home in Britain. For the Irish Jones, these foreign plants were naturalized kindred whose sense of belonging was at the pleasure of the king.[19] By the 1770s, a cosmopolitan nature of this kind, where imperial wealth, influence, and power were on display in massive collections of rare and costly plants, was rapidly being displaced by the view, actively promulgated by Banks, that the true business of empire was not to collect and display natures but to trade in them. Despite the sovereign appeal of being engaged in a symbolic recovery of Eden, and despite the pleasure that an aristocrat might gain from offering friends and visitors a homegrown pineapple or banana, the primary objective of botanic gardens was now to manage the travel and distribution of plants across the globe. Kew was no longer the end point for their journeys; its function was now to serve as the clearinghouse and administrative center of a vast network of botanic gardens, which began to appear in the 1770s and 1780s, allowing Britain to organize and manage the global movement and exchange of plants.[20] Britain sought to administer their travels not only to England but elsewhere. The circulation of plants was to become the very heart and blood of the business of empire. In this context, the botanic garden was less important as a site of display than as a laboratory for remaking colonial natures and for substantially redrawing their boundaries. In order to transform living plants into commodities and to make it possible to redistribute local plants across the globe, Europeans needed to develop the technological means for allowing natures to travel. The botanic garden was this tool. It was an environmental collecting station, the ecological equivalent of Ellis Island, for the immigration and naturalization of plant aliens. As the natural equivalent of Etruria, the botanic garden was a factory for producing cosmopolitan natures, that is, natures that could travel globally, but whose movements would be managed to serve British imperial and commercial needs. Although the gardener king might continue to take an active interest in this activity, his garden as a personal possession was now largely replaced by its position within the globalized commerce in plants. In a very real sense, Darwin's answer to the nature-as-captivity narrative is free trade.

Conclusion

Darwin's commitment to a nature that would no longer be held captive to a place, but would be capable of being transplanted to new parts of the globe, a nature whose cosmopolitanism would be indebted to the governing hand of sovereignty, science, and commerce, was neither unique nor idiosyncratic. Instead, it was a major aspect of the unfolding imperial vision of Britain. It was the same vision that encouraged naval officers and botanists, from James Cook and Banks onward, to explore far-flung parts of the globe in search of useful or curious natures. It also sent Captain Bligh on his fateful breadfruit voyage of 1788. In the Caribbean, this bringing together of natures from elsewhere was being acted out wholesale. By the 1790s, most of the people and plants of Jamaica and Barbados had come from somewhere else. Other islands, as well as the littoral regions of continents, were undergoing similar changes, as traveling natures came to displace indigenous natures everywhere. This heterogeneous mixing of exotic and native natures was not, therefore, just a feature of botanic gardens, English gardens, and colonial plantations. *The Botanic Garden* embodies a vision of the capacity of human beings to change natural ecologies everywhere. The traditional notion of local, native, or endemic natures, whose plant and animal life had always stayed put, occupying the same spots that they had been originally created to occupy, was quickly becoming a thing of the past. "Native" natures were increasingly being displaced by "naturalized" ones, indigenes by immigrants, and the forces that were now producing local natures often lay outside of these localities. The hybrid nature of the botanic garden was a mirror of the commercial and cosmopolitan ideals of the Enlightenment. It was a nature that could be replicated, with appropriate modifications, in other parts of the globe. The ecological boundaries of the natural world were thus being redrawn and a new nature—an expression of "multi-naturalism"— produced in its stead.

The Botanic Garden is an important nature poem of the Romantic period because it presents in the clearest terms this new idea of a modern, cosmopolitan nature, whose character would be determined by science and the marketplace, not by origins. Translation is at the very heart of *The Loves of the Plants*, for the revolutionary transformative capacities of the botanic garden as an imperial technology are formally mirrored in the translational and the transformative capacities of the poem as a *camera obscura*. Mobilized plants are represented by moving pictures and by the poet's self-conscious recognition of the ways in which he is animating these plants poetically. Since Linnaean botanical classification was organized around the study of a plant's reproductive parts, Darwin's emphasis on the gender and sexuality of these plants was in keeping with the concerns of contemporary scientists even as it allowed him to appeal to an audience interested in the sexuality of plants (and their human counterparts). At the same time, by showing that a new kind of nature poetry was possible through the introduction of foreign plants to English literature, Darwin was adopting a stance toward translation that had already been powerfully articulated in Sir William Jones's argument that translation was an essential means by which the stale, conventional imagery of eighteenth-century British poetry might once more be revitalized. In *An Essay on the Poetry of the Eastern Nations*, Jones complains that English poetry has "subsisted too long on the perpetual repetition of the same images, and incessant allusions to the same fables." Since nothing, he believed, compares to the beauty of the "natural images" provided by the East, poetic translation was the means by which a new and vivid set of images might be imported and made available to British poets and readers. He writes that "a new and ample field would be opened for speculation; we should have a more extensive insight into the history of the human mind; we should be furnished with a new set of images and similitudes; and a number of excellent compositions would be brought to light, which future scholars might explain, and future poets might imitate" (*Collected Works* 10:993).

A similar idea of using translation to introduce a vital foreignness into culture is also inherent in Friedrich Schleiermacher's "On the Different

Methods of Translating" (1813). The challenge for the translator in moving the reader closer to the world of a foreign author, he argues, is to "render this feeling of being faced with something foreign to readers to whom he offers a translation in their mother tongue" (53). Schleiermacher did not understand this as a straightforward importation of the foreign, but instead as the "bending" of the language "to a foreign likeness," a process in which a language "should sound foreign in a quite specific way" (53–54). In attempting to explain the benefits of this kind of translation, Schleiermacher has recourse to the idea of plant exchange: "Just as it is perhaps only through the cultivation of foreign plant life that our soil has become richer and more fertile, and our climate more pleasing and milder, so too do we feel that our language, since our Nordic lassitude prevents us from exercising it sufficiently, can most vigorously flourish and develop its own strength only through extensive contact with the foreign" (52). Picking up on Jones's idea of the importance of a translational culture, Schleiermacher argues that the willingness of German culture to adopt foreign languages makes it ideally suited to do for culture what Kew Gardens was doing for global natures: through Germany's "esteem for the foreign and its own mediating nature," it is "destined to unite all the jewels of foreign science and art together with our own in our own tongue, forming, as it were, a great historical whole that will be preserved at the center and heart of Europe, so that now, with the help of our language, everyone will be able to enjoy all the beautiful things that the most different ages have given us as purely and perfectly as possible for one who is foreign to them" (62). This idea of a culture that translates the foreign in order to strengthen itself is contrasted with French culture, which Schleiermacher believes turns everything into itself. "Who would claim," he writes, "that anything has ever been translated, whether from an ancient or a Germanic tongue, into French!" (62). Certainly, Goethe agreed: "Just as the French adapt foreign words to their pronunciation, just so do they treat feelings, thoughts, even objects; for every foreign fruit they demand a counterfeit grown in their own soil" (*West-Easterly Divan* 36).

Such cosmopolitan visions of the translational movement of plants and poems should encourage us not to take for granted what nature writing was during the Romantic period. We should recognize that every writer's nature represented a position taken in a field of contested views about what nature was or should be. The commitment to a cosmopolitan nature remains a strong element in Romantic literature, particularly in the writing of Percy Shelley and, to a lesser extent, in the heavily consumerist natures of Keats and Leigh Hunt. It is not surprising, for instance, that Shelley, a poet who imagined prophecy as a wind blowing the "winged seeds" and "withered leaves" of "dead thoughts over the universe" ("Ode to the West Wind," *Poetry and Prose* 298–301, lines 7, 64, 63), also found pleasure in the idea of mixing English and exotic flowers at home. When Shelley visited Mont Blanc, he bought a packet of seeds at a squalid local natural history museum. "These I mean to colonize in my garden in England," he wrote. "They are companions which the Celandine—the classic Celandine, need not despise; they are as wild and more daring than he, and will tell him tales of things even as touching and sublime as the gaze of a vernal poet" (Shelley and Shelley, *History of a Six Weeks' Tour* 171). Darwin draws a similar comparison between words and plants in his comments on plagiarism. Individual words taken from other authors, he argues, are "lawful game, wild by nature, the property of all who can capture them," but the poet "must not therefore plunder" another writer's "cultivated fruit." Words gleaned from other poets are "like exotic plants, their mixture with the native ones, I hope, adds beauty to my Botanic Garden" (*Loves of the Plants* 132). Whereas Shelley's exotic plants promised to provide the stay-at-home "classic Celandine" with more "daring" stories of new worlds, Darwin's exotic introductions add beauty to both his garden and his poetry.

While Darwin was enthusiastic about the new possibilities opened up by the globalization of local natures, others felt threatened. For the conservative Thomas James Mathias, readers were being "absolutely *debauched* by such poetry as Dr. Darwin's," which marked "the decline of simplicity and true taste in this country" in favor of "the rage of mere novelty" (*Pursuits of Literature* 56). Fearing that the consumerist revolution was leading to a decline in "publick order, regulated government, and polished society" (ix), Mathias uses the distinction between exotic and native plants in order to denigrate poetry that he sees as being foreign to the national spirit of Britain: "I offer the poetry to those who are conversant with the strength, simplicity, and dignity of Dryden and Pope and them alone. . . . There are men, (and women too) who understand. But as to the lovers of exotick poetry, I refer them to the Botanick Garden of Dr. Darwin. My plants and flowers are produced and cherished by the natural invigorating influence of the common sun; I have not raised them by artificial heat" (xi). Here the anti-cosmopolitanism that was becoming increasingly dominant in the post-Burkean political life of the British is being applied to both plants and poems. George Walker also takes a satirical swipe at Darwin in his novel *The Vagabond*, when Doctor Alogos, "a man illuminated with the irradiating principles of the new philosophy" (2:4), distributes to the local townsfolk a pamphlet entitled "The Catechism of Nature," only to find himself on the receiving end of mob violence: "'Oh! curse it,' cried Doctor Alogos, 'they seem to have liberty enough; they are treading down my fine flower garden like an herd of swine: there goes all my exotic shrubs'" (2:98).

Anxiety about the appearance of new plant species in England was also beginning to emerge during this period (Garrard, "Absence of Azaleas" 148–55). In the 1790s, many botanists accepted that English nature would undergo important changes as a result of the naturalization of new plants. John Berkenhout, for instance, commenting in 1789 on the plants listed in his *Synopsis of the Natural History of Great-Britain and Ireland*, declares, "I have described 1600; which I believe is all that have hitherto been discovered growing spontaneously in this country. Some of them, certainly, are not aboriginally indigenous; the present generation however are natives, and propagate their species without cultivation; they are not indeed ancient Britons; but they are Britons nevertheless" (*Synopsis* ix). Sir James E. Smith, president of the Linnean Society, was similarly willing, in his *English Botany*, to grant full rights of citizenship to any new plants that had been naturalized in England; the scientist rather than the king now defined which plants could be considered as having been "naturalized." By the third edition of his *Botanical Arrangement of All the Vegetables Naturally Growing in Great Britain*, retitled as *An Arrangement of British Plants* (1796), however, Withering was already beginning to exclude naturalized plants because they were not indigenous. When William Hooker published the fourth edition of *The British Flora* (1838), naturalists were even more troubled by the problems posed by the inclusion of exotics in their floras. Hooker retained many of them out of respect for Smith, whose works he considered as being "the standard authority for such plants as were deemed indigenous to Britain *at the period of their publication*" (*British Flora* xi). Nevertheless, even though many of these plants could be found growing in "uncultivated ground," there were some that Hooker simply could not bring himself to include in his flora: "The *Martagon Lily* and the American Touch-Me-Not can have no claim to be considered British plants" (xi). Embedded within these debates about a British nature were cultural concerns about whether foreign plants (and by extension foreign peoples) could ever be fully naturalized. Debates over the identities of immigrants were thus being fought out as much in natural history as in British culture.

Whereas both the eighteenth and nineteenth centuries witnessed the extensive planting of foreign tree species on rural estates, for commercial and aesthetic purposes, counter-voices began to be heard during the Romantic period, most notably in Wordsworth's complaints about the planting of larches in the Lake District. Wordsworth's concerns were certainly not without merit. In 1787, for instance, the Bishop of Llandaff, Richard Watson, had arranged for Thomas Clark's nursery in Keswick to plant 85,500 larches at Wansfell, Ambleside, while in 1809, John Curwen, the owner of Belle Isle on Windermere, was awarded a gold medal from the Society of Arts for "having planted in a single year over a million larches and other forest trees" (Harvey, *Early Gardening Catalogues* 117). Extraordinary as these numbers might seem, they were dwarfed by the plantation ambitions of John Murray, the Duke of Atholl, who succeeded in planting over fourteen million larches over a fifty-year period beginning in 1774 (*Account of the Larch Plantations*). Although Wordsworth and his sister, Dorothy, were both avid gardeners, they held a nativist view of the Lake District as a place that was "rich with indigenous produce" (1805 *Prelude* 5:236). Theirs was a local or national nature whose value lay in its not traveling beyond the Lake District, except in the many books and poems that proclaimed its local beauty worldwide. In the Fenwick note to "Love lies bleeding," Wordsworth even goes so far as to argue that commercialism was destroying poetry in England. Citing how "touching & beautiful were in most instances the names" that his English countrymen had given "to our indigenous flowers or any other they were familiarly acquainted with," Wordsworth claims that England is now being inundated by "plants & flowers from all quarters of the globe many of which are spread thro' our gardens & some perhaps likely to be met with on the few commons which we have left. Will their botanical names ever be displaced by plain English appellations which will bring them home to our hearts by connection with our joys & sorrows? It can never be, unless society treads back her steps towards those simplicities which have been banished by the undue influence of Towns spreading & spreading in every direction" (*Poetical Works* 2:495). In the appearance of these plants and in the increasing use of botanical names to identify them, Wordsworth saw the encroachment of urban modernity upon the Lakes and a consequent breakdown of the long-standing relationship between its rural inhabitants and the natural life around them.

In her *Grasmere Journal*, Dorothy Wordsworth writes of frequent excursions into the wild to collect native plants for the garden at Dove Cottage: "I rambled on the hill above the house gathered wild thyme and took up roots of wild Columbine. Just as I was returning with my 'load,' Mr and Miss Simpson called" (*Journals* 24). She also gained other garden plants from friends, purchasing additional penny stock from blind Matthew Newton and obtaining shrubs from Curwen's nursery in Windermere. Dorothy had few qualms about collecting wild plants, because she and her brother sought to incorporate their garden into the native landscapes of the Lake District. Nevertheless, on one occasion, she notes coming upon "a strawberry blossom in a rock. The little slender flower had more courage than the green leaves, for *they* were but half expanded and half grown, but the blossom was spread full out. I uprooted it rashly, and I felt as if I had been committing an outrage, so I planted it again. It will have but a stormy life of it, but let it live if it can" (83). Having initially succumbed to her normal impulse to remove the strawberry plant to her garden for her own use, Dorothy feels an emotional reproof that this "uproot[ing]" breaks a fundamental moral law, that it constitutes an "outrage," and like the young Wordsworth, who guiltily returns the stolen boat, she returns the strawberry to its place, letting "it live if it can." Natures should stay in their place.

Early in her journal, having described the common flowers whose names she knew and many others that she did not, Dorothy wished "we had a book of botany" (*Journals* 16). The next spring, March 1801, she and William finally acquired a copy of the third edition of Withering, *An Arrangement of British Plants*, and on their walks in 1802, as H. J. Jackson observes, they brought these volumes with them "to record places and dates of their own sightings of the plants it describes" (*Romantic Readers* 78).[21] On 28 April 1802, Dorothy writes of a "pretty little waxy-looking Dial-like yellow flower . . . and some others whose names I do not yet know." Two days later, she would identify the plant as *Lysimachia nemorum* (yellow pimpernel) (*Journals* 128n). A similar discovery underlies William's composition of two poems, during the same week, on his favorite flower, the lesser celandine, or common pilewort (named for its traditional use in treating hemorrhoids): "There's a flower that shall be mine, / 'Tis the little Celandine" ("To the

Small Celandine," *Poetical Works* 2:142–44, lines 7–8). Other more showy flowers have "done as worldlings do" (line 54) and have "taken praise that should be thine" (line 55), Wordsworth declares, acclaiming a common flower that he has welcomed for more than thirty years as a harbinger of spring. Mostly an unnoticed plant, "a thing 'beneath our shoon'" ("To the Same Flower," *Poetical Works* 2:144–46, line 50), the lesser celandine is an English commoner, whose native pedigree the poet captures by adopting the archaic Scots ballad form of "shoes." Wordsworth celebrates both the plant and his discovery of it, likening himself to astronomers and explorers who look into the far reaches of space or travel afar in order to find the unknown. "Let the bold Discoverer thrid / In his bark the polar sea" (lines 51–52)—the greatest discoveries are of those things that lie at our feet. In the Fenwick note, Wordsworth commented that "it is remarkable that this flower, coming out so early in the Spring as it does, and so bright and beautiful, and in such profusion, should not have been noticed earlier in English verse" (*Poetical Works* 2:492). The poem is thus not simply about the discovery of a plant, but about its *literary discovery*, and Wordsworth is just as excited about finding this *literary nondescript* as a naturalist would be about coming upon an as-yet-undescribed plant. The poem is certainly a celebration of a plant, but it is also clearly an act of verbal description that incorporates the flower into a written literary tradition, which attaches Wordsworth's name firmly to it. The poem thus demonstrates that the same natural history practices, the same classificatory methods that were making it possible to archive colonial natures, were also making English literary nature visible in new ways. Nevertheless, the most exciting discovery memorialized in "To the Small Celandine" was not the plant, but its name: "I have seen thee, high and low, / Thirty years or more, and yet / 'Twas a face I did not know." The lesser celandine had been a stranger for three decades until Wordsworth finally learned its name—or, at least the Latinate one that he chose to use, since its local name, "pilewort," which he would also have known through the pages of Withering, did not exactly evoke the associations with poetry that he wanted. The poem thus depends on the science that it questions. After Wordsworth's death, the executors of his estate decided that a celandine should be carved on his memorial plaque at Grasmere's Church of Saint Oswald, thus reinforcing, in the same manner as scientific nomenclature often does, the relationship between this flower and the poet who discovered it for literature. Unfortunately, before carving the plaque, they did not consult a copy of Withering, and the greater celandine was used instead! Although in Wordsworth's mind and writings the lesser celandine is not a traveler, the actual plant has displayed far greater ambition. Now spread over twenty states in the northeastern United States and in parts of Canada, it has become one of many plant invaders that threaten the native floodplains flora in these regions. Even if Wordsworth equated this plant with native Englishness, the lesser celandine, perhaps overhearing the tales of "daring" told by Shelley's alpines to its greater namesake, has in turn uprooted itself and ventured to see the world.

CHAPTER TWO

Traveling Natures

Why ... not ... take up your roots and transplant?
(Anon., *Jesuite's Reasons Unreasonable* 20)

The King ought to send a Man of War a Botanist & Gardener for the Plants we want.
(Matthew Wallen to Sir Joseph Banks, 6 May 1785, Banks, *Scientific Correspondence* 3:53)

Colonial history is replete with stories of traveling natures. On 12 October 1806, after a five-month voyage from Calcutta, the East India Company ship *Fortitude* arrived in Trinidad with two living cargoes: a shipment of nutmeg trees to help diversify the island's plantation exports, and 192 indentured Chinese laborers. The historical significance of the latter has not escaped historians' notice, for these workers were the first Chinese to settle in the New World, the first from the Pacific Rim to be transplanted there.[1] That this new diaspora would first set down roots in the West Indies is hardly surprising, for the history of these islands has been inseparably bound up with a variety of forced and inadvertent forms of migration and settlement. Having destroyed most of the indigenous people during the sixteenth century and facing a climate that they believed was inimical to white people, Europeans transported more than eight million black slaves from Africa to the Americas and two million indentured servants as replacement labor between 1492 and 1820.[2] By the beginning of the nineteenth century, the plantation owners of the Caribbean, finding it increasingly difficult to obtain black slaves and anxious about the increased risk of slave rebellions, were looking for an alternate supply of low-cost labor. The planters of the newly ceded island of Trinidad, looking to Ceylon and Penang, where the British had already begun employing indentured Chinese to cultivate pepper, betel nut, nutmeg, and other spices, entered this new venture in the globalized transportation of human beings because Chinese laborers were inexpensive. In outlining the original plan, naval captain William Layman argued that for one-fifth the price of a Demerara slave the planters could import a people who would combine "indefatigable labor with particular skill in the husbandry and preparation of not only every article now produced in the West Indies, but of many other things that if cultivated in our colonies would be greatly beneficial to this country" (Layman, "Hints" 23). He particularly hoped that the Chinese would be used to introduce rice cultivation to the island. Further cost savings would be achieved by using convict ships on their return voyage from Australia. Layman thus envisioned a new maritime trade triangle that would supplement and perhaps supersede the Atlantic triangle, one in which two sides of the trade would be involved in transporting indentured or constrained labor to other parts of the world. Further transplantations would follow, notably indentured Indian laborers after 1860. The resulting *creolization* of races, ethnicities, languages, religions, and cultures, something new, born on the islands from the contact of people who traced their ancestry to different parts of the globe, has been seen as the very prototype of modernity.

The *Fortitude* voyage is an apt emblem of Caribbean history, because these Chinese laborers were also accompanied by nutmeg plants from Penang. The Caribbean islands were places shaped not only by the forced migration and mixing of peoples but also by the transplantation and mixing of plants and animals. Indeed, as Thomas D. Boswell remarks, the West Indies are "a largely imported environment inhabited by imported peoples" (Review of *The West Indies* 363). From 1492 to the beginning of the nineteenth century, the islands were the poster children of European colonialism, providing indisputable proof of the human capacity to alter or remake physical environments to suit human needs. It was here that plantation culture reached its apex, as these islands were seen as being places ideally suited for the agricultural production of natural commodities. The plantocracies that emerged here would celebrate the conquest of nature by human beings. Here too a culture of planters morphed into a culture of trans-planters. Gonzolo, the adviser to King Alonso in *The Tempest*, picks up the theme when he muses, "Had I a plantation of this isle . . . I would with such perfection govern, sir, / T'excel the Golden Age" (*Tempest* 2.1.144, *Norton Shakespeare* 168–69).

If you want to enjoy a paradise, it is not a bad idea to start with one. Columbus had never seen anything like the islands. In October 1492, writing from Cuba, he declared, "The banks of the rivers are embellished with lofty palm trees, whose shade gives a delicious freshness to the air; and the birds and the flowers are uncommon and beautiful. I was so delighted with the scene that I had almost come to the resolution of staying here for the remainder of my days; for believe me, Sire, these countries far surpass all the rest of the world in beauty and conveniency" (quoted in Parry and Sherlock, *Short History* 3). The word "conveniency" already anticipates their commodification. Responding to a landscape of deep forests and lush vegetation and "a loving people, without covetousness, and fit for anything," Columbus would declare that "there is not a better country nor a better people than these" (*Journal* 145). No sooner had he discovered this tropical paradise, however, than he set to work changing it. In the Caribbean, the relation between "natives," "nature," and territory was violently severed, as both the island peoples and their nature were cleared from the land to make way for newcomers.[3] The Arawaks and most of the Caribs, like the huge forests of mahogany, ebony, cottonwoods, and cedar, were razed, and new plants and peoples were forcibly introduced. As a result, few of the plants that now grow in the Caribbean are indigenous. The island environments are thus as much an expression of creolization as the people. In reference to Martinique, for instance, the biologist Clarissa Thérèse Kimber observes that

> interference in the physical and biological processes has been so effective that the term *domesticated* may properly be applied to most of the island landscapes. This domestication has involved changes in the constitution of vegetation by the introduction of exotic species and the selective culling of native species. It has caused changes in the distribution of native vegetation by disturbance, clearing, and burning, and the creation of entirely alien vegetation types. The result is a plant cover composed of a few traditional plantation crops with a cosmopolitan tropical weed population inserted among remnants and modifications of wild New World tropical vegetations. (*Martinique Revisited* 3)

Colonialism produced a new creolized nature, one in which plantation crops and cosmopolitan weeds were rooted in land once occupied by indigenous species. Caribbean history, Edouard Glissant argues, consists of "a polyphony of dramatic shocks," expressive of what he calls the region's "irruption into modernity" (*Caribbean Discourse* 106, 146). The landscapes of the

Caribbean, insofar as they are, in Wilson Harris's words, "saturated by traumas of conquest" (*Whole Armour* 8), are important documents of this history, even as their beauty conceals and erases the violence against indigenous peoples and indigenous natures which brought them into being.

"The way you think and feel about gardens and the things growing in them," writes Jamaica Kincaid, "depend[s] on where you come from" (*My Garden (Book):* 114). As DeLoughrey, Gosson, and Handley have observed, the history of slavery fundamentally shapes how blacks relate to the physical environment in the Caribbean. Whereas American ecocritics "often inscribe an idealized natural landscape that is devoid of human history and labor, the colonization and forced relocation of Caribbean subjects preclude that luxury and beg the question as to what might be considered a natural landscape" (DeLoughrey, Gosson, and Handley, "Introduction," *Caribbean Literature and the Environment* 2). Products of violence, exploitation, and suffering, the "domesticated" natures of the Caribbean islands can only be treated as "natural" by those who ignore the history that produced them. In *My Garden (Book):*, recounting her return to her native Antigua, Kincaid uses her botanical knowledge in order to explain why its landscapes do not and cannot speak to her like those of the Lake District did for Wordsworth. Botany allowed her to recognize that she had grown up in a nature that did not exist for her, but instead denied her very being and her history; she learned that she had grown up on "alien ground" ignorant of the names of the plants around her, which themselves constituted the lingering remnants of a British imperial past.[4]

There is no Romantic "return to nature" for Kincaid, because the nature in Antigua has no place for her. "What did the botanical life of Antigua consist of at the time . . . [Christopher Columbus] first saw it?" she asks.

> To see a garden in Antigua now will not supply a clue. I made a visit to Antigua this spring and most of the plants I saw there came from somewhere else. The bougainvillea . . . is a native to tropical South America: the plumbago is from Southern Africa; the croton (genus *Codiaeum*) is from Malay Peninsula; the *Hibiscus rosa-sinensis* is from Asia and the *Hibiscus schizopetalus* is from East Africa; the *allamanda* is from Brazil; the poinsettia . . . is from Mexico; the bird of paradise flower is from Southern Africa; the Bermuda lily is from Japan; the flamboyant tree is from Madagascar; the casuarina comes from Australia; the Norfolk Pine is from Norfolk Island; the tamarind tree is from Africa; the mango is from Asia. The breadfruit, that most Antiguan

(to me) and starchy food, the bane of every Antiguan child's palate, is from the East Indies.... It's as though the Antiguan child senses intuitively the part this food has played in the history of injustice and so will not eat it. But, unfortunately for her, it grows readily, bears fruit abundantly, and is impervious to drought. ("Alien Soil" 1017–18).

Kincaid realizes that the nature of Antigua exists by virtue of its having erased another nature that antedated it. Her relationship to the land is thus very different from Wordsworth's, for whereas the English poet's nature was the ground and anchor of his being, Kincaid's child seems to know already that the nature around her is not for her, and she spits it out. Instead of using botany to discover her connection to a place, Kincaid uses it to understand her ungroundedness. Kincaid is thus able to come to terms with a land that, despite its beauty, speaks to her of violence, loss, powerlessness, and dispossession. To love nature, you must first be able to afford to love it, Kincaid suggests. "The ignorance of the botany of the place I am from (and am of)" she writes, "really only reflects the fact that when I lived there, I was of the conquered class and living in a conquered place" (*My Garden (Book)*: 120). Instead of allowing Kincaid to engage in a Wordsworthian return to origins, botany allows her to understand critically why as a child she felt alienated from the nature around her. Nevertheless, she still associates nature with remembrance, particularly through the personal associations that she has with the Antigua botanic garden, especially the rubber tree "from Malaysia (or somewhere)" under which she and her father used to sit during a year when they were both sick, "he with heart disease, I with hookworms," and the bamboo grove where she used to rendezvous with lovers. For Kincaid, however, the botanical garden is a symbol not of an inclusive cosmopolitan nature, but of the power of those who established it, reducing her to insignificance: "The botanical garden reinforced for me how powerful were the people who had conquered me; they could bring to me the botany of the world they owned. It wouldn't at all surprise me to learn that in Malaysia (or somewhere) was a botanical garden with no plants native to that place" (120).

Eighteenth-century British settlers in the Caribbean also used their knowledge of the natural world in order to define themselves. Like Kincaid, they were just as aware that they were living in a transplanted nature; however, rather than producing feelings of alienation and disempowerment, this recognition gave rise to tremendous national pride in their mastery and control of tropical (and other) natures across the globe. In this regard, as Richard

Drayton remarks, "Botany had become the symbol of an 'improving' plantocracy" (*Nature's Government* 115). It provided the British with a way of seeing their power manifested in the landscapes around them. Shortly after arriving in Jamaica in 1801 with her husband, the newly appointed lieutenant governor of the island, Maria Skinner Nugent visited the home of the planter Lewis Cuthbert at Clifton, in the hills of Liguanea. Her excitement about the scenery is manifest, as is her satisfaction in her new role at the center of the political and cultural life of the colony: "the road beautiful and romantic, overhung with bamboos, and different picturesque trees and shrubs. . . . The palms and cotton trees on each side of it were quite majestic. It was all singularly beautiful, and my delight was increased upon arriving at Clifton (Mr. Cuthbert's seat), which is indeed indescribably lovely. The views from it are quite enchanting" (*Lady Nugent's Journal* 25). Here, Nugent employs what Elizabeth Bohls terms "the cultural power of the aesthetic to legitimize colonial plantation culture" (*Women Travel Writers* 47). The beauty of the landscape, expressive of the ability of the planters to select and organize a nature that fully meets their economic and aesthetic needs, conceals the violence that underpins this society. Nugent's use of the language of scenic tourism, with its emphasis on the capacity to move from one view of the landscape to another, enacts the privilege of mobility and of power. Anticipating the formal tour of the entire island which would take place between 5 March and 24 April 1802, this first visit to a neighboring estate is inherently a symbolic action of taking in, that is, possessing with her eye, the island that her husband now governs. Seeing is thus a means of asserting and enjoying one's control over this nature.

At Clifton, Nugent can hardly wait to leave the party to go to her room, "the better to enjoy the landscape, as from my windows it is enchanting indeed" (*Lady Nugent's Journal* 25). A somewhat lengthy prospect description follows, which merits some analysis:

> Imagine an immense amphitheatre of mountains, irregular in their shape and various in their verdure; some steep and rugged, others sloping gently, and presenting the thickest foliage, and the most varied tints of green, interspersed with the gardens of little settlements, some of which are tottering on the very brinks of precipices, others just peep out from the midst of cocoa-nut trees and bamboos, the latter looking really like large plumes of green feathers. The buildings are like little Chinese pavilions, and have a most picturesque effect. In front is a view of the sea, and the harbours of Kingston,

Port Royal, Port Henderson, &c. full of ships of war and vessels great and small; the whole affording an exceedingly busy and interesting scene. The plain, from the Liguanea mountains, covered with sugar estates, *penns*, negro settlements, &c. and then the city of Kingston, the town of Port Royal, all so new to an European eye, that it seemed like a paradise; and Clifton, where I stood, the centre of the blissful garden. . . . The garden contains a great variety of flowering shrubs and fruit trees, and the hedge round it is of lime trees, kept constantly cut, which makes it thick and bushy. The limes were ripe, and the yellow tint mixed with the bright green had a beautiful effect. Here and there the logwood was seen, which is something like our hawthorn. In other places are seen rows of orange trees, the fruit just turning yellow; mangoes, red and purple; forbidden and grape fruit, in clusters; the acqui, a tree that bears a large scarlet fruit, the inside of which, they say, when dressed is like a sweetbread; and the avocado pear, or real vegetable marrow, which poor Lord Hugh told me he ate for his breakfast on his toast, instead of butter. There were also pomegranates, shaddocks, &c. in abundance, and a tree, that looks like the cherry at a distance, but is redder and much larger. Coffee, too, is a very pretty shrub, bearing a bright red berry. Besides these, there are several trees from which perfumes are made, but I forget their names. One had a narrow very green leaf, and a very bright pink flower, which looks at a distance like a large full blown rose. Another tree has small dark green leaves, and tufts of scarlet flowers, something like the geranium. But it is quite impossible to describe the great variety of beautiful plants, trees and shrubs, that at this moment delight my eyes, and regale my nose. General N. and I spent the whole morning, looking about and admiring every thing, as far as the scorching sun would permit. (26)

Whereas male Romantic poets tend to seek out a hilltop prospect, Nugent looks out upon the world from her chamber window. Her eye takes in the surrounding scene, dwelling less on the wild tropical verdure of the mountainsides (a nature as yet uncolonized) than on the many signs of British presence in the landscape—"the gardens of little settlements," buildings, the harbor with its warships and trade vessels, the sugar plantations, the cattle pens, slave quarters, and the city of Kingston and town of Port Royal. Discordant elements are subsumed within the aesthetics of what Bohls elsewhere has called the "planter picturesque," the most notable elements being the buildings that look like "Chinese pavilions" peeking from behind plumed clumps of bamboo (*Slavery and the Politics of Place* 15–53). Yet to notice the *chinoiserie* in the landscape is also to see its fashionable novelty, the manner in which this place, like Kew Gardens, with its plants and architectural elements, brings together East and West by translating the Orient to the islands, creating a "west" Indies. This scene finds its economic and political center in the estate garden where Nugent, as the lieutenant governor's consort, now stands. For her, it is a recovery of paradise: "all so new to an European eye, that it seemed like a paradise; and Clifton, where I stood, the centre of the blissful garden."

Strikingly, much of Nugent's appreciation of this garden, of what she both sees and smells, comes from her knowledge that most of the plants that she is encountering have come from elsewhere. They are as much new *arrivants* settling in Jamaica as she is. Along with the bamboo, which Matthew Wallen brought to the island from the Philippines early in the 1770s, Nugent recognizes an abundance of introduced fruits—oranges, ripe limes ("the yellow tint mixed with the bright green had a beautiful effect"), forbidden fruit, mangoes, shaddocks, grapefruits, and pomegranates—along with logwood (introduced around 1715 for dye), coffee (which arrived in the 1720s), and ackee, only recently arrived from Africa.[5] Three days later, on a visit to the estate of Mr. Hutchinson, in Saint Andrews, she would marvel at its bamboo walk ("Every ten or twelve feet there is a cocoa-nut tree, as a pillar to support the feathering bamboo. Nothing could well be more beautiful"), its breadfruit trees ("here in great perfection"), and its jackfruit ("Like an enormous pumpkin, growing on the trunk") (28).

One of the highlights of her subsequent official tour of the island was her visit to the public botanic garden at Bath, Saint Thomas. Established in 1779 by the island's first appointed botanist, Dr. Thomas Clarke, but now under the supervision of Dr. Thomas Dancer, the garden was the means by which most of the plants that Nugent encountered had been introduced to Jamaica.[6] Here Nugent was fascinated by their sheer diversity:

> We ... were really much gratified, in seeing the variety of plants, shrubs, and trees, all so new to an European eye. The bread-fruit, cabbage tree, jack-fruit, cinnamon, &c. were in great perfection; as likewise were the sago [palm], and in short a number of beautiful shrubs I can't describe, and some of them as curious and extraordinary as they are beautiful.—The leaf of the star-apple tree is like gold on one side, and bright green on the other. Another tree, the name of which I can't recollect, was purple on one side, and also green on

the other. The Otaheite apple is a beautiful tree, bearing a bright pink blossom, like a tassel; but it is impossible for me to describe all the beautiful plants I saw (*Lady Nugent's Journal* 67–68).

A decade after arriving in the Caribbean via Captain Bligh's second voyage, the breadfruit had already become a common sight on plantations. The aesthetic order pictured by the eye thus replicates the physical order embodied by the landscape; seeing mirrors doing. For Wordsworth, one of the primary reasons why people enjoy seeing new and exotic plants in their gardens is that "the circumstance of their not being native, may, by their very looks, remind us that they owe their existence to our hands" (*Guide* 2:218). Jamaican landscapes produced the same pleasure on a grander scale, for these newly introduced plants, by their very existence, also remind Nugent that she belongs to a society whose power lies in its capacity to transplant and resettle people and natures across the globe. The transports of such landscapes are colonial and imperial as they speak of a nation whose claim to the future lies in its peoples' capacity to root themselves in transported natures. This is truly the naturalization of the social order in its profoundest sense. Nugent's experience of nature would suggest that European colonists learned how to control the natures with which they came into contact, often by replacing them. Increasingly over the course of the nineteenth century, they also read their destiny as imperial powers in their capacity to translate plants to new locations. Given her preoccupation with portable natures, it is not surprising that Nugent's eye ignored the wilderness around her, because her interest was not in preserving Jamaica's indigenous nature, but in envisioning the new imperial nature that was being ushered into existence there—Jamaica's future nature, not the indigenous nature associated with its past.

Britons during the eighteenth century increasingly viewed themselves as being a people whose fundamental identity lay in their capacity to mobilize people and things. Any number of nineteenth-century writers came to celebrate the idea of a stationary nature, grounded in place and removed from modernity, yet we should not overlook the degree to which British society at the time was also fundamentally engaged in defining Britain as the nation whose genius lay in its capacity for translating natures, people, ideas, and laws to new places. The genius of hybrid Britain, unlike that of England, was not rooted in place, like a *genius loci*, but instead resided in a *genius translatio*.

Transplanting Breadfruit

One of the most famous and well-documented stories of a traveling nature is that of William Bligh's two voyages to transfer breadfruit from Tahiti to the West Indian plantations of Saint Vincent and Jamaica. It was the subject of Thomas Gosse's *Transplanting of the Bread-Fruit-Trees from Otaheite*, a mezzotint produced on 1 September 1796. Bligh and a member of the Arreoy, representing two societies, Britain and Tahiti, are shown sharing a common goal as they direct the sailors and natives who together load the potted breadfruit into the ship's launch. The engraving is about the translation of an insular plant into a globalized commodity. It memorializes that moment when a distinctive part of Tahitian nature picked up its roots, packed its trunk, and crossed the beach, leaving its island nature and embarking on a journey to new parts of the globe. In the engraving cultural contact is centered on and understood in terms of the exchange of a plant that now can cross the seas. Two kinds of bounty, two kinds of providence, are also linked together. On one side, there is the rich vegetal exuberance of Tahiti, a tropical forest of palms, bananas, and, unmistakably, breadfruit loaded with produce. On the far right side lies a different kind of providence—HMS *Providence*, waiting to receive its living cargo. Two natures are thus ostensibly portrayed in the engraving: a local insular nature that up until this time has remained in place, and a new nature that can be traded, transported, and transplanted. By virtue of the journey of the *Providence*, other island natures would also be brought into this picture—Tasmania, Java, Saint Helena, Saint Vincent, Jamaica, and England. A maritime nation, which sees all natures as being eminently transportable across the seas, is juxtaposed against another island nature, which is going global.

Gosse stages the action on a beach, this being the space, remarks Greg Dening, where most colonial cultural contact in the Pacific took place. "Every islander has had to cross a beach to construct a new society," he writes. "Across those beaches every intrusive artifact, material and cultural, has had to pass. Every living thing on an island has been a traveler. Every species of tree, plant and animal on an island has crossed the beach." These were, indeed, places of cultural contact, but just as importantly, beaches were also the spaces where colonial natures met and mixed, where local natures encountered and/or became traveling natures, where island insularity was either affirmed or rejected. Dening reminds us that "crossing beaches is always dramatic. From land to sea and from sea to land is a long journey

Thomas Gosse, *Transplanting of the Bread-Fruit-Trees from Otaheite*, National Library of Australia, nla.pic-an6016209

and either way the voyager is left a foreigner and an outsider" (*Islands and Beaches* 31–32). This happened to the breadfruit. In Jamaica, it became something very different from what it had previously been. A transplanted plant is, in all senses of the word, a translated nature. Natures, like words, do not travel alone; these biotic translations also brought new histories and meanings to their new homes. Consequently, the new ecologies brought into being at this time by these transplants, largely forged of an assemblage of transnational populations inserted into indigenous environments, were inherently heterogeneous, as much riven by disjunctions and conflict as the colonial societies that brought them into being.

It took two tries for the breadfruit to reach Jamaica and Saint Vincent. The first boatload of seedlings traveled only as far as the Friendly Islands, where they found a watery grave, thrown overboard by Fletcher Christian and his fellow mutineers on the HMS *Bounty*. Bligh and the nineteen men who joined him on the ship's launch fared better. The story of the first voyage—of Bligh's tyrannical authority, of the tropical attractions of Tahiti and its people, and of his astonishing thirty-six-hundred-mile journey across almost uncharted seas—has overshadowed what is even more extraordinary about this expedition: the decision, made twice by the British government, David Mackay reminds us, to send a ship on a thirty-thousand-mile expedition just to transport a collection of plants from one part of the world to another.[7] Plants are rarely accorded a significant place in world history, yet the breadfruit, like the sugarcane, might easily claim to be as contradictorily Romantic as any of the period's charismatic personalities—Burke, Napoleon, or Byron. European primitivists idealized Tahiti as an arcadia, in which health and pleasure freely coexisted in happy simplicity. They represented the breadfruit as being as noble as the island people who cultivated it. This was the plant that was believed to have freed Tahitian society from the necessity of labor. Joseph Banks extolled its benefits: "In the article of food these happy people may almost be said to be exempt from the curse of our forefathers. Scarcely can it be said that they earn their bread with the sweat of their brow when their chiefest substance Breadfruit is procurd with no more trouble than that of climbing a tree and pulling it down" (*Endeavour Journal* 1:341). Bligh agreed, quoting Hawkesworth: "if a man plants ten of them in his life-time, which he may do in about an hour, he will as completely fulfil his duty to his own and future generations as the native of our less temperate climate can do by ploughing in the cold winter, and reaping in the summer's heat, as often as these seasons return" (*Voyage to the South Sea* 12).[8] This was, indeed, the food of paradise, and like the mutineers, who chose Tahiti over the navy, the plant constituted an "imaginary counterweight to the European society of consumption, manners, and crime" (Spary and White, "Food of Paradise" 75).[9]

Banks was the driving force behind George III's decision to organize the expedition. He saw it first and foremost as an opportunity to promote in a highly visible manner the public utility of natural history as a science and to display on a grand scale the capacity of the newly emerging field of botany to improve the condition of humankind. By increasing the productive powers of the earth through transplants, botanical science would benefit human beings across the globe. The spread of the breadfruit was thus intrinsically bound up with spreading the word about the value of European scientific Enlightenment. Transformed from a *local* into a *cosmopolitan* plant, breadfruit promoted a cosmopolitan ideal of nature as much as of society. Yet it is also the case that the immediate purpose of the transfer had little to do with Enlightenment cosmopolitanism, for it was to serve the needs of the Caribbean sugar economies. From the 1770s onward, West Indian planters, who were attempting to convert most of the land into cash crops, had been seeking ways to reduce the cost of food for their slaves and, as Bryan Edwards commented, to lessen "the dependence of the Sugar Islands on North America for food and necessaries" (*History* 1:xli). The American War of Independence and its aftermath only increased this concern, so they offered a prize and a substantial reward for anyone who could successfully bring the breadfruit to Jamaica. In 1784, the Jamaican planter Hinton East wrote to Banks that "the Acquisition of the best kind of the Bread Fruit woud be of infinite Importance to the West India Islands in affording exclusive of variety, a wholesome & pleasant Food to our Negroes, which w$^{d.}$ have this great Advantage over the Plantain Trees from whence our Slaves derive a great part of their Subsistence, that the former wou'd be raisd with infinitely less labor and not be subject to be destroyd by evry smart Gale of Wind as the latter are" (Banks, *Indian and Pacific Correspondence* 2:62). The noble fruit of Tahiti was thus never intended for universal consumption, nor did it express a desire to extend the ideal of Tahitian brotherhood to other parts of the world. Instead, it was to be the food of slaves, providing subsistence to a people who had themselves been transplanted to the islands to provide the intense labor required by the sugar trade. Less labor spent on growing slave food meant more time to produce sugar. Whereas in Tahiti breadfruit had made possible, as described in Byron's *The Island*, a society "Where all partake the earth without dispute, / And bread itself is gathered as a fruit" (1.213–14), "bak[ing] its unadulterated loaves / Without a furnace in unpurchased groves" (2.262–63), in Jamaica it was cheap food for a people who had been uprooted, turned into property, and denied the fruits of their labor. A symbol of Edenic freedom in Tahiti, the transplanted Jamaican breadfruit was the fruit of exploitation, "Pacific manna for Britain's empire" (Fulford, Lee, and Kitson, *Literature, Science and Exploration* 115), a plant embodying the absolute commodification of nature and of human beings. Perhaps not surprisingly, black slaves did not willingly take to the breadfruit, and throughout most of the nineteenth century, it was a crop primarily used to feed pigs—and a young girl growing up in Antigua in the 1950s, Jamaica Kincaid, would spit it out. The story of the global translation of this plant was thus interpreted in radically different ways.

Natural History and the "Insular Empire"

The transplantation of breadfruit from Tahiti to the West Indies was only part of a far more extensive vision of the role that Britain might play in the commercial globalization of natures. Whereas earlier natural history expeditions (Cook's, for instance) had engaged in the worldwide collection and scientific identification and description of plants, Bligh's voyage was different because its objective was to collect *living plants* along the route and to redistribute them in new locations. It was inherently an activity of mobilizing natures, which would lead to what Yvonne Baskin has called the "rearrangement of the earth's living heritage" (*Plague of Rats* 20). "The Object of all former voyages to the South Seas, undertaken by command of his present majesty," remarked Bligh, "has been the advancement of science, and the increase of knowledge. This voyage may be reckoned the first, the intention of which has been to derive benefit from these distant discoveries" (*Voyage to the South Sea* 5). The Admiralty's instructions for the first voyage stressed that after leaving Tahiti Bligh was to visit Java in order to replace "any breadfruit trees which may have been injured, or have died" with "mangosteens, duriens, jacks, nancas . . . and other fine fruit trees of that quarter, as well as the rice plant which grows upon dry land" (7). Banks confirmed this objective in a letter written to a member of the Jamaican Assembly which appeared in the *Royal Gazette* on 25 July 1791. By this time, his plans for introducing the globalized control and administration of nature were more concrete, detailed, and expansive. En route, Banks writes, Bligh was to stop at French-controlled Mauritius in order to procure their spices. It had, indeed, been the French, through the enterprising spice-smuggling skill of its chief administrator Peter Poivre, of "picked-a-peck-of-pickled-peppers" fame, who had shown the English the commercial opportunities to be had by getting into the business of colonial transplantations. Banks indicates that Bligh "will have orders to procure *all the fruits and useful plants of the East*, wherever he may touch; so that the cargo will be far more valuable than a cargo of Bread-fruit trees alone" ("Extract of a Letter"; my emphasis). Banks is here envisioning the wholesale transfer of the exotic fruits, vegetables, medicines, and woods of the East to the West Indian isles under British control, and he declares that he cannot imagine "an undertaking really replete with more benevolence . . . than that of transporting useful vegetables from one part of the earth to another where they do not exist." Sugar and coffee had already traveled from "the East to the West," he noted, and "that all the remaining valuables of the East may follow them is my ardent wish, as they will all equally succeed under a tropical climate." That the geographical difference between East and West could be dissolved, that each could be translated into the other, was further indicated by the fact that other plants, such as custard apples, cashews, papaya, and pineapple, had gone the other direction. The only obstacle was how to transfer this nature, for Banks was sure that "if the plants once arrive" they would thrive ("Extract of a Letter").

Such views were shared by West Indian planters, who saw the islands as industrial gardens to be planted with useful or economically valuable plants. In a letter to Hinton East (which was then forwarded to Banks in 1785), Matthew Wallen, who introduced at least 115 plants into Jamaica (among these *Cannabis sativa*, from India, watercress, chickweed, wild pansy, violets, and the English oak), provided an alphabetical list (he stopped at "N") of sixty-eight plants, many of them ornamentals, which he hoped to obtain for East's garden. The letter proceeds as if all the world were at East's disposal, urging him to obtain valuable fruits and spices, domestic animals, birds, and rabbits:

> When you see any of your Brother Planters, I think it would not be amiss to spur them on to get the Bread Fruit as well as other Fruits, Roots, & Curious Plants from Otaheite as the new Zealand hemp. The Mangoustan, Nutmeg, Clove, Jack, Teak & walking Cane from the E. Indies. The Cows with large Tails, the Sheep with the Wool of which the Shawls are made from Thibet. The Broad-tail'd Sheep very different from the Barbary & much finer, the Angora Goat of the Hair or Wool of which the finest Camblets are made from Smirna. They would make all our dry rocky S. Side Hills the most valuable Part of the Island. The large domesticated red legg'd Partridge from the Island of Scio, or Chios, would be an Acquisition. Blackbirds, Thrushes, Larks, Goldfinches, Linnets, Nightingales & Canary Birds to turn out at your Mountain. Hares too, I would bring out for the same Purpose. Bring out 4 or 5 Alderny Cows for yourself. (Matthew Wallen to Joseph Banks, 1785, Banks, *Scientific Correspondence* 3:2)

In thinking about Bligh's voyage, then, we need to widen our perspective to see it as the expression of a fundamentally new attitude toward the natural world, one that saw nature as being eminently transportable and transplantable. Against the traditional view that the geographical distribution of plants expressed a divine purpose, the majority of colonial naturalists believed that the designed earth, structured by natural boundaries, could be redesigned by human ingenuity. The idea of global transplantation was first suggested by Thomas Sprat, when he urged fellow members of the Royal Society to experiment with "*Transplanting*, and *Communicating* of the several *Natural Commodities* of all *Nations*, to other *Airs*, and other *Soils*." Sprat clearly had the early successes of the West Indian plantation economies in mind, for he argues for "transplanting the Eastern spices and other useful vegetables into our Western Plantations" (*History of the Royal-Society* 385). The success of the West Indian plantation economies demonstrated that such a goal could be achieved, at least selectively.

With the rise of Linnaean botany in the 1730s, the notion of redistributing the earth's natural productions to benefit European nations seemed a fully realizable goal. In 1746, as Linnaeus began developing the idea of "economic botany," he explained to the Swedish Academy of Science that although "nature has arranged itself in such a way that each country produces something especially useful; the task of economics is to collect [plants] from other places and cultivate [at home] such things that don't want to grow here but can" (quoted in Koerner, *Linnaeus* 2). Although Sweden lacked an overseas empire, he hoped that botany would enable it to reduce its reliance on others for natural resources by introducing exotic crops and animals—cotton, tea, mulberries for silk production, merino sheep, angora goats, and American bison—into Lapland. As one Swedish noble caught up in the enthusiasm of the moment put it, "In Lapland, we have our West Indies" (quoted in Koerner, "Linnaeus' Floral Transplantation" 159). If "the Public" were "to plant a little garden up there [in Lapland], and to entrust it to the care of a skilled person, one who understands how to tend these foreign plants" (127), these floral newcomers would soon become self-seeding ferals that would invade the tundra. Linnaean botanists would be the agents of a wholesale transformation of the ecology of Lapland by colonizing its indigenous nature with one that would better suit Sweden's material needs. "Scandinavia's tundra," in Lisbet Koerner's words, "would flourish with tea plantations, saffron meadows, and cedar forests, all tended by nomads turned farm laborers" (*Linnaeus* 139).[10] In keeping with this grandiose Laputan plan, Linnaeus sent his botanical disciples on far-flung expeditions to gather knowledge and importable plants from across the globe. To his great disappointment, however, he learned that he had been far too sanguine about the capacity of plants to adapt to a subarctic environment.

Nevertheless, the idea of global plant translation increasingly gained traction. In 1769, John Ellis, a naturalist and merchant in the linen trade and royal agent for West Florida, published in the *Transactions of the American Philosophical Society* a list of eighty-two plant species that might usefully be cultivated in North America. A year later, he provided much-needed instructions on how to transport living plants on ships. "Many valuable trees and plants yet unknown to us, grow in the Northern Provinces of China," he writes, "which would thrive well in North-America. . . . But as the distance is great, the manner of preserving the seeds properly, so as to keep them in a state of vegetation, is an affair of considerable consequence" (*Directions* 2).[11] European countries with overseas possessions, such as England and France, recognized that rather than attempting to grow tropical plants at home, they could transplant them to the places that they controlled, which were more suitable for their cultivation. Natural history was thus reconceptualized as a science intrinsically concerned with learning how to make natures travel, its goal being to refashion the natural world in terms of the commercial production and exchange of natural products. Although colonizing Lapland with exotic feral plants proved to be an outright failure, for a person like Joseph Banks, the potential for translating the natures of the East and West Indies and of Africa had no limits. As director of Kew Gardens and as the president of the Royal Society, he placed the globalization of natures at the heart of the imperial enterprise. Henri Lefebvre has commented that "a revolution that does not produce a new space has not realized its full potential" (*Production of Space* 54). For late eighteenth-century British naturalists, this revolutionary new space was nature itself. When, for instance, Banks's protégé Mungo Park considered the interior of Africa, he envisioned a "whole catalogue of exportable commodities" that might be derived from it, but then added that "all the rich and valuable productions, both of the East and West Indies, might easily be naturalized, and brought to the utmost perfection, in the tropical parts of this immense continent" (*Travels* 272). Bligh's ship, HMS *Providence*, was thus aptly named, and Bligh loved his symbolic place in this new scheme of things. He wrote to Banks, "I hope that Great Providence which has hitherto protected me will send your little Providence back to you with success in due time" (24 and 26 Nov. 1791, *Papers of Sir Joseph Banks*, sec. 9, ser. 50.16).

Rev. John Walker, who outmaneuvered William Smellie in 1779 to become the first Regius Professor of Natural History at Edinburgh, provides a powerful statement of this new imperial vision of plant transfer in his *Essay... of the Translation of Plants from the East to the West Indies*.[12] Walker was a knowledgeable botanist, closely connected with the international Linnaean circle and with the British natural history correspondence networks established during the 1760s and 1770s by Lord Bute, Pennant, Banks, and the botanist Richard Pulteney. He too believed in the utility of natural history. "My leading idea in Natural History is to render it subservient to the Purposes of Life," he commented (quoted in Withers, "Neglected Scottish Agriculturalist" 135). This close connection between economics and natural history led to his being appointed in 1764 (and then again in 1771) by the Scottish Commissioners of Annexed Estates to do a natural history survey of the Hebrides and the Highlands, his instructions being "to examine the natural histories of these countries, their population, and state of their agriculture, manufactures and fisheries" (J. Walker, *Report of the Hebrides* 6). The goal of this kind of natural history, in which the description and cataloging of regional flora and fauna were part of an overall assessment of a region's human and natural resources for the purposes of economic management and improvement, would be outlined in his 1788 lecture "On the Utility and Progress of Natural History, and Manner of Philosophizing": "When the productions of a country have been scientifically examined, when their species are investigated, and the natural order known to which they belong, opportunities must then occur to the naturalist, of discovering in them many useful properties; and that, not by accident, which is usually the case, but upon principles" (*Essays on Natural History* 328). It was during this survey that Walker conceived the idea of producing a complete natural history of Scotland, which, if it had been completed, "would have placed its author among the first botanists of the eighteenth century" (quoted in Taylor, "John Walker" 178). Unfortunately, the project suffered delays, and Walker finally shelved it when John Lightfoot's book on the subject appeared in 1777.

Probably around this time, or shortly thereafter, his botanical interests took on the more imperial focus expressed in the *Essay... of the Translation of Plants from the East to the West Indies*, which passed through the hands of John Dalrymple and Lord Bute, ending up at Banks's house in Soho Square. The manuscript, which provides both a list of plants and botanical advice on the transplantation of tropical plants from the East to

Jamaica, was originally conceived as the first installment in a much larger project that would identify and provide information on all the plants across the globe which could be valuably transplanted to areas under British control. Walker presents the reasoning behind the essay as a syllogism:

> That as the Territories now belonging to Britain possess every Climate from the Line to the Pole; they are thereby enabled to produce in perfection, every Plant which inhabits our Globe: And,
>
> As the chief part of the British Imports from other countries consists of vegetable Substances, which the British Dominions can be made to afford, in equal perfection:
>
> It is therefore an Object worthy the attention, not only of Individuals, but of the Public, to translate to the Territories of Britain, the Plants, from which those vegetable Articles of commerce are derived, which we import from Foreign Nations. (*Essay . . . of the Translation* 1)

Like Banks, Walker sees plant transference as a primary state concern and botany as the science focused on the commerce in plants. Preliminary to achieving this goal, however, it was necessary "to form accurate Catalogues of the Plants, thus to be translated, containing their Names and Botanical History, the Places where they are to be found, and the Countries to which they ought to be transported." Walker spoke for many when he suggested that "this is . . . the most important purpose, to which Botany can be applied; and few Sciences can promise any discovery or improvement of more public Emolument" (1).

Underlying Walker's and Banks's understanding of the economic importance of botany is the idea that commerce could serve as the remedy for the unequal distribution of natural products across the earth. "As every climate has its peculiar produce," writes Gilbert White, "our natural wants bring on a mutual intercourse; so that by means of trade each distant part is supplied with the growth of every latitude" (*Natural History of Selborne* 195). For the parson of Selborne, the inequities in the apportionment of natural products across the earth showed the wisdom of the Creator, who had organized the earth to encourage global trade, and from it mutual enlightenment and shared material benefits. Globalization and the exchange of natures were thus part of God's plan. For this process to work, however, it was necessary to know what kinds of natural commodities were available, where they were to be found across the globe, and the uses to which they could be put. For the American William Stork, whose natural history of Florida sought to promote European settlement of the region, its capacity to manage the global movement of plants gave botany a particularly important role: "Here therefore is a field in which the naturalist may make his science peculiarly useful. His knowledge extending throughout the vegetable world, informs him where every valuable plant, grain, or tree is to be found, and also in what country it is wanting, and may be propagated to advantage" (*Description of East-Florida* 2:2).

Behind this vision of the planetary management and mobilization of natures lay an ecological experiment that for more than a century had been playing out in the plantation economies of the Caribbean. In 1695, a similar, if more brutally economic, view was expressed by the eminent Royal Society member John Houghton when he argued that "in order to improve this *West-Indian* Trade, I believe it would be well worth while to have it some body's Business to make a good Natural History as well as can be, and to study how every thing there may be improved, and what useful known matters grow in other Countries, that in Probability might grow here, and also to settle the *Guinea-Trade* for *Blacks*, which are the usefullest Merchandize can be carried thither, except *White-Men*: for according to their plenty is the product" (*Collection for Improvement* 1:457). For Houghton, every element of West Indian production and trade, from African slaves to plants, was a commodity, a form of "Merchandize." The rise of the plantation economies in the Caribbean expressed a new conception of global natures and of the role that natural history might play in furthering the commercial objectives of England. Instead of just exploiting the local natural resources of these regions, Europeans had discovered that enormous profits could be made by transplanting a race of people from one part of the globe to another in order to grow plants that were themselves also often imports from other places. Colonization was like gardening, and the earth provided places where it could be done cheaply and profitably: all that was needed was to control and manage profitably the transportation of people and plants.

Islands played a pivotal role in this new economy of globalized transplantation. Throughout history islands have been seen as embodiments of isolation and inwardness, an association clearly registered by the fact that the word "insularity" is derived from the Latin *insula*, or "island." Islandedness has long played a prominent role in English exceptionalist self-understanding, a view most powerfully expressed in John of Gaunt's famous celebration of England's insularity in *Richard II*:

> This royal throne of kings, this sceptred isle.
> This earth of majesty, this seat of Mars,
> This other Eden, demi-paradise,
> This fortress built by nature for herself,
> Against infection and the hand of war,
> This happy breed of men, this little world,
> This precious stone set in the silver sea,
> Which serves it in the office of a wall,
> Or as a moat defensive to a house,
> Against the envy of less happier lands.
> This blessed plot, this earth, this realme, this England. (2.1.40–50)

Here Gaunt brings together an arsenal of symbols that have shaped England's cultural imaginary: the idea that England is an island kingdom (which is actually not the case), that it is an "other Eden," a "little world," a "precious stone set in a silver sea," protected from the "envie" of its "less happier" continental neighbors. Against this isolationist model of Englishness, however, should be set a contrary view, which rose in tandem with England's increasing naval supremacy, of seeing England as a maritime nation. In the opening lines of the *History of England*, Thomas Macaulay saw the beginnings of this new national identity in the thirteenth century: "Then it was that the great English people was formed, that the national character began to exhibit those peculiarities which it has ever since retained, and that our fathers became emphatically islanders, islanders not merely in geographical position, but in their politics, their feelings, and their manners" (1:13).[13] By the latter part of the eighteenth century, and particularly after the unprecedented success of the Seven Years' War, the British increasingly saw their insularity not as what separated them from the world, but as a distinctive geopolitical advantage they held over their continental neighbors. The British Empire was to become an "insular empire." Through its mastery of the seas, Great Britain would become an island that ruled a global network of islands joined by commerce, ships, and the sea, that "great high road of communication to the different nations of the earth," as Adam Smith remarked (*Theory of Moral Sentiments* 349). Who needed roads, when all seas led to London?

Britain's understanding of empire not as the acquisition of large landmasses but instead as the strategic control of islands across the globe was particularly suited to a period when shipping was the primary means of transportation, communication, and commerce. "In the era of commercial capitalism," observes John R. Gillis, "coasts and islands were the core and the continents the periphery of geographical transfers of capital, people, and knowledge" ("Taking History Offshore" 29). Because of their size, boundedness, and accessibility, islands could be more easily controlled and exploited by a maritime power. During the Napoleonic Wars, the idea of an insular empire took on geopolitical importance, particularly in the Mediterranean, where Gould Francis Leckie, in *An Historical Survey, of the Foreign Affairs of Great Britain* (1810), argued that "an empire of the sea will always balance that of the land" (vi). He believed that "a system of insular aggrandizement" (202) was of paramount strategic importance if Britain was to counter the continental power of Napoleon. "We must therefore sometimes conquer; and, if we are excluded for a time from the continent of Europe form ourselves an insular empire, complete in its parts, and sufficient to itself" (8), he argued. Although Leckie emphasizes the geopolitical advantages of controlling the islands in the Mediterranean, his views developed out of a decidedly more far-reaching contemporary vision of insular colonization, which saw islands as the ideal sites of economic exploitation. Particularly through the agency of Banks, the British were increasingly focused on using their maritime supremacy to acquire and integrate islands and littorals within a broader global network. Through its maritime control of India, Britain was able to establish "the strategic centre of a vast commercial network that included the Straits Settlements, the Indonesian archipelago, China, and Australia on the one hand, and on the other, Persia, Arabia, Egypt, and East Africa... the Indian Ocean had become... for all practical purposes... a 'British lake'" (Graham, *Tides of Empire* 45–47). In the new globalized maritime economy, islands were to serve as the sites of new plantation economies, as naval bases, trading posts, and shipping depots. They would be the primary distributive points in the circulation of goods, people, and natures.

Gosse's engraving commemorates the integration of another island into the insular empire, marking the moment when Tahitian nature went global, losing its status as a local or islanded ecology to become a stopping point in an emerging globalized trade in natures. The physical ecology of present-day Tahiti has changed substantially from the nature that first greeted Samuel Wallis when he arrived on the island in 1767. Although the Polynesians had introduced about thirty domestic and fifty accidental species to the island, the island at that time is estimated to have had about seven hundred indigenous species of plants, of which roughly 60 percent were endemic to French Polynesia (Whittaker and Fernández-Palacios, *Island Biogeography* 318). Most of the island's nature, in other words, existed nowhere else in the world. Currently, with 467 native flora, that is, having lost one-third of its indigenous species, and having gained more than fifteen hundred exotic species, Tahiti is essentially a cosmopolitan nature, with three times as many immigrants as native plants. Of these introduced species, more than two hundred "have established themselves beyond control." So much has changed, Stanley L. Welsh observes, that it is "difficult to know at present which of the plants are indigenous and which are alien" (*Flora Societensis* 2). By the same token, certain plants that were once endemic to Tahiti and the Polynesian islands have now spread across the globe.

Bligh's Voyages

Both of Bligh's voyages required extensive planning and organization. Each included two botanist-gardeners with responsibility for procuring and preserving the plants during the voyage. Banks provided them with extensive instructions on what plants to collect and how to care for them while at sea. Following the tenets of Baconian science, he emphasized the collection of "useful" plants, those that could be used for food or other commercial purposes. However, Banks also encouraged them to collect any "particularly beautiful or curious" plants, which were to be brought back to Kew, "provided however that the stock of breadfruit trees and useful Plants is never diminished by the admission of curious ones" (Banks to Wiles and Smith, 25 June 1791, *Papers of Sir Joseph Banks*, sec. 9, ser. 49.09). Arrangements were also made for the collection and reception of plants at various points on the journey. Here Banks drew on a newly established globalized network of botanic gardens, at Cape Town, Fiji, Saint Helena, Saint Vincent, and Jamaica, which he had been instrumental in developing over the previous decades.

Bank's 15 May 1787 letter to Sir George Yonge, secretary for war, concerning the proposal by Colonel Robert Kyd to establish a botanic garden at Calcutta indicates the important role that botanic gardens played in his plans for the global transplantation and exchange of plants. "Ceres was deified for introducing wheat among a barbarous people," he writes. "Surely then the natives of the two great Continents who will in the prosecution of this Excellent work, mutualy receive From each other numerous products of the

Earth as valuable as wheat will look up with veneration to the Monarch who protected & the ministers who carried into execution a plan the benefits of which are above appreciation to the present generation & will extend their beneficent influence to the Latest posterity of those who receive them" (*Indian and Pacific Correspondence* 2:190). With spice plantations on the Isle de France capable of supplying "before the end of the present Century . . . as large a quantity of these Spices as the whole Consumption of Europe will demand," it was a matter of some strategic importance that Britain "retreive the advantage which our active neighbors have obtaind over us in point of time" (191). Some might think, Banks observes, that these plans were not motivated by "general benevolence," but instead were a "mere attempt of Filching From another country its commercial advantages" (192). However, Banks stresses the "nobler prospects" of Enlightenment universal improvement:

> The Exchange between the East & the west Indies the productions of nature usefull for the support of mankind that are at present confind to one or the other of them to increase by adding [to] this variety the real Quantity of the produce of both Countrys, & by this means their population Furnishing to the inhabitants new resources against the dreadfull effects of Hurricanes & Droughts to one or the other of which all intertropical Countries are subject are the more immediate objects of his majesties present intentions the disinterested humanity of which seems alone sufficient to inspire diligence & activity into the minds of all those who may be Fortunate enough to be allowd a Share in the honor necessarily consequent in having carried Ideas of such Exalted benevolence into execution. (192)

Much work remained to be done by parties in both countries "to know with accuracy the things each wants & the other posesses" (192), but Banks provided a preliminary list of eleven plants that he hoped might be obtained from Bengal, along with thirteen others that he thought might usefully be introduced from the West Indies.

The *Bounty* and the *Providence* can be seen as technological advances in the European management of nature, for both ships underwent substantial remodeling to allow them to serve as mobile botanic gardens or floating islands. The *Bounty*, a 215-ton West Indiaman, was designed to carry 533 plant pots, the majority being six inches in diameter, with some of eight or eight and a half inches (Bligh, *Voyage to the South Sea* 1). Bligh described its refitting thus: "The great cabin was appropriated for the preservation of the plants and extended as far forward as the after hatchway. It had two large skylights, and on each side three scuttles for air, and was fitted with a false floor cut full of holes to contain the garden-pots in which the plants were to be brought home. The deck was covered with lead, and at the foremost corners of the cabin were fixed pipes to carry off the water that drained from the plants into tubs placed below to save it for future use" (Mackaness, *Life* 1:52). The 420-ton *Providence*, almost twice the size of the *Bounty*, was accompanied by a 110-ton tender ship, the *Assistant*, reflecting the much grander plans for plant exchange and transfer of the second voyage. The garden on this ship contained more than nine hundred pots, and these were of a larger size (640 nine-inch pots and 266 seven-inch ones).

On both ships, the comfort of the officers and sailors was entirely subordinated to the requirements of the plants. Matthew Flinders, who would later command the *Investigator* in its exploratory circumnavigation of Australia, was a young midshipman on the second voyage. He recorded that water was frequently in such short supply that "he and others would lie on the steps and lick the drops of the precious liquid from the buckets as they were conveyed by the gardener to the plants" (cited in Mackaness, *Life* 316). Since the plants on both vessels were understood as being the property of King George III, no additional plants were allowed on board.

The second voyage provides a clear picture of the procedures of globalized transplantation. The two ships left Spithead on 3 August 1791 already stocked with pineapples (the plants being sold by London nurseries for a price somewhere between 2 s. 6 d. and 4 s. each), nectarines, and a variety of seeds that were to be distributed at various points on the journey.[14] Bligh was allocated £500 for the purchase of plants, and he also brought an assortment of goods—wearing apparel, iron, beads, and trinkets—to be used for trade (Banks, "Memorandum"). At Cape Town, he picked up about 240 additional plants, including figs, pomegranates, quinces, and grape vines. In addition to providing the botanist-gardeners with important experience in tending plants at sea, these would later be distributed to various islands in the South Pacific. In exchange, Bligh left two nectarine trees, which he later remarked "were the only ones ever in that country."[15] On 9 February 1792, Bligh anchored in Adventure Bay in Tasmania to obtain wood and water, to collect plants, and to check on the status of the trees and vegetables that he had planted there four years earlier on the first voyage. He knew the place well, for he had first visited it in 1777, as sailing master of the *Resolution* on Captain Cook's third voyage. The plants had not fared well. Of the

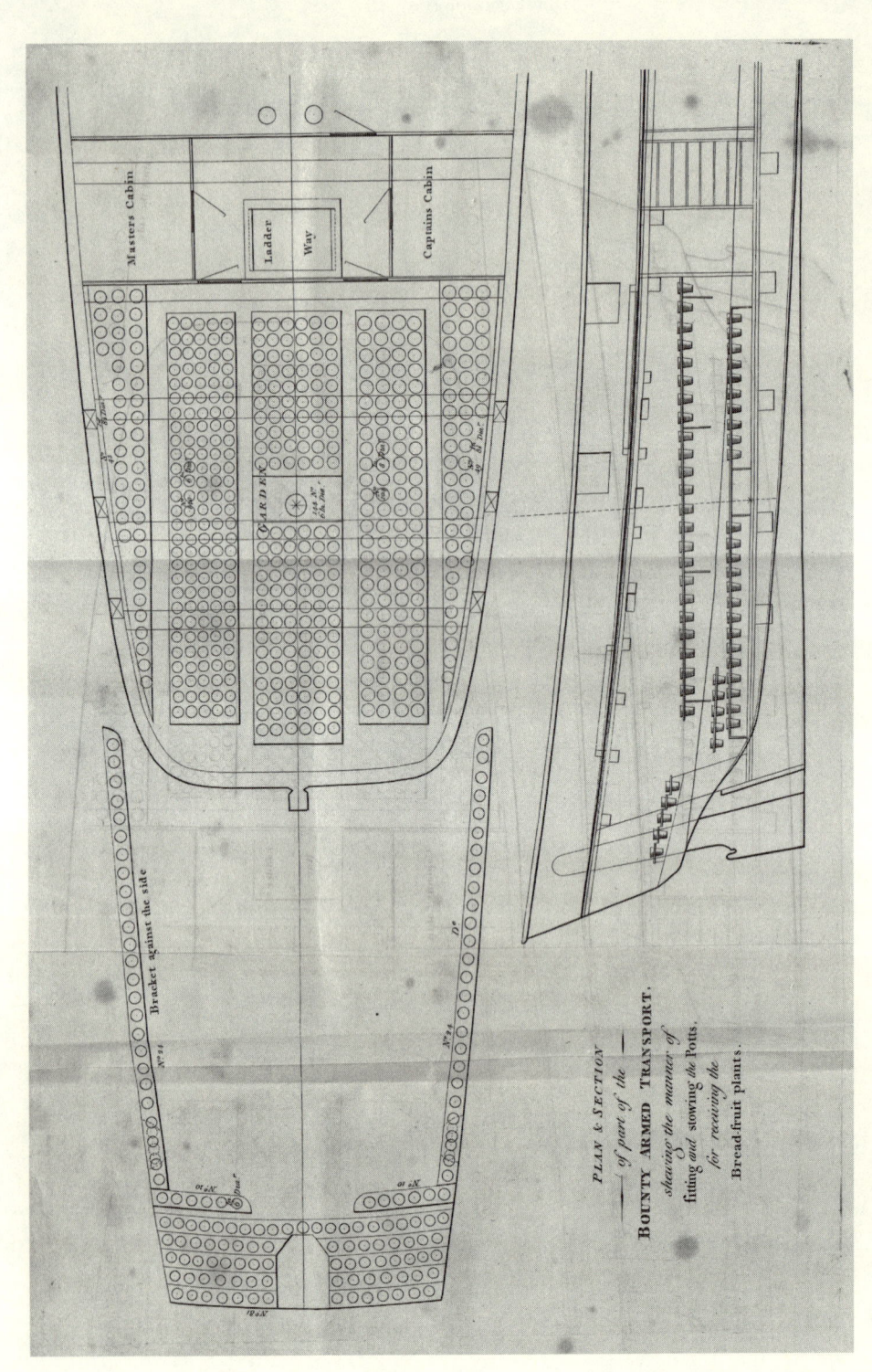

Anon., *Plan and Section of Part of the Bounty Armed Transport*; William Bligh, *Voyage to the South Sea*, National Library of Australia, ANL

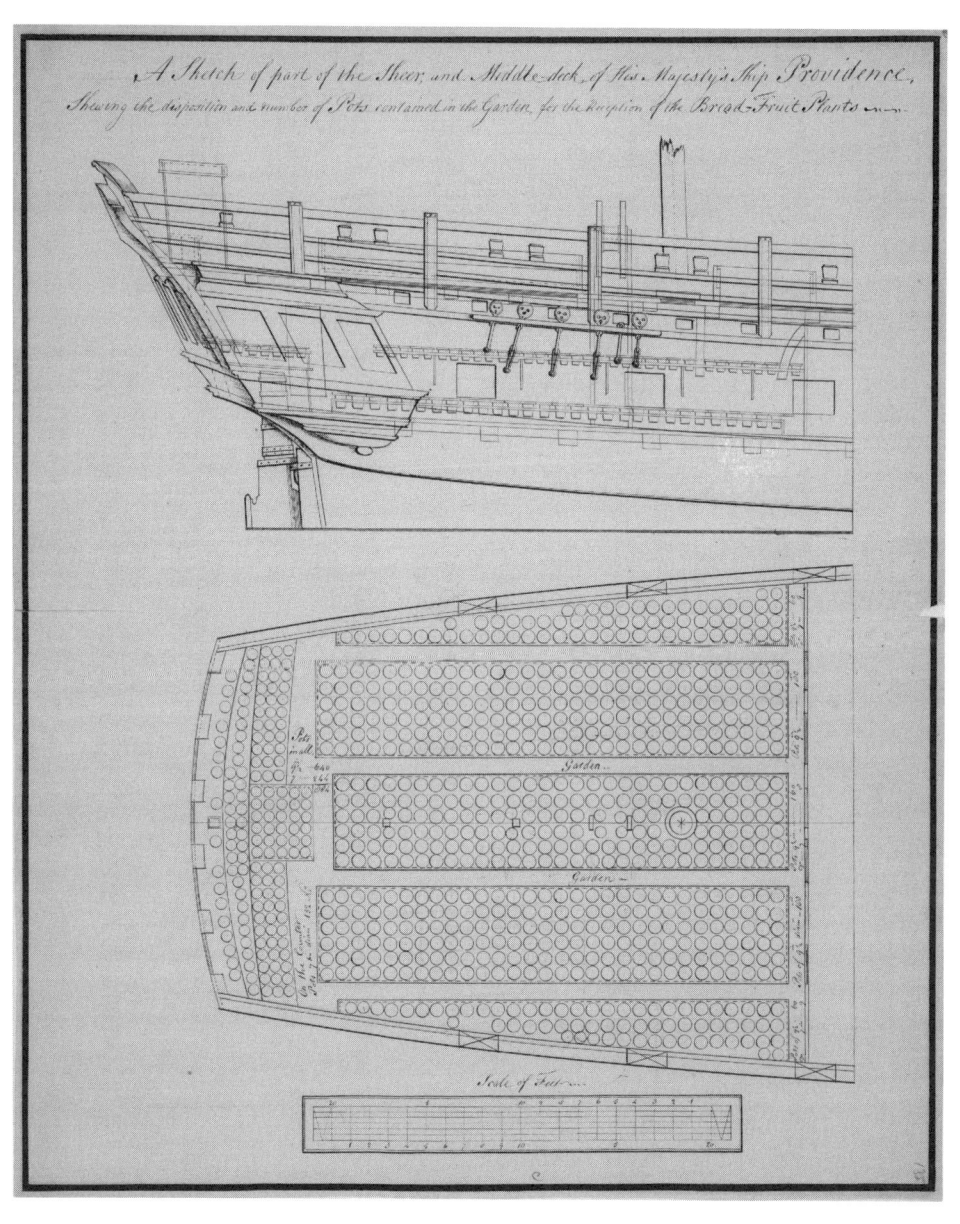

Anon., *A Sketch of Part of the Sheer and Middle-deck of His Majesty's Ship Providence*, State Library of New South Wales "three fine young apple-trees, nine [grape]vines, six plantain-trees, a number of orange and lemon-seed, cherry-stones, plum, peach, and apricot-stones, pumpkins, . . . two sorts of Indian corn, and apple and pear kernels" that he had originally planted, only one apple tree had survived, and though it was still healthy, it had hardly grown at all. At other spots, where Bligh had planted "onions, cabbage-roots, and potatoes," there was no sign of these plants, although he would later regret not having searched for the latter by digging (*Voyage to the South Sea* 49). Given how the first voyage had ended, Bligh could not hide his disappointment and wrote in his logbook, "I had hoped that my last voyage might have been productive of some good" (Lee, *William Bligh's Second Voyage* 10). Undeterred, he set his two botanist-gardeners, James Wiles and Christopher Smith, to work putting in more plants: figs, pomegranates, quinces, oaks, and a Spanish chestnut, along with twenty strawberry plants, watercress, and a rosemary. He also seeded Penguin Island, in the bay, with firs, almonds, apricots, and plums. Before leaving Tasmania, Bligh marked his territory. On the tree where Cook had earlier carved his own memorial inscription, he added his own: "Near this tree, Captain William Bligh planted seven fruit trees, 1792:—Mssrs. S. And W., botanists" (23). We do not normally tend to think of eighteenth-century explorers as being interested in gardening, yet Bligh's actions powerfully express the new manner in which Europeans asserted territorial claims by the end of the eighteenth century. No longer appropriating a place by planting a wooden cross on a beach, Bligh made his mark with a pen, a spade, and a hoe. Through maps, surveys, and natural histories, Europeans translated foreign natures into their own knowledge framework. By planting trees, Bligh was translating foreign natures in a more palpable manner, not simply incorporating them into European knowledge, but naturalizing Europe elsewhere.

La Pérouse displayed a similar understanding of the role of the explorer when he landed in Maui in 1787. First, he mocked the idea that land could be claimed by raising a flag. "The practice of Europeans, in this regard, is too utterly ridiculous," he wrote. "Without a doubt it must grieve philosophers to reflect that for these people, just because they have guns and bayonettes, 60,000 of their fellow men count for nothing; ignoring the most sacred rights, they regard as an object of conquest a land watered by the sweat of its inhabitants and containing the tombs of their ancestors." La Pérouse's modern explorer is no longer concerned with conquest, but seeks instead to observe the customs of others to complete the natural history of man, to achieve the reconnaissance of the globe, and to contribute to the happiness of the islanders whom he encounters by "augmenting their means of subsistence." "It is on these principles," he remarks, that modern explorers "have transported to these islands bulls, cows, goats, ewes, and rams and have planted trees and sown different grains" (de Galaup, *Voyage de Lapérouse* 89–90; my translation). Here one sees a classic version of what Pratt has called the "anti-conquest" strategy, a means "of representation whereby European bourgeois subjects seek to secure their innocence in the same moment as they assert European hegemony" (*Imperial Eyes* 7). The appeal to science and to ideas of improvement provided, as many have noted, the ideological underpinnings of an imperialism that sought to distance itself from the blood and violence of earlier modes of territorial acquisition. Yet can we not also see in such statements the emergence of a new form of territorial possession, where claims on a place were made in gestures toward remaking its nature? Planting was an assertion of control, as it pointed toward future settlements of plants and people. In this sense, the voyage of HMS *Providence* was not simply about global reconnaissance and the transfer of plants. It was also ultimately about producing the nature that the British would eventually claim, and about using Britain's role in producing this transplanted nature as the legal basis for that claim, replacing a *terra nullius* with one that was cultivated.[16] Bligh's inscription on Cook's tree marks, in the way that writing only can, not only a naming but also the beginning of a new natural history for Tasmania, the first stage in the replacement of its indigenous nature. In the future, the very trees themselves would speak of a European presence and of the "improvements" that had already been made. Perhaps this is why, in his logbook, Bligh wrote that "the most valuable articles I have planted this time are nine fine young oak plants about eight inches high" (Lee, *William Bligh's Second Voyage* 23–24). No doubt, he envisioned a time when Tasmania would be able to provide the navy with the oaks that were fast disappearing from the English landscape.

Less than fifty years later, when the indigenous people had been forcibly removed from Tasmania, the biological presence of the British would have been obvious to any visitor. When Bligh visited the island, however, the new plant colony that he sought to establish was little more than a biotic beachhead. There is no record of how the native Tasmanians greeted the new plants that had suddenly arrived on their shores, but it can hardly be doubted that they recognized that their nature was changing. Whatever the Tasmanians might have thought about the new alien plants that were beginning to take root on Bruni Island, the marks of Bligh's visit did not go unnoticed by the French naturalist Jacques-Julien Labillardière when he arrived there a year later as part of d'Entrecasteaux's expedition. Ostensibly searching for signs of the missing expedition of La Pérouse, he too was prowling the Pacific, surveying, collecting plants, and in quest of breadfruit for the Jardin des Plantes in Paris (Spary and White, "Food of Paradise" 79). Labillardière had visited Banks in 1785, and in preparation for the voyage he had received advice from him about where to collect and how to preserve specimens. He must have been disappointed to learn that another of Banks's protégés had reached the island before him. Knowing something of the circumstances surrounding the mutiny on Bligh's first voyage, and concluding that "an abuse of authority on the part of this captain had been the cause of his misfortunes" (*Voyage in Search of La Pérouse* 1:80), the French revolutionary naturalist vented his anger on the mark of that precedence, the inscription itself, by defending the science of botany (2:76–77). He noted that, after writing his name in full, Bligh had supplied only the last initials of the two botanists on the voyage, Smith and Wiles. Labillardière easily recognized the biological newcomers that the British had introduced to the island, and he carefully recorded the presence of oaks, figs, pomegranates, quinces, celery, and cresses. Having only just arrived in Tasmania, these plants were thus already documented aliens, carefully registered by both the British and the French. The French naturalist even appears to have derived a little satisfaction in noting that one of the seven fruit trees was already dead.

Bligh was not the first explorer to introduce a new species of plant or animal to Tasmania. Following the precedent of the Spanish and Portuguese, who introduced pigs, goats, and citrus trees to many of the islands that they visited for the purpose of revictualing their ships, Cook also brought vegetables and domestic animals to the Pacific islands. On the third voyage, in keeping with the imperial objectives of Banks and George III, this activity was substantially increased. The *Resolution* left England crammed with domestic livestock: bulls and heifers, horses, sheep, goats, pigs, rabbits, turkeys, geese, ducks, and even a peacock and peahen supplied by the Earl of Bessborough, all of them "intended for New Zealand, Otaheite and the neighbouring islands, or any other place we might meet, where there was a prospect that the leaving of some of them might prove useful to posterity" (Cook, *Journals* 3:23). At Cape Town, Cook took on more animal passengers. He wrote somewhat sardonically to Banks, "We are now ready to proceed on our Voyage, and nothing is wanting but a few females of our own species to make the Resolution a complete ark for I have added considerably to the Number of Animals I took onboard in England" (cited in Beaglehole, *Life* 511). At most stops along the way, Cook also put in vegetable gardens, employing the assistance of David Nelson, who would serve eleven years later as the gardener on Bligh's first voyage.[17] At Adventure Bay, Cook set a boar and sow loose. Originally, he had planned to include "a young bull and a cow, and some sheep and goats ... as an additional present to Van Diemen's Land. But I soon laid aside all thought of this, from a persuasion that the natives, incapable of entering into my views of improving their country, would destroy them." For Cook, the animals were a "gift" that would "improve" the nature of the island, yet he was also not certain that the native Tasmanians would willingly enter into his plans, so when he was sure that they were not "near enough to observe what was passing, I ordered the two pigs, being a boar and a sow, to be carried about a mile into the woods, at the head of the bay." He noted that since "that race of animals soon becomes wild, and is fond of the thickest cover of the woods, there is great probability of their being preserved" (Cook and King, *Voyage to the Pacific Ocean* 1:98). In 1792, Bligh looked for signs that "the hogs left here by Captain Cook or any breed of them were alive," but he saw none. Nevertheless, following his mentor's lead, he left "a cock and two hens" in Tasmania, hoping that they "will breed and get wild" (Lee, *William Bligh's Second Voyage* 23).

One could easily show that Cook's and Bligh's actions were not unusual for late eighteenth-century explorers, who not only mapped coastlines, observed native customs, and archived natural resources, plants, and animals but also were active in promoting the globalization of useful plants. On the beaches of the Pacific, it was seeds, not crosses, that were planted. The beaches thus became the sites of new biotic colonies, of invaders and settlers, which would fundamentally reshape the very nature of the colonial world. Settling plants in new places was not that different from settling people there, yet the former could be seen as a form of *benevolence*, as Anna Seward suggests in her "Elegy on Captain Cook," helping to ease the conscience of a people that was spreading across the globe faster than weeds. "BENEVOLENCE," she argues, bade Cook "on each inclement shore / [to] Plant the rich seeds of her exhaustless store" (*Poetical Works* 2:34). In New Zealand, he

> Pours new wonders on th' uncultur'd shore
> The silky fleece, fair fruit, and golden grain;
> And future herds and harvests bless the plain.

O'er the green soil the kids exulting play,
And sounds his clarion loud the bird of day;
The downy goose her ruffled bosom laves,
Trims her white wing, and wantons in the waves;
Stern moves the bull along th'affrighted shores,
And countless nations tremble as he roars.
. . .
Then Wisdom's Goddess plants the embryon seed,
And bids new foliage shade the sultry mead;
'Mid the pale green the tawny olive shine,
And famish'd thousands bless the hand divine. (38–39)

One might easily assume that when Bligh first arrived in Tahiti in 1788, he was visiting a pristine insular environment as yet unaffected by the emerging network of global nature exchange which I have been outlining. This was hardly the case. Cook had introduced sheep and goats to the island on his second voyage, and most of the animals that he had picked up in Cape Town on the third voyage disembarked at Matavai and Pare. Cook was pleased to bid farewell to his brute passengers. He felt "lightened of a very heavy burden, the trouble and vexation that attended the bringing these Animals thus far is hardly to be conceived. But the satisfaction I felt in having been so fortunate as to fulfill His Majestys design in sending such usefull Animals to two worthy Nations sufficiently recompenced me for the many anxious hours I had on their account" (*Journals* 3:194). As he did on other South Pacific islands, such as Nomucka and Huahine, Cook also started a garden, growing melons, potatoes, pineapples, and shaddocks (Cook and King, *Voyage to the Pacific Ocean* 1:23). All the animals that Cook brought to the islands would not have been so welcomed. Black and Norway rats also arrived with the Europeans. Bligh recalled that in 1777 they had become outright pests: "They flocked round the people at their meals for the offals." It must have been very difficult to enjoy a feast in those early years after European contact, and no doubt at that time the islanders must have questioned the benefits of European contact. By 1788, however, the rats were no longer so common, a change that Bligh attributed to "the industry of a breed of cats left here by European ships" (*Voyage to the South Sea* 121).[18] He did not give any thought to what these feral cats might also be doing to the smaller species inhabiting the island.

During his three months on the island, Bligh seems to have been less interested in its indigenous nature than in preserving the biotic legacy of Cook. Almost his first question upon arriving there on his first voyage had to do with the whereabouts of the cattle "that had been left by Captain Cook" (63). He then went to great lengths to locate them. On his first walk on the island, Bligh was intent on discovering whether there remained any biological signs of previous European visits: "In this walk I had the satisfaction to see that the island had received some benefit from our former visits. Two shaddocks were brought to me, a fruit which they had not, till we introduced it. And among the articles which they brought off to the ship and offered for sale were capsicums, pumpkins, and two young goats" (64). Bligh's interest was not in native but in transplanted natures. He had brought seeds with him, and four days after arriving on the island, even before beginning to collect breadfruit, he was at work on a garden. "I planted melon, cucumber, and sallad-seeds," he writes. "I told them many other things should be sown for their use; and they appeared much pleased when they understood I intended to plant such things as would grow to be trees and produce fruit" (68). A month later, not fully satisfied with this garden, he started another at Point Venus. On 31 January 1789, he describes a walk that he made "to see the cattle and the gardens." He was pleased to see that the cow and bull he had recovered were now grazing "in a very fine pasture. I was informed that the cow had taken the bull; so that, if no untoward accident happens, there is a fair chance of the breed being established." The garden was also in good shape: "The Indian corn was in a fine state and I have no doubt but they will cultivate it all over the country. A fig-tree was in a very thriving way, as were two vines, a pine-apple plant, and some slips of a shaddock-tree." The garden at Point Venus, however, was almost destroyed. "I had the mortification to find almost everything there destroyed by the hogs," writes Bligh. "Some underground peas and Indian corn had escaped, and likewise the caliloo greens and ocra of Jamaica" (121). Bligh's list of the plants in this garden makes it clear that even on the first voyage, Tahiti was not just the origin of plants headed for the West Indies but also already a destination for many others.

The list that Wiles and Smith sent to Banks indicates that the number of plants introduced to Tahiti was substantially increased by the second voyage: "18 Pine[apple] Plants, 14 Figs, 2 Pomgranates, 4 Lemons, 1 Guava, 1 Almond (raised in the passage) 4 Oranges, 3 Oaks, 1 Myrtle, 2 Aloes, 3 Banksia, 3 Metrocedera [a eucalyptus from Tasmania], a creeping umbelliferous

Plant we call'd Adventure Bay Parsley, between 200 and 300 seedling Orange, Limon and Citron Plants, a great number of young Palma Christi and seedling Firs, besides Indian Corn and Kitchen Garden Plants" (James Wiles and Christopher Smith to Joseph Banks, 17 Dec. 1792, *Papers of Sir Joseph Banks*, sec. 9, ser. 52.09). Nature during this period had become exchangeable, and Tahiti was already playing its part in the trade in plants. When the *Providence* left Tahiti, it carried 1,686 breadfruit plants, 132 Malay apples, 55 Otaheite apples, 67 Tahiti chestnuts, 21 kou trees (used for furniture), 8 dye figs (for dying clothing), 20 plantains, Tahitian arrowroot (to supplement slave diets), Tahitian coconut, and 32 "curious" plants destined for Kew.

In the stifling heat of the Torres Strait, Bligh lost about one-fifth of the breadfruit plants, but he quickly made up for these at the Dutch-controlled island of Timor, where he obtained ninety-one pots of "the best plants of this place," many of these being fruit trees (Lee, *William Bligh's Second Voyage* 207). These were new and unfamiliar plants with strange and exotic names: mango (*Mangiferi indica*); nanka or jackfruit (*Artocarpus heterophyllus*); jambo armarvah, the rose apple (*Syzygium jambos*); jambo marree, the Malay apple (*Syzygium malaccense*); balumbeng, the cucumber tree (*Averrhoa bilimbi*); cherimalah or Malay gooseberry (*Phyllanthus acidus*); bintalloo or the betel-nut tree (*Areca catechu*); seeree boah or betel leaf (*Piper betle*); cattahpas; nam nam (*Cynometra cauliflora*); teak (*Tectona grandis*); and bughnah-kanangah (the tree whose flowers now provide the scent for Chanel No. 5). When the ship docked at Saint Helena, it was thus carrying plants from Tasmania, Tahiti, and Fiji. Any plants that could be spared were passed on to Henry Porteus, who was in charge of the East India Company's botanic garden on the island. In exchange, he contributed another 830 plants to the ship's living cargo, most of these from the island, but also some plantains, almonds, and coffee that had been brought earlier to the botanic garden from elsewhere and were now ready to renew their travels. Banks had specifically requested that Wiles and Smith "dig up one or two Fern trees with large Balls of Earth adhering to their Roots and Plant them in large baskets for his Majesties use" (Banks, "Instructions for Mr. James Wiles"). They succeeded in collecting six of these magnificent trees, along "with every different Plant on the Island worth notice" (James Wiles and Christopher Smith to Banks, 24 Dec. 1792, *Papers of Sir Joseph Banks*, sec. 9, ser. 52.11). Bligh thus sailed to Saint Vincent with almost the entire flora of an island on board, many of these in very large pots that would be taken out on the deck during the day for air and sunlight. No wonder the third lieutenant, George Tobin, would describe the ship as "our floating forest" (quoted in Powell, *Voyage of the Plant Nursery* 26). When the boat arrived at Saint Vincent on 22 January 1793, another wholesale exchange of plants took place. Dr. Alexander Anderson, who was the superintendent of the botanic garden at Saint Vincent, established in 1765, was allotted 544 plants—a major share of breadfruit and half of the others, "particularly the Fruits and useful Plants." In return, he supplied another 465 pots of plants, some to be dropped off in Jamaica, others headed for Kew.[19] In preparation for the arrival of the ship, Anderson had also undertaken a collecting expedition to Demerara, Guyana, and Trinidad, so the plants that eventually embarked on the *Providence* would have also included many from South America. When the *Providence* arrived in Jamaica, on 5 February 1793, more plants were exchanged. A total of 346 breadfruit plants were brought ashore, along with 244 additional plants that Bligh hoped might prove well adapted to thrive in the West Indies. Among the new arrivals were Otaheite apples, Malay apples, Tahitian chestnuts and arrowroots, different varieties of banana, coconuts, Kou trees, jackfruit, mangos, dwarf peaches, coffee, almonds, and nutmeg, to name just a few.

Although the heterogeneity of Caribbean culture has long been recognized, what has been less readily grasped is the degree to which the physical natures of these islands were as much the product of a mixing of natures as of peoples. Drawing on the work of Crosby and Grove, Elizabeth M. De-Loughrey remarks that "perhaps there is no other region in the world that has been more radically altered in terms of flora and fauna than the Caribbean islands" ("Island Ecologies" 298). Beginning with Columbus's second voyage, plants introduced from South America, Europe, Africa, South Asia, and (through Bligh) the South Pacific steadily displaced the indigenous plants of the Caribbean. Most of the plants that now commonly grow on the islands came from somewhere else, and these transplanted natures were as creole as the people that depended on them. The Caribbean islands can thus be said to have provided the paradigm for modern natures, where the role of immigration, transplantation, migration, and naturalization in their formation has been dramatic. The primary factors governing the ultimate character of these new environments had less to do with climatic factors than with the location that these islands occupied in the global networks of plant transference managed by the English and the French. Economics was inseparably bound up with these ecologies, for the natures that arrived on the shores of the Caribbean had been routed there for commercial purposes, as cash crops supported by the labor of slaves and indentured labor. The conditions for the emergence of these modern natures thus lay to a great extent outside of the control of the localities within which they appeared. To Arjun Appadurai's list of the transnational flows of people, capital, technologies, media, and ideas that produce "localities" nowadays and to the imagined world landscapes that produce them—ethnoscapes, financescapes, technoscapes, mediascapes, and ideoscapes—we need to add another category that his analysis overlooks, that is, the *bioscapes* produced by the movement of natures ("Disjuncture and Difference" 1–23). Also alongside the colonial formation of multicultural societies, we need to consider natures that were inherently "multi-natural" in character, products of the coming together—the naturalization in the original legal sense—of biota from many different parts of the world.

Having introduced a large swathe of the flora of the Pacific to the Caribbean, having fulfilled Joseph Banks's most "ardent wish" that "all the remaining valuables of the East" might follow the journey of sugar and coffee to the West, the *Providence* returned to England with 1,283 plants, of which 876 were from Jamaica. Another 338 came from the botanic garden at Saint Vincent, while the remaining 69 plants were from Saint Helena, Timor, New Guinea, Tahiti, and Tasmania. Fifteen years before Bligh arrived in the West Indies, in 1778, Jamaica's national fruit, also an immigrant, came on a slave ship from West Africa, where it was called "Akye fufo." Over the course of the next century, it was not the breadfruit but this plant, commonly called "ackee," that would play a major role in Jamaican black culture, so much so that contemporary anthropologists now look for its presence as "a strong indicator of the location of abandoned slave village sites throughout Jamaica" (D. V. Armstrong, "Afro-Jamaican Community" 1). No sooner had ackee arrived in Jamaica than it was booked as a passenger on the *Providence*, now destined for Kew. There it was given its scientific name, *Blighia sapida*, in honor of the captain who had played such an important role in changing Jamaica's nature. To his contemporaries, the despotic captain and subsequent governor of New South Wales would be remembered as "Breadfruit Bligh." For botanists, this harsh taskmaster's memory is preserved by an African newcomer to Jamaica, its name "sapida" referring to the pleasant flavor of its fruit. Like the breadfruit, this plant embodies the contradictory historical forces that underlay the hybrid natures that began to emerge with such clarity during the eighteenth century.

CHAPTER THREE

Translating Early Australian Natural History

> The Whites did not . . . possess the Aborigine's country any more than they spoke his language. They possessed a country of which the Aborigine was unaware.
>
> (P. Carter, *Road to Botany Bay* 64)

The dream of botanizing runs deep in the Australian psyche, so deep that it even appears to have been a frequent preoccupation of the most notorious and callous of the Tasmanian bushrangers, Michael Howe. In the first piece of general literature printed in Australia, Thomas Wells's pamphlet *Michael Howe, the Last and Worst of the Bushrangers*, published in Hobart in 1818, we learn that during the last year of Howe's life, he found time, while seeking to evade the authorities, to indulge in his dream of gardening in Tasmania. Writing in blood in a little kangaroo-skin notebook, Howe recorded his dream-visions. Most of these had to do with an intense foreboding that he was about to be murdered by the Aboriginal people or captured and killed by the soldiers, but sometimes he also fantasized about better things, such as seeing again his old friends and his sister. His favorite dream, however, was about gardening in the bush, for the now-lost journal contained, as Wells notes, "long lists of such seeds as he wished to have, of vegetables, fruits, and even flowers!" (*Michael Howe* 37). Even a violent outlaw on the run from the authorities seems to have found it hard to resist the temptation of compiling lists of plants that could be introduced to the Australian bush. Such botanic dreams, however, rarely coincided with the harsh realities of this environment: instead of seeds, it was Howe who was planted somewhere in the woods near the River Shannon, and instead of plants it was his head that was shipped for display in Hobart, in a strange colonial recasting of Keats's *Isabella, or a Pot of Basil*. Australia was all about plants and transportation.

The dream of using botanical knowledge to transform an alien landscape into a world-class nature, global in every sense, seems to have been there from the beginning, registered symbolically by Captain Cook's renaming of the place where he initially landed on the Eastern coast of Australia in April 1770 from Sting Ray Harbour to Botany Bay in recognition of "the great quantity of New Plants &c Mr. Banks & Dr. Solander collected in this place" (*Journals* 1:310). Later, first in 1779 and then in 1785, when Sir Joseph Banks was questioned by committees of the House of Commons about the best place in the world to establish a penal colony, he proposed Botany Bay. The connections between the international globalizing ambitions of eighteenth-century botany and this new penal settlement were thus forged in the public mind from the beginning. There can be little question that the settlement at Botany Bay was an experiment in using transported convicts as colonists, yet on the basis of the publications it produced one might easily conclude that it held more interest for the public as a project in natural history. J. M. Powell remarks, for instance, that "the chief importance of early Australia in intellectual and cultural terms was as a new field for science: the discovery of the new continent greatly excited botanists, geographers, zoologists and geologists everywhere, and the main preoccupation of the best-educated settlers was also with matters of scientific interest" (*Environmental Management in Australia* 13).

Banks played a primary role in the establishment of the colony and in the acquisition of natural history knowledge. As Ann Moyal observes, "With the despatch of the First Fleet to Sydney, Banks corresponded with governors, issued clear directions to ships' captains for collecting and preserving specimens, forwarded useful plants for cultivation from Kew and set in motion that traffic in botanical and zoological specimens that laid the foundation of the study of Australian natural science" (*Scientists in Nineteenth Century Australia* 11). Natural history and highly centralized governmental power worked hand in hand. Many of the major figures connected with the early settlement were naturalists: the first collaborative work published on the settlement, the government-sponsored *Voyage of Governor Phillip to Botany Bay* (1789), was delayed in order to include twenty-nine natural history plates. John White, who was appointed chief surgeon for the First Fleet and later took up the role of surgeon general of New South Wales, authored *Journal of a Voyage to New South Wales* (1790), which included sixty-four natural history plates and scientific descriptions by eminent metropolitan naturalists. He planned a more extensive project and put together a collection of drawings, many by the convicted forger Thomas Watling, but he appears to have abandoned this project as he became increasingly disillusioned with the convict settlement. Captain John Hunter, second-in-command of the First Fleet, and second governor of the colony from 1795 to 1800, published *Historical Journal of the Transactions at Port Jackson and Norfolk Island* (1793), which included Blake's famous illustration *A Family of New South Wales*. Hunter was also a serious naturalist who produced over one hundred watercolors of the flora and fauna of the region. By the urging of Banks, in recognition of his role in introducing the flora of the East Indies to the Caribbean, Bligh was also appointed as the fourth governor of New South Wales; in the wake of the Rum Rebellion of 1808, which put an end to his governorship, he spent the remainder of his time on the continent collecting shells and drawing birds in Tasmania.[1]

I follow Deirdre Coleman in understanding colonization "as a leap of the imagination as well as a leap in geographical space and time" (*Romantic Colonization* 2). Following Britain's success in transforming the physical natures of the Caribbean islands, and on the heels of its loss of the American colonies, Australia briefly held the promise in the late 1780s of being the new poster child for the capacity of Europeans to build a cosmopolitan nature, brought from the four corners of the earth, that would serve the needs not only of the settlement but also of people living in Britain. Realities on the ground soon dispelled such visions, but it was an idea of a nature that would be created across distances, through the imperial technologies of communication, transportation, resource management, and trade. Responding in August 1771 to the news of Cook's first voyage, Benjamin Franklin remarked that "Britain is said to have produced originally nothing but *Sloes*. What vast advantages have been communicated to her by the Fruits, Seeds, Roots, Herbage, Animals, and Arts of other Countries! We are by their means become a wealthy and a mighty Nation, abounding in all good Things. Does not some *Duty* hence arise from us towards other Countries still remaining in our former State?" ("Introduction to a Plan for Benefiting the New Zealanders," *Papers* 18:214). Since Britain's originally inferior physical nature had been remade through technology and the importation of other natures, why could not other peoples benefit just as much from this translational activity? Franklin proposed that a ship under the command of Alexander Dalrymple be equipped for this purpose, and the third voyage of Cook suggests that his ideas took hold. In the film *Field of Dreams*, while walking through an Iowa cornfield, Ray Kinsella hears a voice that tells him, "If you build it, he will come," so Kinsella plows under his corn to build a baseball field, and Shoeless Joe Jackson arrives, followed soon after by seven other baseball players. Australia during the last decade of the eighteenth century was, indeed, a British field of dreams. Colonial natural history was seen as the means by which a local indigenous nature, characterized as inferior, was to be replaced by another nature built for settlers—"If you build it, [they] will come." The British possessed Australia *in* and *through* translation, as linguistic translation proceeded in tandem with more material economic and biological forms of transference and exchange.

Bernard Smith, in his groundbreaking study *European Vision and the South Pacific, 1768–1850*, argues that "the opening of the Pacific is . . . to be numbered among those factors contributing to the triumph of romanticism and science in the nineteenth-century world of values" (1). Nineteenth-century Romanticism would not have been what it was in the absence of the natures and cultures of the Pacific which began flooding into Britain and France from the 1770s onward. Smith's work is important because he recognized the degree to which Europeans, when faced with the radically different physical environments and cultures of the Pacific, initially transplanted their own aesthetics there. Recent work, notably by Paul Carter and Nicholas Thomas, has complicated this model by integrating it more strongly within contexts of translation and exchange. This chapter seeks to extend these approaches by emphasizing the degree to which the commitment to translation, exchange, and mobility fundamentally shaped the nature that came into being with the European settlement of the continent.[2] What differentiated the nature that the Aboriginal people knew from the one that began to appear at the end of the eighteenth century is that whereas the former was intrinsically *local* and *rooted* in place, existing in the Aboriginal peoples' detailed knowledge and material and cultural interaction with native plants and animals over generations, colonial Australian nature was essentially a product of European natural history and the globalizing technologies of empire. Relying on highly centralized modes of decision making, the new settlers set out to change and improve this nature, as J. M. Powell observes, "long before there had been time to achieve more than a rudimentary grasp of the continent's most crucial ecological characteristics" (*Environmental Management in Australia* 33). Thus, as Geoffrey Bolton suggests, the environmental history of Australia can be seen as "a conflict between those who exploited the country to serve preconceived economic goals and imported attitudes of mind, and those on the other hand who sought to create a civilisation where human use of resources was compatible with a sense of identity with the land" (*Spoils and Spoilers* 23). The nature that the settlers sought to produce was fundamentally translocal, that is, it came into existence between places and peoples, a nature that traveled because it was valued.

The history of natural history in Australia is a history of renamings and rewritings mapped on erasures, of new natures standing where others stood, of new possessions standing in the place of the dispossessed and having little in common with them, and thus of presences that are haunted by ghostly absences. In thinking about Australian natural history, therefore, we should not think of it simply as an origin, a thing, or an idea, but instead as the fractured product of a history of positions taken in relation to the natural world of the region. Although for those who occupied positions of power within the settlement the strange and exotic natural history of Australia initially seemed to represent an extraordinary opportunity, their confidence in their capacity to manage and control this environment was soon severely tested. The early documents suggest that they also suffered feelings of intense alienation and anxiety. Robert Hughes sees a substantial difference in how Americans and Australians perceived the frontier environment around them. Whereas Americans associated space with freedom and the possible recovery of paradise, "in colonial Australia, space itself was a prison. You walk across the country, find nothing, then die" (Hughes, *Beyond the Fatal Shore*). Where the British separated themselves from the nature around them, seeing the bush as something to be collected, conquered, and controlled, Aboriginal people had developed a comfortable long-standing relationship to the "more-than-human nature" around them. As Martin Mulligan and Stuart Hill write, the Aboriginals inhabited an ancient place that they had lived in and known for thousands of years, so "the recent history of those places soon merges into ancient stories that incorporate creation myths . . . every place has its stories" (*Ecological Pioneers* 234). Australian colonial nature must be understood, therefore, as being structured by conflict and riven by struggle. It was not a homogeneous entity, but instead an arena where the conflict between old and new natures was often as intense as the conflict between peoples. If we examine its history carefully, it speaks of nature as a site of colonial struggle, of possessions and dispossessions, of natures made and lost through translation.

Thomas Gosse's mezzotint print *Founding of the Settlement of Port-Jackson* (1799) provides a powerful illustration of this sense of Australia as a divided nature, structured by competing local and translocal elements. Gosse never visited Australia, so the print is essentially a product of remediation, indicating what he had gleaned of Australia from the abundant literature of Australian settlement and, perhaps, from seeing watercolors, prints, or specimens of Australian animals at the Leverian Museum. The print followed on Gosse's extremely successful print entitled *Transplanting of the Bread-Fruit-Trees from Otaheite* (1796), which I discussed in the previous chapter as inaugurating artistically the era of imperial globalized plant transfer epitomized by the Bligh breadfruit expedition. Echoing something of the structure of Christian nativity scenes and certainly drawing on images of Noah's Ark, the engraving displays the gathering together of a people and their animals at a place of new beginnings. This beginning is depicted as the replacement of the natural world and people of precontact Australia with a new society that brings its nature with it. At the center of the print, in front of the tents that were initially used to house the convicts and sailors, stands Captain Arthur Phillip, the first governor of New South Wales, directing the establishment of the penal settlement. Like Ray Kinsella standing in his cornfield, Phillip points to their dream city (as yet unbuilt in 1788), which appears in the distant background flying the Saint George's Cross. He has turned his back on a naked Aboriginal male who stands cowering before a marine, on the right side of the print, while deep in the distance, under darkness and clouds, three other Aboriginal people pass through a gloomy and barren landscape. The need to use violence to bring the indigenous people and their nature under British rule is a central theme of the print. In the foreground, the convicts are busy cutting stakes, building fences, and flipping over sea turtles to be later used for food. Another man shoots a parakeet from a tree filled with these birds, and it is shown plummeting headfirst from the top center of the painting.

Gosse's print is formally divided into two different natures: the indigenous nature of Australia is being literally pushed aside or crowded out of the picture not only by the people in the foreground but also by the new domestic animals that are being introduced to the continent—a pig and two cattle that are being unpinned from their transportation pens on the right-hand side of the painting and seem ready to take possession of this new land. It is also in the process of being turned into a *nature past* by the new nature that is replacing it. As Sir Joseph Banks had made clear in his testimony before the Bunbury and the Beauchamp Committees of Parliament, a major attraction of Botany Bay was that it was sparsely settled by indigenous people

Thomas Gosse, *Founding of the Settlement of Port-Jackson*, National Library of Australia, nla.pic-an6016205

and animals, so they could easily be replaced by British settlers and livestock. Speaking of the Aboriginal people, Banks declared that "from the Experience I have had of the Natives of another part of the same Coast I am inclined to believe they wo^d speedily abandon the Country to New Comers" and that "they were armed with Spears headed with Fish Bones but none of them we saw in Botany Bay appeared at all formidable" (*Indian and Pacific Correspondence* 2:94). At the same time, "there were no tame Animals, and he saw no wild Ones during his Stay of Ten Days"; furthermore, "there were no Beasts of Prey, [so] he did not doubt but our Oxen and Sheep, if carried there, would thrive and increase." He also stressed that colonization required the transport not only of people but also of plants and animals: "Black

Cattle, Sheep, Hogs, and Poultry; with Seeds of all Kinds of *European* Corn and Pulse" along "with Garden Seeds (10 Apr. 1779, 1:251–52).

Although Gosse hoped to profit from the public interest in the events taking place in the new penal colony at Port Jackson, the print did not sell well. His focus on the violence of colonial contact must have been a major factor in its lack of success with a British audience. Smoke seems to have spread across and darkened the entire scene—both from the gun and from the fire at the center of the print being used to cook the turtles and birds. The landscape, based on a restricted palette of browns and grays, seems to be suffering from smoke damage, despite the efforts of the colonists to drape a white tent over it to protect it. Also, the convicts in the print avoid our gaze, either because they are too busy working to build the settlement or, perhaps, because they have something to hide. The marine interrogating the Aboriginal man has turned his back completely on us, while most of the other figures look down or away from the viewer. The print presents a troubling image of the future of settlement, of its silences and uncertainties. When Gosse produced this engraving, he strongly believed, like many millennialists of the time, that the world would come to an end in 1800. *Founding of the Settlement of Port-Jackson* is a print about one world coming into being at the expense of another. If his goal was to create a nativity scene, it is one that seems darkened by apocalyptic foreboding.

More successful in promoting a seamless idea of the inauguration of a new nature in New South Wales is Erasmus Darwin's poem "Visit of Hope to Sydney Cove." This twenty-six-line poem appeared among the prefatory materials of the official publication of Arthur Phillip's *The Voyage of Governor Phillip to Botany Bay* (1789) and was based on a medallion produced at Wedgwood's Etruria factory to commemorate the new settlement (iv). As its title suggests, Wedgwood's medallion portrays in allegorical form "Hope encouraging Art and Labour under the influence of Peace to pursue the employments necessary to give security and happiness to the infant colony." This medallion epitomizes the processes of remediation by which Australian nature was remodeled for a British marketplace. Starting with the actual soil of Australia, which was then transported to the Wedgwood factory in Stoke-on-Trent, Staffordshire, to create a medallion, this object, in turn, appears both as an engraved vignette on the title page of a government-sponsored narrative of the new settlement and as the basis of Darwin's verbal ekphrasis, which also appears as part of this volume. From soil, to medallion, to engraved medallion, to poem, to travel narrative, this is a nature whose value lies in its capacity to travel and to undergo multiple transformations and remediations.

Darwin's poem "Visit of Hope to Sydney Cove" celebrates the benefits that colonial natures will derive from European exploration, travel, and settlement, as Art and Labour remake nature "under the influence of Peace" (in Phillip, *Voyage of Governor Phillip* iv). Surprisingly, Hope does not directly address the convicts, who probably needed all the encouragement they could get, but instead the land itself. Standing "high on a rock," "'Hear me,' she crie[s], 'ye rising realms! Record / Time's opening scenes, and Truth's unerring word,'" as she makes her address "To each wild plain," "High-waving wood and sea-encircled strand" (v). No longer hopeless, these "rising realms" of plains, woods, and sea are asked to become historic, to "record" the new beginning that is being marked out *by* and *through* them at Sydney Cove. Nature, like the soil that Wedgwood used in creating the medallion, is thus the document that records its own historic transformation—not by remaining in place, but through its remediations and travels. "Time's opening scenes" will be registered in the changes in the land. As if Hope were Governor Phillip, giving directions about where and how to build this new settlement, her address maps out the lineaments of the future city:

> There shall broad streets their stately walls extend,
> The circus widen, and the crescent bend;
> There, rayed from cities o'er the cultured land,
> Shall bright canals and solid roads expand;
> There the proud arch colossus-like bestride
> Yon glitt'ring streams, and bound the chasing tide;
> Embellished villas crown the landscape scene,
> Farms wave with gold, and orchards blush between:—
> There shall tall spires and dome-capped towers ascend,
> And piers and quays their massy structures blend;
> While with each breeze approaching vessels glide,
> And northern treasures dance on every tide! (v)

Darwin envisions the future Australia as a "cultured land," of farms, orchards, and villas linked by transportation and trade arteries ("bright canals and solid roads" and bridges "bestrid[ing] / Yon glitt'ring streams, and... chasing tide") to what he believes will be the first of many cosmopolitan cities. The city itself will be linked to the cities of England by trade: "with each breeze approaching vessels glide,/ And northern treasures dance on every tide!" Darwin's Sydney is an integral participant in global trade.

One might argue that Darwin is not describing nature, but instead the founding of a city. My point, however, is that he does not see nature and the city as being antithetical, but instead as mutually supporting each other through histories that are fundamentally linked together. Without the coming of this city, nature, for Darwin, is without hope. The different technologies of mobility to which he refers—roads, bridges, and canals—spread like rays "from cities o'er the cultured land," linking the country with cities in a profound unity, and the city he envisions, in turn, "Courts her young navies" (in Phillip, *Voyage of Governor Phillip* v), as the ocean-going vessels connect this place to others across the globe. No wonder Hope waxes Shakespearean as she recalls Prospero's spectacular vision of the mighty "fabric" of theater, which can create "The cloud-capp'd towers, the gorgeous palaces,/ The solemn temples, the great globe itself" (*Tempest* 4.1.152–53). With its "tall spires and dome-capped towers," the city of Sydney and the nature that it encompasses are just as much a product of dreams, but rather than being produced by theater, this new "cultured land" is woven from the fabric of European transportation technologies and trade. As William Cronon argues, the city may also ultimately be based on theft, for "a sizable share of the new city's wealth was the wealth of nature stolen, consumed, and converted to human ends" (*Nature's Metropolis* 206). Darwin powerfully articulates the concept of a nature that does not exist apart from human values and needs, instead finding its meaning through its integration within the networks of circulation and exchange provided by labor and art. It is worth stressing that although the roads and canals joining nature and the city might seem like lines connecting distant points together, what really links them is the movement and communication of goods along these routes. They make possible inflows and outflows, imports and exports, arrivals and departures. Thus, it can be said that Darwin is conceiving of a nature that is new because it depends for its existence on its capacity to travel; it is a nature that moves on land and water along the routes that have been engineered for its transport. Darwin draws attention to this relationship when he puns on the "High-waving wood," probably echoing Pope's famous description of the oaks of Windsor Forest: "With joyful Pride survey'st our lofty Woods,/ Where tow'ring Oaks their growing Honours rear,/ And future Navies on thy Shores appear" (*Poems* 1:144–94, lines 220–22). Darwin liked the pun so much that he used it again, later in the poem, declaring that "Farms wave with gold," once again linking the value of farms to maritime commerce. Australia was built on the legal practice of transportation, so it is perhaps fitting that the nature that Darwin envisions is just as much a product of transport.

As I suggested earlier, every new nature comes into being by erasing or transforming the natures that preceded it, and Darwin's vision of Sydney Cove is no different. That is why Hope's allusion to *The Tempest* is so relevant, because the play is deeply focused on the struggle between Prospero and the cannibal Caliban for the island. Darwin's poem works within the binaries that structured the British understanding of the colonial natural world. Hope is racially and economically associated with Europe through her "golden hair" and "snowy hand," while the Calibans of Australia are simply erased from the landscape, as if they did not exist. The nature and people of precontact Australia are all but ignored because they are understood as belonging to the past. They are on the way to becoming ghosts, even as the future is reserved for the new and useful nature that is being ushered into being by transportation and settlement. Utility and mobility shape all. Darwin's poem emblematizes the ways in which translations call communities into being. In the same manner as the Australian soil carried across the waters to Wedgwood's factory is transformed into a medallion, Darwin's poem returns across the waters, using print to surmount time and space, in order to build a new place and a new people shaped by this Hope. Hope's address, like Prospero's, creates the social order that she prophesies, and the peals of applause that greet her speech allegorize the new "imagined community" that is being created by this translation:

> Then ceased the nymph—tumultuous echoes roar,
> And Joy's loud voice was heard from shore to shore—
> Her graceful steps descending pressed the plain,
> And Peace, and Art, and Labour joined her train. (in Phillip, *Voyage of Governor Phillip* v)

Amid the echoing applause of this now joyful Australian nature, a critical eye can certainly recognize the silences—of both Aboriginal people and convicts—that in this new time of nature occupy the edges of such public demonstrations of a new beginning.

The Biological Translation of a Continent

If, instead of adopting biblical allegory, Gosse had sought to portray the true dimensions of colonial nature transport opened up by Australian settlement, he would have needed a much larger space than a small mezzotint engraving. When the First Fleet landed at Botany Bay in 1788, it brought more than four hundred animals and numerous seeds and plants. Included in the livestock were a stallion, 3 mares, 3 colts, 2 bulls, 5 cows, 29 sheep, 19 goats, 49 hogs, 5 pigs, 5 rabbits, 18 turkeys, 29 geese, 35 ducks, 122 fowls, and 87 chickens, along with Governor Phillip's greyhounds and Reverend Richard Johnson's cats (Colbey, *Sydney Cove* 154). Shortly thereafter, a bull and four cows strayed into the bush and were not found for seven years, until November 1795. The plants transported to Australia included bamboos, grains, grasses, coffee, cocoa, cotton, indigo, bananas, oranges, lemons, guava, tamarind, the prickly-pear cactus (complete with cochineal insects to produce dye), a fig tree, oaks, Spanish reed, *Eugenia*, ipecacuanha, sugarcane, vines, quince, apple, pear, strawberries, and myrtle. Such a number of biotic fellow travelers, not to mention the weeds and vermin that the British were also accidentally bringing with them, should remind us that even if the primary objective of England was to reproduce British culture in a faraway part of the world, the reproduction of that culture was understood as depending on their capacity to introduce to this continent a physical nature meeting their needs—a cosmopolitan nature selected first for its economic utility. The British were not seeking to adapt themselves to a new environment, but instead responding to its difference by changing it.

The biotic dimensions of colonization required careful planning, and it was in this capacity that Sir Joseph Banks, as director of Kew Gardens, entered the imperial limelight. Much of the imported biota was brought from England, but Arthur Phillip also picked up additional plants and livestock in his stopovers in Rio de Janeiro and the Cape of Good Hope. In a letter to Banks from Rio, Phillip writes, "I have procured every Seed I think likely to grow in NSW & likewise fruits & plants. particularly the Indigo, Coaco, Coffee & Jambo or Pomona Rose, Cotton, Tobacco, Vines, &c. with the Cochineal which I hope to cultivate" (31 Aug. 1787, Banks, *Indian and Pacific Correspondence* 2:231). Since plant exchange was not simply a global activity among the members of Banks's circle but also the primary medium of social exchange, Phillip did not neglect to include a shipment of plants with his letter, along with promises of more to come, from Brazil and New South

Wales. Nature went global during the late eighteenth century, and it primarily did so as part of an expansive British vision of imperial plant and animal transfer. In Cape Town, Phillip's primary concern was with obtaining livestock for introduction into New South Wales. On 10 July 1788, he kept his side of the bargain and sent Banks four cases of seeds and plants, along with animal skins and other natural history specimens. Global plant transfer was a two-way street; plants and animals were going in all directions across the globalized networks established by Banks, centered at Kew. Thus, it was not where the plants came from that defined their modernity, but instead their capacity to travel. Shortly after arriving, Phillip planted food plants, but these failed miserably because they were planted in the wrong season. He then established the Sydney Domain to grow plants and crops. In 1807, William Bligh created the first Australian botanic garden on this site, which is now the Royal Botanic Garden. Australia was thus already taking steps to become a leading participant in the creation of the globalized nature that emerged over the eighteenth and nineteenth centuries. The nature that was brought into being was consequently not simply a substitution of one nature for another, but instead a uniquely hybrid nature, of colonizers and native species. It was a nature built by virtue of its communication, as Darwin suggests, with cities both within and beyond Australia. It was also a nature whose justification lay in the assumption that it was replacing a *terra nullius*, an empty land belonging to no one. This way of seeing Australian nature, as Mulligan and Hill remark, "not only obliterated the indigenous people as active agents, it also led to a profound devaluing of most of Australia's phenomenally rich flora and fauna, which were commonly regarded as having no utility. Most plant communities in Australia were perceived, for example, as 'rubbish vegetation' that had to be cleared to make way for the more familiar 'useful' exotics" (*Ecological Pioneers* 304).

Many of these introductions proved invasive, among the first of these being rabbits and the prickly pear cactus.[3] When Robert Brown, the botanist on the Matthew Flinders expedition, visited Sydney in 1802, he could already list twenty-nine nonindigenous weeds growing there wild; the plantain (*Plantago major*), nettle (*Urtica urens*), scarlet pimpernel (*Anagallis arvensis*), shivery grass (*Briza minor*), petty spurge (*Euphorbia peplus*), mouse-eared chickweed (*Cerastium glomeratum*), and catchfly (*Silene gallica*) probably came from England, while lesser swinecress (*Coronopus didymus*) and cape gooseberry (*Physalisperuviana*) arrived from South America, and swan plant (*Gomphocarpus fruticosus*) had traveled from South Africa (Swarbrick,

"Weeds of Sydney Town" 42).[4] Flinders's expedition, modeled on the scientific voyages of Cook, was sent out to map the Australian coastline and to collect specimens of its natural history.[5] At each successive anchorage, following the practice of Captain Cook, the gardener on the voyage, Peter Good, sowed European grains and vegetables (P. I. Edwards, "Botany of the Flinders Voyage"). Many of these plants soon took root. Forty years later, large regions of Australia were overrun with alien plants, among these the blackberry introduced to Victoria by the Baron von Mueller in 1843 and Spanish flag (*Lantana camara*), which was colonizing the north. The Bathurst burr probably arrived in 1840, carried on the tails of horses imported from Chile, while the Noogoora burr came mixed with cotton seed (Burt and Williams, "Plant Introduction in Australia" 255–56). By the 1850s, Joseph Hooker counted more than 130 species that "have run wild, and assumed the positions and importance of native plants" in Melbourne alone (*Flora Novae-Zelandiae* cv). He commented to Charles Darwin that almost all of these were "social or roadside or cultivated-field plants of England, that must have been introduced over & over again into Australia & by hundreds of people" (Hooker to Charles Darwin, 25 Jan. 1859, *Correspondence of Charles Darwin* 7:238–39). Those transportation arteries radiating from cities envisioned by his grandfather had proven to be just as useful for weeds as for natural commodities.

As if the deliberate and accidental reseeding of Australia had not done enough damage, in the 1850s, under the auspices of the Acclimatisation Society of New South Wales, white Australians sought to expand their biological colonization of the continent on an even greater scale. The "great object of this Society," founder Dr. George Bennett declared, "is that of stocking our waste waters, woods, and plains with choice animals, making that which was dull and lifeless become animated by creatures in the full enjoyment of existence, and Lands before useless, become fertile with rare and valuable trees and plants, teeming with excellent fruits, variety of foliage, and gay and brilliant flowers" (cited in Low, *Feral Future* 30). As a result, over the past two hundred years much of Australia's indigenous nature has been shifted, displaced, and obliterated to suit the economic, aesthetic, and cultural needs of Europeans, and the impact of these ecological changes on the Aboriginal people and their culture has been enormous. With a contemporary insouciance that is frightening, Marianne North would remark in 1880, "It is curious, how we have introduced all our weeds, vices, and prejudices into Australia, and turned the natives (even the fish) out of it"

(*Vision of Eden* 171). As R. L. Burt and W. T. Williams note, "Australia is probably the only country in the world in which, with a single exception (*Macadamia*, the Queensland nut), every edible cultivar and every crop plant is, or has been derived from, an introduction, as indeed is true of most of the ornamentals, lawn grasses and sown pasture species" ("Plant Introduction in Australia" 252).

Australian Natural History as Translation

I have so far been discussing Australia as part of a broader global remaking of natures made possible by the managed transfer of plants and animals across a worldwide network of trade that radiated out from Kew Gardens with a view to suggesting that we recognize in this process a kind of biological translation, which transformed diverse local natures by transforming the physical contexts within which they existed and integrating them within a globalized system of trade and knowledge exchange. In thinking about the historical changes that took place in Australia and other colonial environments, it is clear that we should jettison the notion that these were homogeneous, stable environments composed of a permanent assemblage of unmoving plants and animals, each guaranteed their place in Creation since time immemorial, or that the plants and animals that were introduced into these natures did not also change these environments. Instead, we need to think in terms of natures in motion—of natures transplanted, transported, and displaced—and of environments undergoing major transformations, often at lightning speed. Australia's colonial nature was the site of an intense life-and-death struggle among different groups of human beings and between different biota and/or ecologies. What I have said about the mobility of physical natures during this period is even more obviously the case in regard to the natural history of the continent. Although this literature represented itself as a neutral activity of describing, classifying, and making available to a scientific and popular audience the natural history of Australia, it was symbolically engaged in taking possession of that nature on paper. Like Hope's address to Australia, colonial natural history was a *performative* language that took possession of the natures that it described, claiming these natures *in* and *through* translation.

One of the more interesting natural histories that appeared at this time is John White's *Journal of a Voyage to New South Wales with Sixty-Five Plates of Non-Descript Animals, Birds, Lizards, Serpents, Curious Cones of Trees and Other Natural Productions* (1790).[6] The title of the book indicates its hybrid generic character, for it seeks to provide a day-to-day journal account of the early settlement of Australia while also capitalizing on the public interest in natural history by providing a portfolio of high-quality natural history plates of some of the as-yet-undescribed animals. Whereas the journal recounts both the violence and anxiety that underpinned the colonial settlement of New South Wales, as the British acquired this new territory in the usual manner—through guns, disease, trade, coercion, and occupation—it also allows us to recognize the way in which the British claimed to possess a new and unfamiliar nature and on what basis that claim was made. Rather than being fought on the ground, the British claim on Australian nature was made textually, in the books, taxonomic lists, illustrations, and commentaries produced by naturalists. In the same way that early explorers claimed ownership of territory through maps and landscape studies, scientific naming and describing were inherently a claim of ownership, based on the idea that the British were the first people to truly understand this nature. Unlike the indigenous people, whom the British mistakenly believed had interacted with this nature while leaving it unchanged, they thought that through natural history they could improve on and remake this nature and that they were in the best position to speak for it. In this sense, colonial Australian natural history conforms to Michel Callon and Bruno Latour's definition of translation as "the set of negotiations, intrigues, acts of persuasion, calculations, acts of violence by which an actor or a force accords or allows itself to be accorded the authority to speak or to act in the name of another actor or force" ("Unscrewing the Big Leviathon" 279).

Natural history texts constituted a new relationship with that nature, one in which the naturalist, as a spokesperson whose authority derived from his participation in an international community of like-minded scientists, replaced the native inhabitants as the person best suited to speak and act for that nature, which otherwise remained silent. The relationships with plants and animals which most mattered, therefore, were not those that happened on the ground, but those that occurred in books. To experience nature was not enough; it had to be written down so that it could be transported back to England, where it could be used as the basis of comparisons and subsequent observations. Australian nature, in other words, happened elsewhere. It was brought into being as a repetition. That is why the possession of Australian nature did not operate solely within physical space, but instead was achieved symbolically in the world of print; although acted out on local ground, it was made possible by carrying this nature across the oceans, through the intervention of European metropolitan naturalists. And yet, because colonial natural history was produced collaboratively, by a community that extended across vast distances, occupying different social and geographical locations, it was itself contradictory and heterogeneous, riven by very different attitudes that its practitioners had toward the nature that they were translating. What we call "Australian nature" was ultimately not the product of a local understanding operating in a local context—as if John White were the single author of a book that recorded his observations of nature in situ—but instead something that was fabricated collaboratively across a complex network of writers, editors, scientists, and illustrators, a collaboration made possible by the fact that this nature had already left Sydney's shores. It emerged in an in-between space, out of what Homi Bhabha has called a "poetics of relocation and reinscription" (*Location of Culture* 323), by virtue of its being translated into something that could be circulated, exchanged, and, in turn, further transformed, by other people, many of them living in England.

The story of this collaborative translation began when White decided to keep a journal of the voyage. Encouraged by his London naturalist friend Thomas Wilson, he provided additional observations in natural history and collected a substantial number of specimens of flora and fauna, among these the skins of many birds, reptiles, and mammals, for which he apparently supplied notes, which are now lost. The journal manuscript, covering the first ten months of the settlement, and the natural history specimens, along with some of his own drawings, were sent to Wilson from Australia on the *Golden Grove* or *Fishburn*, which left Port Jackson on 19 November 1788 (Nelson, "John White's *Journal*" 110). In the final paragraph of his journal, White indicates that he was planning to continue the journal and to provide considerable "additions to the Natural History" with the "next dispatch" (*Journal* 217). Given the public interest in the new colony, Wilson immediately set to work preparing the manuscript (which is also now lost) for publication. First, he arranged for the zoological skins to be prepared, stuffed, and mounted. This work would have been mostly completed by 6 July 1789, when he displayed many of the botanical specimens and birds at a meeting of the Society for Promoting Natural History. The specimens were then advertised, 26 August to 11 November 1789, and displayed at the Leverian Museum (C. E. Jackson, *Sarah Stone* 52). Around the same time, he commissioned watercolor illustrations of the mounted specimens. Most of these were done by illustrators connected with the Leverian Museum, notably Sarah Stone, who painted watercolors of the majority of the birds; Frederick Polydore Nodder, who painted the plants; Edward Kennion, who contributed a plate on the root of the yellow gum tree; and Charles Catton Jr. and Mortimer, who illustrated the mammals. These watercolors were then displayed on the premises of Debrett's, a bookseller in Piccadilly, and the exhibit was used to publicize and obtain subscribers for the book. Wilson also brought several of them to a 5 October meeting of the Society for Promoting Natural History, of which he was a member. At the same time, he had arranged to have the drawings engraved by Thomas Milton, a descendant of the brother of the poet, who also produced the vignette that appears on the title page of the volume. The engravings were completed by 29 December 1789, and the book appeared sometime early in 1790.

In his advertisement to the volume, Wilson drew attention to the accuracy of the drawings and engravings, writing that "the Public may rely, with the most perfect confidence, on the care and accuracy with which the Drawings have been copied from nature, by Miss Stone, Mr. Catton, Mr. Nodder, and other artists; and the Editor flatters himself the Engravings are all executed with equal correctness, by, or under the immediate inspection of Mr. Milton. The Birds, &c. from which the drawings were taken are deposited in the Leverian Museum" (*Journal* vi). Wilson also used his connections with the Society for Promoting Natural History in order to enlist the assistance of notable metropolitan naturalists in identifying and scientifically describing the specimens. The founder of the Linnean Society, the preeminent botanist Sir James Edward Smith, was responsible for the plants, while the anatomist John Hunter dealt with the mammals, and Sir George Shaw, the keeper of zoology at the Museum of Natural History, examined the birds, reptiles, and fish. White appears to have originally provided the Aboriginal names for the specimens whenever he could, and John Hunter stressed their importance, writing that "it is much to be wished that those gentlemen who are desirous of . . . promoting the study of Natural History . . . would endeavour to procure all the information they can relating to such specimens as they may collect. . . . In collecting animals, even the name given by the natives, if possible, should be known; for a name, to a Naturalist, should mean nothing but that to which it is annexed" (269–70). Noting, for instance, that the *potoroo*, which, following White, he called the "Kangaroo rat," was, properly speaking, neither a "kangaroo" nor a "rat," Hunter preferred to use the Aboriginal names for these species. Thus, he refers to the *Wha Tapoua Roo* (i.e., the common brushtail possum); the *dingo*, or dog (now *Canis dingo*); and the *Tapoa Tafa*, or *Tapha* (now referred to as the brush-tailed phascogale). He also emphasized the need to provide as many observations as possible about the habits and life of the animals collected. Smith and Shaw, however, excised all traces of Aboriginal culture, replacing any indigenous names that they might have had with new European scientific ones.

We should understand the resulting text, then, as a collaborative work, the product of the translation of a nature from New South Wales to England, in which materials first provided by White were reworked by an editor and supplemented by metropolitan artists and scientists. Originally drawing on the Aboriginal knowledge and names for this nature, White sent his observations and specimens to London, where they were subsequently adapted to conform to a British scientific understanding of what this nature was and to confirm British hegemony in the science of natural history. With each new stage in this process a new version of that nature appeared. Already lifted out of the living biological and cultural contexts in which that nature had existed for millennia, the nature that appeared in White's *Journal* was a product of collaborative translation adapted to a metropolitan audience, a "carrying across" that drew its character as much from European science and readers as from its original Australian context.

In preparing the *Journal* for publication, Wilson faced a predicament, for he had in his hands basically two very different kinds of material: first, there was White's narrative of the early months of the colony; second, he had sixty-four superbly illustrated plates of the largely unknown flora and fauna of Australia (which he incorrectly numbered as sixty-five), along with verbal descriptions and taxonomic identifications by the leading naturalists of the day. Unfortunately, given the extreme hardships faced by the colonists, with little available food, rampant scurvy, ongoing thefts, public lashings and hangings, and the daily threat of violence from the Aboriginal people, taking time out to study the natural world, beyond the objective of gaining knowledge about edible and medicinal plants, does not appear to have been a major priority for White, at least not on the evidence provided by the journal. For both the settlers and the Aboriginal people, at least as they are portrayed by White, the nature that seems to have mattered most was the nature that they could eat. With the exception of about four pages on the kangaroo (*Journal* 179–83), most of White's comments about the animals that he encountered have less to do with their appearance or habits than with their suitability as food. When animals and birds are mentioned in the narrative, it is less often as part of a natural history observation than in reference to their having been killed for food, as in the following passage, which may have influenced Gosse: "The trees around us were immensely large, and the tops of them filled with loraquets and paroquets of exquisite beauty, which chattered to such a degree that we could scarcely hear each other speak. We fired several times at them, but the trees were so very high that we killed but few" (151). We learn from White, for instance, that a crested cockatoo when combined with two crows makes "a kettle of excellent soup" (148), and that even though he was impressed by the speed of the kangaroo, he felt that "in any other country I am sure that such food would be thrown to the dogs; for it has very little or no fat about it, and, when skinned, the flesh bears some likeness to that of a fox or lean dog" (183–84). White undertook some anatomical work on an emu brought to camp, but probably his most important observation was that "the flesh of this bird was very good, and tasted not unlike young tender beef" (131). Nature is here deeply connected with utility and survival, so it is perhaps not surprising that many of the major conflicts between the British and the Aboriginal people happened in situations where they were fishing or gathering food. In fact, White's first account of possible conflict occurs when the British were seining for fish:

> Some of them were present, and expressed great surprise at what they saw, giving a shout expressive of astonishment and joy, when they perceived the quantity that was caught. No sooner were the fish out of the water, than they began to lay hold of them, as if they had a right to them, or that they were their own; upon which the officer of the boat, I think very properly, restrained them; giving, however, to each of them a part. They did not at first seem very well pleased with this mode of procedure, but on observing with what justice the fish was distributed, they appeared content. (116–17)

This priority given to nature as food in the journal must have caused Wilson some difficulty because such a view of nature ran counter to the notion of scientific disinterest shaping natural history as a professional field. Although Captain Bligh's well-known sketch of an echidna was done just before he ate it, and although Charles Darwin is famous for having eaten a lesser rhea before he realized it was a new species, naturalists were not supposed to be preoccupied with eating the birds they studied or with considering their characteristics on the end of a fork.

Even more to the point, White rarely even mentions in his *Journal* the animals that he collected. Wilson's answer to this problem, which I would essentially characterize as there being literally no place in the *Journal* for the nature that he had on hand, was to find whatever places he could in the narrative to insert the plates and the accompanying verbal descriptions, while reserving the rest for the appendix. He also studiously avoided inserting birds when they were mentioned as having been shot for food, as in the case of the parakeets, mentioned above. Usually he inserted a plate whenever a species was first mentioned by White, as in the following: "The *Pennantian Parrot* [i.e., crimson rosella] (of which see plate annexed) was about this time first noticed" (174); or "We this day discovered the Banksian Cockatoo [i.e., black cockatoo]" (139); or "We discovered the *New Holland Creeper* [i.e., New Holland honeyeater]; (See plate annexed)" (186). At other times, the plate follows directly on White's mention of having collected a specimen, as in the following instance: "A *New Holland Cassowary* [emu] was brought into camp" (129). Occasionally, he decided against inserting a plate in the main text, for instance, when White mentions the red gum tree; at other times, he seems to have missed an opportunity, for instance, when White discusses the white feluca (i.e., white gallinule) of Lord Howe Island, a flightless bird that quickly became extinct, because of the ease with which it could be taken. In cases where Wilson inserted a bird illustration into the text, he also included Shaw's scientific description, often making it appear as if these observations were actually being made by White. Here is the description of the wattled bee-eater (i.e., little wattlebird):

> April 17: ... We, accordingly, after an expeditious walk, reached the stream from whence we had set out in the morning, and, taking up the tents and provisions which we had left, proceeded a little farther down, to the flowing of the tide, and there pitched our tents for the night; during which it rained very heavily, with thunder and lightning. The Wattled Bee-eater, of which a plate is annexed, fell in our way during the course of the day. This bird is the size of a *missel thrush* but much larger in proportion, its total length being about fourteen inches. The feathers on the upper part of the head, longer than the rest, give the appearance of a crest; those of the underpart are smooth; the plumage for the most part is brown, the feathers long and pointed, and each feather has a streak of white down the middle; under the eye, on each side, is a kind of *wattle*, of an orange colour; the middle of the

belly is yellow; the tail is wedge-shaped, similar to that of the magpie, and the feathers tipped with white; the bill and legs are brown.

This bird seems to be peculiar to New Holland, and is undoubtedly a species which has not hitherto been described (144).

The polyvocal aspects of the *Journal* can clearly be recognized here, for two authorial voices, existing in two different parts of the globe, are here being brought together in a single text. At other times, especially with birds that had already been described scientifically, Wilson makes it seem like White is drawing on John Latham's *Synopsis of Birds*, which was based on specimens obtained through Cook's voyages. Thus, for the blue-bellied parrot (i.e., rainbow lorikeet), the text reads, "We likewise saw several Blue-bellied Parrots. This is a very beautiful bird; and Mr. *Latham*, whose leave we have to copy the account of it, from his *Syn.* vol. i., p. 213, N° 14. B., describes it thus: 'The length is fifteen inches; ....'" (140).

The collaborative polyvocality produced by multiple authors also extends to the illustrations, for they retain the traces of the process by which this nature was brought into being. Sarah Stone is a wonderfully accomplished artist, whose skill in drawing and coloring these specimens was well appreciated by her contemporaries. Although Wilson insisted that "the Drawings have been copied from nature," it must also be admitted that her birds look less like creatures living in a physical environment than specimens of eighteenth-century taxidermic display. In plate 2, her image of the kookaburra, labeled the "Great Brown King's Fisher," Stone transforms the twig upon which the specimen was mounted into a miniature tree, which has the effect of making either an eighteen-inch bird gigantic or a large tree into a bonsai specimen. At other times, the mounting twig becomes a branch on a ruined stump, complete with lichens. Since Stone knew little about the actual flora of Australia, her birds inhabit a nondescript nature. Also, as Christine E. Jackson has observed, Stone was intent on recording "precisely what was in front of her," so she often reproduces "all the inadequacies of eighteenth-century taxidermy" (*Sarah Stone* 10). Many of her birds, most notably the white feluca, have extraordinarily long necks, not because that is how they looked in life, but because their necks had been stretched when they were hung from the hunter's belt. Hanged birds would seem particularly appropriate representatives of life in the new convict settlement. The birds in Stone's illustrations also often look either overstuffed or understuffed, but never like creatures that actually have musculature just beneath

Sarah Stone, *Great Brown King's Fisher*, National Library of Australia, ANL PET—FRM F97 / Phillip's voyage 1789. NK3799

their feathers. Also, because the colors of birds' legs, eyes, and soft parts usually fade soon after death, in portraying these birds as she saw them, Stone's paintings recorded how they appeared in death. This is a dead nature. To recognize it as such, we only need to see where the paintings fail to translate the specimen into the living bird. Despite these shortcomings,

Stone cannot be blamed for more egregious failures, such as the emu that looks like a character on *Sesame Street* (she was copying an illustration sent by White), or the disastrous illustration of a kangaroo standing at attention like a trained dog. No one who had ever seen a living kangaroo could ever have drawn it in this manner.

Because White had little occasion to discuss the wildlife around him in his journal, Wilson often had to resort to inserting plates and descriptions wherever he could find room for them in the narrative. The effect of these insertions is often very troubling. For instance, on 21 May White describes the wounding of William Ayres and the murder of Peter Burn. Midway through his account (which appears on pp. 156 and 158 of the *Journal*), Wilson inserted a description of the Port Jackson thrush (i.e., grey shrike-thrush). The narrative thus reads,

> After the operation, he informed us that he received his wound from three of the natives, who came behind him at a time when he suspected no person to be near him except Peter Burn, whom he had met a little before, employed on the same business as himself. He added, that after they had wounded him, they beat him in a cruel manner, and stripping the cloaths from his/back, carried them off; making signs to him (as he interpreted them) to return to the camp. He further related, that after they had left him, he saw Burn in the possession of another party of the natives, who were dragging him along, with his head bleeding, and seemingly in great distress, while he himself was so exhausted with loss of blood that, instead of being able to assist his companion, he was happy to escape with his life.
>
> The *Port Jackson thrush*, of which a plate is annexed, inhabits the neighbourhood of Port Jackson. The top of the head in this species is blueish-grey; from thence down the hind part of the neck and the back the colour is a fine chocolate brown; the wings and tail are lead colour, the edges of the feathers pale; the tail itself pretty long, and even at the end; all the under parts from chin to vent are dusky-white, except the middle of the neck, just above the breast, which inclines to chocolate. The bill is of a dull yellow; legs brown.

On 30 May, White described in gruesome detail the deaths of William Okey and Samuel Davis, who were cutting rushes used for thatching. The account covers two pages of text (*Journal* 160, 162) and provides the occasion for a plate and description of the yellow-eared flycatcher (i.e., yellow-eared honeyeater), which is inserted in the middle of the account (161).

Sarah Stone, *Port Jackson Thrush*, National Library of Australia, ANL PET—FRM F97 / Phillip's voyage 1789. NK3799

Captain Campbell of the marines, who had been up the harbour to procure some rushes for thatch, brought to the hospital the bodies of William Okey and Samuel Davis, two rush-cutters, whom he had found murdered by the natives in a shocking manner. Okey was transfixed through the breast with one of their spears, which with great difficulty and force was pulled out. He had two other spears sticking in him to a depth which must have proved mortal. His skull was divided and comminuted so much that his brains easily found a passage through. His eyes were out, but these might have been picked away by birds. Davis was a youth, and had only some trifling marks of violence about him. This lad could not have been many hours dead, for when Captain Campbell found him, which was among some mangrove-trees, and at a considerable distance from the place where the other man lay, he was not stiff nor very cold; nor was he perfectly so when brought to the hospital. From these circumstances we have been led to think that while they were dispatching Okey he had crept to the / trees among which he was found, and that fear, united with the cold and wet, in a great degree contributed to his death. What was the motive or cause of this melancholy catastrophe we have not been able to discover, but from the civility shewn on all occasions to the officers by the natives, whenever any of them were met, I am strongly inclined to think that they must have been provoked and injured by the convicts. We this day caught a Yellow-eared Flycatcher (see annexed plate). This bird is a native of *New Holland*, the size of a martin, and nearly seven inches in length; the bill is broad at the bottom and of a pale colour; the legs dusky; the plumage is mostly brown, mottled with paler brown; the edges of the wing feathers yellowish; the under part of the body white, inclining to dusky about the chin and throat; the tail is pretty long and, when spread, seems hollowed out at the tip; beneath the eye, on each side, is an irregular streak, growing wider and finishing on the ears, of a yellow or gold colour (160–61).

In such moments, a reader cannot help but ask, "Why on earth is a bird appearing at this point in the narrative? What is the relationship between the nature in this illustration and the journal in which it is found?" Here the seamlessness of the text breaks down, and different voices and values shaping the translation of this nature become apparent. While White was deeply troubled by the deaths of Burn, Okey, and Davis, the editor Wilson was not, and, of course, the metropolitan scientists, illustrators, and engravers were able to play their roles in the production of this translation without even being aware of this history. As Susan Bassnet has remarked, "translation . . . is a primary method of imposing meaning while concealing the power relations that lie behind the production of that meaning." Citing censorship, she notes how easy it is "to see how translation can impose censorship while simultaneously purporting to be a free and open rendering of the source text. By comparing the translated version with the original, the evidence of such censorship is easy to see where written texts are concerned" ("Transla-

Sarah Stone, *Yellow Eared Fly Catcher*, National Library of Australia, ANL
PET—FRM F97 / Phillip's voyage 1789. NK3799

tion Turn" 136). White's *Journal* allows the reader to see the selective aspects of the Australian nature that appeared in England in the 1790s. The contradiction between the plates and the journal makes it clear that this nature achieves its heightened aesthetic calm by virtue of its separation from its original local context, the realities of colonial settlement. The plates present a nature that is oblivious to all that stands beyond it. White came to feel that this indifference extended to the colonists themselves, and in December 1790, he published a scathing letter critical of the entire colonial enterprise, describing Australia as "a country and place so forbidding and so hateful as only to merit execration and curses" (Britton, *Historical Records* vol. 1, pt. 2, 332). By this time, whatever plans he had for continuing his journal and natural history studies were over.

One might conclude, on the basis of the serenity of the dead birds portrayed in these plates, that Australian nature was removed from the human struggles that the journal reminds us were taking place all around them, but the narrative suggests otherwise. The encounter between the British and the animal world seems normally to have been punctuated by gunshots, which ring out on almost every page of the journal: "This day, for the first time, a Kangaroo was shot and brought into camp" (*Journal* 126); "we found nine birds, that, whilst swimming, most perfectly resembled the *rara avis* of the ancients—a *black* swan. We discharged several shot at them" (137); "we saw a great many ducks and teal; three of which we shot" (149); "we saw a kangaroo, which had come to drink at an adjacent pool of stagnated water, but we could not get within shot of it" (150); "We this day shot a *Knob-fronted Bee-eater*; (See plate annexed)" (190); "We this day shot the *Sacred Kings-Fisher* (See Plate annexed)" (193); and so on. Nothing in White's journal speaks of human beings being part of a greater living society, because he was not seeking to portray the life of nature. Instead, he was collecting the natural world to be transported elsewhere. Whereas the convicts, instead of dying, were transported, Australia's nature suffered both sentences. Australian nature was isolated from human life, and, no doubt, much of the theoretical innovativeness of natural history emerged from its separation of nature from culture. Yet in the modern world, where nature is constantly affected by human beings, the pretense of separation has left nature unprotected. And, of course, in colonial contexts, where specific natures were associated with specific peoples, this separation placed the cultural factors underlying the transformation of colonial natures outside of consideration.

In his *Journal* White notes that in his reconnaissance of the area around Port Jackson he observed that the Aboriginal people had produced rock engravings, "proofs of their ingenuity" in "figures cut on the smooth surface of some large stones. They consisted chiefly of representations of themselves in different attitudes, of their canoes, of several sorts of fish and animals" (141). In their pictographs, the Aboriginal people usually drew images of hunting

William Westall, *Chasm Island, Native Cave Painting*, National Library of Australia, call number 585900

and fishing scenes and also depicted the animals that they most depended on for food, frequently kangaroo and wallabies, though bandicoot, opossum, echidna, and dingos were also portrayed.[7] Fish were also depicted, notably sharks, sawfish, stingrays, and sunfish, as well as other marine animals, such as dolphins, dugongs, and turtles. Birds appear less frequently, though jabiru, herons, scrub hens, emu, pigeons, and geese can be found. Snakes and lizards are also pictured, but they are uncommon. That the Aboriginal people had a developed knowledge of the natural world was obvious, even to British settlers. The marine William Dawes, in the dictionary that he compiled in 1790 of the "Grammatical Forms of the Language of the N. W. Wales in the neighbourhood of Sydney," notes fifteen names for birds, fourteen for animals, seventeen for fish, and eighteen for plants. If he had been allowed to remain in the settlement, this number probably would have increased. Through these representations, the Aboriginal people were marking their place in nature and linking that nature to a place. Rock etchings and paintings constituted a cultural act of claiming this nature as their own, and, as such, they were similar to the representations being produced by British naturalists, cartographers, and landscape artists. The primary difference was that the British laid claim to this nature from a distance, by carrying it back to England and integrating it within the globalizing networks of science and trade. It was not nature itself but its mobilized inscription that mattered most. William Westall, who was a close friend of Robert Southey and became one of Wordsworth's favorite Lake District landscape artists, learned his craft as the appointed landscape artist for Matthew Flinders's *Investigator* expedition. In 1803, he came upon a substantial number of cave drawings on Chasm Island in the Gulf of Carpentaria. In what can be seen as an enactment of the fundamental gestures of British colonial natural history, Westall painted the drawings. Whereas Aboriginal natural history illustration, like the nature that it represented, was fundamentally linked to place, Westall lifted these paintings out of the context in which he found them and brought them back to England.

CHAPTER FOUR

An England of the Mind
Gilbert White and the Black-Bobs of Selborne

> England is London says one, England is Parliament says another, England is the Empire says still another; but if I be not much mistaken, this stretch of green fields, these hills and valleys, these hedges and fruit trees, this soft landscape is the England men love.
>
> (Collier, *England and the English* 316 17).

Gilbert White could not hold back his disappointment when he learned in the spring of 1768 that his new acquaintance, Sir Joseph Banks, who had recently gained a reputation in British natural history circles by virtue of his six-month expedition to Newfoundland, was now about to leave England again, this time to accompany Captain Cook on his first voyage to the Pacific. The rural parson had hoped that Banks and the fellow naturalist and antiquarian Thomas Pennant would visit the small Hampshire village of Selborne on their way to Goodwood, where two living moose had recently arrived from Canada. White had promised Banks a specimen of the rare parasitic plant toothwort (*Lathraea squamaria*), and he was certain that "some rare matters may be discovered" on the adjacent moors. Birds abounded in the chalk hills, bogs, and woods of the area, he wrote, and to add a further enticement, White dropped the hint that there were "vast large bats" in the area, the as-yet-unidentified noctule bats that White would eventually be credited with discovering. In light of this new sea voyage, the plans changed, and White broke his customary reserve in acknowledging his frustration, writing, "I must plod on by myself with few books and no soul to communicate my doubts and discoveries to" (12 Apr. 1768, *Natural History and Antiquities of Selborne* 2:242). Crusoe on his island could hardly have been more isolated from the scientific world than this provincial clergyman in this out-of-the-way village accessible only by narrow cart paths and lanes. When two years later White responded to Daines Barrington's proposal that he write "an account of the animals in [his] neighbourhood" (Foster, *Gilbert White* 129), his isolation from a scientific community surfaces again. "It is no small undertaking for a man unsupported and alone to begin a natural history from his own autopsia," he complained.[1] White would demonstrate how much could be learned by "seeing with one's own eyes," but he was acutely aware that science was a collective venture and that he was at its periphery.

Both Banks and White shaped British natural history in important ways, yet the differences between the two men and what they meant by "nature" could hardly have been greater. For the young, wealthy, ambitious, globe-trotting Banks, botany was in the vanguard of modern science, and its scope was international and imperial. The engine of the agricultural revolution that was transforming British rural life, botany was a science that promised to change the very nature of things. Like most cosmopolitan naturalists, Banks understood natural history as a global science that mobilized local natures for the benefit of both Britain and humankind. Except for a short stint at Oriel College, Oxford, the fifty-year-old bachelor White lived most of his life in Selborne. In an age of travel, he preferred to stay at home. Where Banks would circumnavigate the earth searching for plants and setting an example for the many traveling naturalists who followed him, White's constitution prevented him from even taking a coach ride, so on the few occasions that he traveled, for instance, between Selborne and his curacy in Farringdon, he did so by horse (Holt-White, *Life and Letters* 1:168). Unlike Banks, whose London home on New Burlington Street served as the center of a global correspondence network of naturalists bent on cataloging and exchanging plants, prints, and knowledge, White communicated with a small circle of friends, most of them family members, and he lived in what was seen as one of its most backward rural parishes. Nevertheless, White's letters, once they were published as a book, would become the first English "classic" natural history book and over the course of the next two centuries would go through more than two hundred editions. Charlotte Smith, John Clare, and Wordsworth were all deeply influenced by him, and in 1833 Wordsworth asked Henry Crabb Robinson to purchase a copy for Dorothy, who had long wished for one.[2] It was also Parson White who would become, over the course of the nineteenth century, "Saint Gilbert," the father of English ecology (Mabey, *Gilbert White* 9). White's world was small and provincial, a static fragment of a way of life and of a nature that for most people in England was rapidly becoming a memory—soon even a myth—of its past. Yet this smallness also magnified things and allowed White to see things as a whole. By considering in detail the complex interrelations that the creatures of Selborne had with each other, White would establish a new kind of ecological writing. In his hands, natural history became the study of a living nature, focused on capturing "the life and conversation of animals" (*Selborne* 140), not on reproducing the textual equivalent of a library or museum.

For White, natural history was preeminently a science of the local. The deep relationship between nativity and place and the multifarious ways in which creatures inhabited his locality meant everything to him. Natural history was thus grounded in place; it was a form of "place-making," what Martha Bohrer has called "a tale of locale," reflecting the increasing importance that place would play in English conceptions of the nation during the nineteenth century. In the advertisement to *Selborne*, White asserts, with "proper deference," but with apparently little sense of irony, that his book represents a new kind of "*parochial history*," focused not only on the antiquities but also on the natural history of a parish. Such histories, he suggests, are best accomplished by "stationary men," who "pay some attention to the districts on which they reside" (*Selborne* 7). At a time when natural history was a traveling science, written by innumerable "traveling naturalists" who sought out exotic and undiscovered species of plants and animals across the globe, White stands out for his insistence that it could just as easily be done by those who stayed at home.[3] "Every kingdom, every province, should have its own monographer," he writes, because "men that undertake only one district are much more likely to advance natural knowledge than those that grasp at more than they can possibly be acquainted with" (115). Furthermore, "all nature is so full, that that district produces the greatest variety which is the most examined" (51). The natural world described in *Selborne* is as stationary and centered as its observer. It embodies what Lisa H. Malkki has termed a "sedentarist metaphysics," a world in which the identity of every person and creature is rooted in its place or "station" in the natural and social order ("National Geographic" 24–44). In representing very different conceptions of nature and what natural history should be and do, White and Banks can be said, then, to constitute the systolic poles of British natural history. Alongside the emergence of a natural history aimed at collecting, mobilizing, and changing nature on a worldwide scale, an imperial natural history that would be embodied in the foreign and exotic natures that were increasingly appearing in museums, zoos, gardens, and books, there emerged an alternate nativist, localist, domestic natural history that valued the familiar and the everyday inhabitants of the English countryside over the exotic and the foreign. Against a nature shaped by travel and the globalized exchange of plants, animals, and knowledge, White produced a nature that was valued for being rooted in place, a nature that could not move, except in books.

Both men brought into view new natures that had not been seen before, and they did so by having recourse to different forms of print media. Banks was catapulted into the cultural center of British life by his participation in the first of Cook's voyages to the South Pacific. Through the publication of the lavishly illustrated accounts of the voyages and the material display of exotic specimens and cultural artifacts, these expeditions set the standard by which other scientific expeditions would be measured. Cook's expeditions were not the first to include a team of naturalists and artists to collect plants and animals, make observations, and produce scientific illustrations, since this precedent had already been established by the French, but they did firmly establish in the public mind for the next fifty years the value of such voyages. Alan Frost speaks of the period from 1764 to 1806 as "a second great age of European exploration," distinguished from its predecessor by the important role that natural history played within these state-sponsored expeditions ("Science for Political Purposes" 27). In the ambitious, state-authorized naval scientific voyages of La Pérouse, Vancouver, Baudin, and Flinders, culminating in Darwin's voyage on the *Beagle*, and in the land expeditions of Pallas in Russia, Napoleon in Egypt, Franklin in northern Canada, Lewis and Clarke in the western frontier of the United States, and Ruiz and Pavón in South America, cartographical surveys were no longer enough: the goal was to produce complete inventories of physical natures, thoroughly detailed natural histories of a region's plants, animals, and people. The French expedition of Nicolas Baudin, for instance, set out for Australia in 1800 in two ships, the *Géographe* and *Naturaliste*, with a scientific company of four astronomers and hydrographers, three botanists, five zoologists, two mineralogists, four artists, and five gardeners. Over a four-year period, the expedition would collect over one hundred thousand animal specimens, twenty-five hundred of which were new species. Flinders's expedition to Australia on the *Investigator* (1801–4) was overseen by Banks and was similarly equipped with the botanist Robert Brown (who returned to England with over three thousand specimens, half of which were new), the botanical illustrator Ferdinand Bauer, the landscape painter William Westall, a gardener, a miner, and an astronomer. As Bernard Smith has suggested, the empirically based art and science of these voyages was of momentous importance in demonstrating how science, working with writers and artists, could collect and organize the materials for a concrete and coherent representation of a new world. "The opening of the Pacific," he writes, "is ... to be numbered among those factors contributing to the triumph of romanticism and science in the nineteenth-century world of values" (*European Vision and the South Pacific* 1). The cultural reproduction of the Pacific and its translation to England were triumphs of print technology and material culture.

Cook and Banks returned to England with a massive quantity of cultural artifacts, natural history specimens (including the young Tahitian Omai), paintings, cartographic surveys, geographies and ethnographies, and a wealth of stories, and from this abundance of objects and representations emerged the idea of a vast Pacific world, composed of an extraordinary diversity of islanded peoples and cultures, which took hold of the British public imagination. Banks and Cook directed Britain's gaze outward, to encompass global possibilities. Certainly, it was the sheer amount of stuff that would most have impressed anyone visiting Banks's residence in Soho. In a 1772 letter to White, William Sheffield, keeper of the Ashmolean, conveyed his astonishment: "Had I not been an eyewitness of this immense magazine of curiosities, I could not have thought it possible for him to have made a twentieth part of the collection" (Sheffield to White, 2 Dec. 1772, *Natural History and Antiquities of Selborne* 2:98). In one room alone, there was

> a large collection of insects; several fine specimens of the bread & other fruits preserved in spirits; together with a compleat *hortus siccus* of all the plants collected in the course of the Voyage. The number of plants is about 3000; 110 of which are new genera and 1300 new species, which were never seen or heard of before in Europe. What raptures must they have felt to land upon countries where everything was new to them! Whole forests of nondescript [i.e., undescribed] trees clothed with the most beautiful flowers and foliage; and these too inhabited by several curious species of birds, equally strangers to them. I could be extravagant upon the topic. (99)

White's impact was less immediate, but it was in many ways just as far-reaching. Although White successfully distinguished the three species of willow warbler and was the first to describe the lesser whitethroat, the harvest mouse, and, extraordinarily, the noctule bat, the largest bat in England, his major importance derives from *The Natural History of Selborne* (1788), which took him almost twenty years to write. This book essentially produced what the British meant when they referred to "English nature" during the nineteenth century. *Selborne* came to embody textually the national myth of England as a domestic rural idyll. The natural world portrayed in *Selborne* was, nevertheless, just as much a product of scientific representation as the natures introduced into England through colonial expeditions and surveys. In fact, as Tobias Menely has observed, "its distinctiveness was in fact made visible and compelling as a result of the global accumulation of natural knowledge that marks the latter half of the eighteenth century" ("Traveling in Place" 48), for White transferred the techniques of field observation which had been developed in colonial settings to the description of his birthplace. Rather than opening outward onto the seas, however, White's was a landlocked nature, mostly composed of insiders. It was in every sense English, social, and domestic. White's great achievement was to make the creatures that composed this domestic rural nature as important as its people, and some would say even more important. The American poet James Russell Lowell captured its mythic dimensions when he described the book as "the journal of Adam in Paradise" (*My Study Windows* 1). Since Eden has always been too foreign and too tropical for English taste, White recovered his own temperate and seasonal version of it in a rural village in Hampshire. Less interested in naming species than in learning how they lived, treating them as if they were neighbors and he a somewhat "nosy neighbour," White's descriptions performed a truly *naturalizing* function in the original political sense of the term: they accorded the creatures of Selborne a status akin to citizenship by giving them a place and a home. Despite whatever migratory wanderings they may have engaged in throughout the year, his natural history made them English through and through. In this regard, *Selborne* is the literary precursor of a long tradition of books about English domestic nature, such as *Wind in the Willows*, *Alice in Wonderland*, or *Winnie the Pooh*. Like Wordsworth's Lake District, the cultural influence of the nature of *Selborne* on nineteenth-century readers was inseparably bound up with its formal textual qualities. Instead of being an expensive edition with detailed scientific illustrations, White produced a nature that was primarily embodied in conversational prose. The intimacies of the epistolary form and the room that it gave for gossipy anecdotes allowed White to provide a heightened sense of day-to-day discoveries and to speak personally of the impact that this nature had had on him over the decades. By virtue of the place-making power of his letters, White created a nature that was a comfortable fit, as parochial as its author and as English as house sparrows and hedgehogs.

White is the great mythologist of English nature as a *nature present*, that is, a nature that exists for us in a permanent present tense. Despite the change in the seasons, White's nature remains the same, whether he is talking about the past, present, or future, and his book makes us believe that this nature will continue for as long as England celebrates village life. In this regard, White does not really display any nostalgia for the natural world (he is not a Wordsworth), because nature for him is not associated with the past. As the nineteenth century wore on, however, the stable rural nature portrayed in *Selborne* became more and more a thing of the past, and the book became the popular locus of nostalgia for England's past, as if its author, like an indigenous informer, were living in the past without knowing it. Instead of declining, the cultural importance of the book increased as it became a textual equivalent of what Cannon Schmitt has called a *"lieu de mémoire,"* that is, "a place that by virtue of its own perceived archaism returns those who traverse it to the past" (*Darwin and the Memory of the Human* 40). This situation has much to do with the increasingly important role that *Selborne*, the book, rather than Selborne, the place, came to play in the cultural imaginary of an increasingly mobile nation. Benedict Anderson, in *Imagined Communities*, has argued that nationalism was an outgrowth of print culture, as the common activity of reading newspapers brought otherwise isolated and widely separated individuals together, allowing them to imagine themselves as parts of a larger national community. Local insularities, he suggests, were forged into common identities through the process of sharing the news. *Selborne* contributed to the emergence of the idea of a nation in a somewhat different manner, not by giving its readers directly a shared political milieu, but by providing them with a powerful myth of what true English nature was and how they should relate to it: it gave them a nature that they could share, if not as villagers, then as readers of this book. It grounded English nature in an actual place, in the rural Hampshire hamlet of Selborne. By virtue of print, this local nature, which few people would have actually known firsthand, was given national, and then international, distribution. *Selborne* provided its readers with something that the news could not, that is, access to an experience of a natural world that was to a great degree quite different from the actual places in which they lived.

The embodiment of traditional rural English nature, *Selborne* appealed to modern readers, seeking a nature that they believed was fast disappearing and being replaced by industry and empire. It allowed them to stand beside White at the center of a world that primarily existed on the margins of English cultural life. Empire required worldly knowledge and globalized forms of networking and exchange, and late eighteenth- and early nineteenth-century British society was undergoing political and social revolutions of many kinds. *Selborne* provided its readers—both at home and abroad—with an anchoring counter-nature: through its hyper-Englishness and its anti-cosmopolitanism, this science of a locality in all its limitedness was everything that imperial Britain was not. In a world where identities were no longer certain, stable, or homogeneous, *Selborne* provided the British with a national myth of a *native* nature whose essence lay in its insularity, partialness, and unwillingness to change. As an anonymous reviewer of the 1830s commented, Selborne was removed from "the improved knowledge and refinement which belong to these enlightened and virtuous times. It has been excluded from the blessings of increasing commerce and population, from factories and filiations, manufactures and Methodism, genius and gin, prosperity and pauperism" ("Selborne" 565–66).

British national identity in the age of empire was, indeed, the product of reading, but it was fractured—proudly expansionary yet also nervously agoraphobic, supremely confident about the nation's capacity to rule others while simultaneously worried that painting the globe in pink was not making it any less foreign, strange, or dangerous. Global rule could also lead to global dependency. *Selborne* and *Robinson Crusoe* are thus the two great *island narratives* of the eighteenth century: through Defoe, a British reader could act out the contradictory fantasy and anxiety of possessing an island all to himself (despite the threat of others); in White, s/he was given the equally ambiguous opportunity of retiring from the world of global commerce to the original native nature of the island, to enjoy the security of living simply, naively, and even a little complacently, life at the very center of a very small world. In a society increasingly characterized by mobility, change, and the integration of localities into global commercial and informational networks, White's stationary, isolated, domestic nature appealed to those who felt that the modern world, in Wordsworth's phrasing, was simply "too much with us." "His world," commented James Fisher, "is round and simple and complete; the British country; the perfect escape. No breath of the outside world enters in; no politics; no ambition; no care or cost" (*Natural History of Selborne*, ed. Fisher, x). As Banks's dream of the imperial management of a globalized nature faded into the past, White's *Selborne* continued to function as a spiritual anchor, as it confirmed that despite Britain's involvement in changing everybody else's natures, its own native nature, that of the rural English parish, could still be found, if only between the covers of a book. Amid all the changes wrought by empire and settlement, *Selborne* gave the British the security of knowing that as long as their villages still had sparrows, swallows, and martins, they were never far removed from their national nature.

The reach of *Selborne* went far beyond England. Richard Mabey notes that "when English settlers emigrated to the colonies in the last century, *Selborne* was packed alongside the family Bibles and sprigs of heather" (*Gilbert White* 6). *Selborne* provided them with a sense of the place that they had lost and with a guarantee that, in spite of migration, the rural values that had defined the island kingdom still remained intact. Even though White was rooted in place, he succeeded in turning the local nature around him into a portable text that could go anywhere, making it possible for generations of readers, spread across the far-flung reaches of empire, to feel that they could still drink at the fount of English nature. Paradise lost could be paradise regained. To buy a copy of *Selborne* was thus equivalent to purchasing a little piece of England (or a piece of "little England"), a place in an "imagined community" that existed apart from change and dislocation. In *Portable Property*, John Plotz has examined the important role played by the many portable objects that Victorian travelers cherished as the bearers of cultural value and identity in new places. *Selborne* was intrinsically a contradiction, for it represented a place that, like its author, refused to move, and yet the impact of the book lay in its capacity to go anywhere.

Salman Rushdie reminds us that even as all exiles and emigrants are haunted by a sense of loss and the urge to reclaim their homeland, they can never actually reclaim "precisely the thing that was lost." Instead, they "create fictions, not actual cities or villages, but invisible ones, imaginary homelands, Indias of the mind" ("Imaginary Homelands" 10). For the British who settled abroad, and for those who stayed at home, *Selborne* was very much this kind of "imaginary homeland," a "dream-England" (18) as Rushdie suggests, a place whose very partiality and limitations—its fully lived insiderism— made its readers feel that it was their symbolic home. The letters that composed *Selborne* could thus function as letters *from* Selborne—as letters from their imagined "home." When in 1880 the American James Russell Lowell visited England, he, like many others, made an obligatory pilgrimage to Selborne. His response powerfully expresses the impact of seeing for the first time a place that he had inhabited for many years in his imagination:

> My eye a scene familiar sees,
> And Home! is whispered by the breeze.
> My English blood its right reclaims;
> In vain the sea its barrier rears.
> Our pride is fed by England's fame,
> Ours is her glorious length of years;
> Ours, too, her triumphs and her tears. (cited in Holt-White, *Life and Letters* 1:275)

Lowell's concluding allusion to Wordsworth's "Intimations Ode" provides a fitting conclusion to a poem that seeks to bridge Lowell's nostalgia over the loss of a home that he never had. *Selborne* made it possible for members of what James Belich calls the "Anglo-world," "a transcontinental, transnational entity . . . politically divided and sub-global, yet transnational, intercontinental, and far-flung" (*Replenishing the Earth* 49), to imagine themselves as having a true nature of origin, that despite time, space, and migration they nevertheless still belonged to this special spot on the earth. As the embodiment of what was seen as its ancestral nature, *Selborne* was fundamental in fostering what Benedict Anderson has called "long-distance nationalism" (*Spectre of Comparisons* 58–76). White taught his colonial readers to love a certain kind of nature and to see it as an important part of their being, and thus the book became a "prominent model of settler nature writing" (Dunlap, *Nature and the English Diaspora* 35), and it played a significant role in fostering the desire of British settlers to transplant the nature of England to the colonies wherever they could. At the same time, it taught them to love a nature that, for most of them, neither was nor ever had been theirs.

From Local Informant to Sentimental Naturalist

Virginia Woolf shrewdly characterized *Selborne* as "one of those ambiguous books that seem to tell a plain story" (*Collected Essays* 3:122). Over most of its history, the book has not been seen as a crafted epistolary narrative, but as an unmediated response to a historical reality, as the simple, plain-speaking expression of one man's deep, if somewhat quirky, love for his rural birthplace. For example, David Elliston Allen speaks of *Selborne* as "the testament of Static Man: at peace with the world and with himself, content with deepening his knowledge of his one small corner of the earth, a being suspended in perfect mental balance" (*Naturalist in Britain* 51). Recent criticism, however, has sought to recover a more sophisticated and worldly author behind the text.[4] The circumstances leading to the composition of *Selborne* are well known but worth repeating for they help clarify the emergence of this anti-cosmopolitan, nativist strand of natural history. White's interest in natural history followed a trajectory not unlike that of Erasmus Darwin. Having taken possession of the family home, the Wakes, in 1749, he became an avid gardener, and keeping a detailed gardening journal led by the mid-1760s to a more serious interest in natural history. The year 1766 saw his first venture into natural history with a manuscript entitled "Flora Selborniensis," which recorded the blooming times of plants in his garden and in the surrounding area, along with other natural occurrences for that year. From the beginning, White had good publishing connections. His brother Benjamin, seeing an opportunity in the rising public interest in natural history, had established a publishing house on Fleet Street, where, as Edmond Gosse comments, he "issued most of the standard works on natural history which appeared in London during the second half of the [eighteenth] century" (*Gossip in a Library* 198).[5] Gilbert would comment that "anything in the naturalist way now sells well" (White to John White, 9 Mar. 1775, Holt-White, *Life and Letters* 1:279). White was also quite familiar with the most important natural history books of the period.[6] When Benjamin "hired a foreigner" (the naturalist on Cook's second voyage, Johann Reinhold Forster) to translate Pehr Osbeck's *Voyage to China and the East Indies* (1771), White had the responsibility of correcting the translation (Foster, "Gibraltar Correspondence" 235, 231).

In 1767, through Benjamin, White met the Welsh naturalist Thomas Pennant, who had just lost money with the luxury folio edition of the *British Zoology* (1766) and was now seeking to recoup his losses with a cheaper, and ultimately more successful, octavo edition. As the idea of Britain as a unified political nation was being forged, the time seemed right for a "British" natural history. He and other naturalists, such as William Withering, in the *Botanical Arrangement of all the Vegetables Naturally Growing in Great Britain* (1776), and John Walcott, in *Flora Britannica Indigena* (1778), set out to provide this new nation with a common nature. Confirming Linda Colley's argument that British nationalism was produced in opposition to France, Pennant justified his book by referring to continental naturalist history. "At a time, when the study of natural history seems to revive in *Europe*; and the pens of several illustrious foreigners have been employed in enumerating the productions of their respective countries," he writes, "we are unwilling that our own island should remain insensible to its particular advantage; we are desirous of diverting the astonishment of our countrymen at the gifts of nature bestowed on other kingdoms, to a contemplation of those with which (at lest [sic] with equal bounty) she has enriched our own" (*British Zoology* 1:i). Pennant would go on to write numerous other publications, among these Indian and Arctic zoologies, tours, and world geographies, much of his success lying in his capacity for developing networks of "local informants," whose information he then compiled.[7] Pennant was happy to enlist the help of White as a local informant, because the rural curate could provide him with information on the fauna of "the most southerly county" (*Selborne* 37) of England. The arrangement was intended to be mutually rewarding for both parties, with letters, specimens, and books being the essential medium of exchange. In return for specimens and local information, White would gain access to an informal scientific community (Pennant was scrupulous about acknowledging his debts to others in his books) and would be given general advice and assistance in identifying species. For White, it was a real opportunity. In his first letter to Pennant, he writes, "It has been my misfortune never to have had any neighbours whose studies have led them towards the pursuit of natural knowledge. For want of a companion to quicken my industry and sharpen my attention, I have made but slender progress in a kind of information to which I have been attached from my childhood." He then goes on to provide information about migratory species in the neighborhood, apologizes for not sending specimens of a new species of mouse that he had discovered (*Micromys minutus*), and includes a hawk (a peregrine falcon) that he was having trouble identifying: "you will excuse me if it should appear as familiar to you as it is strange to me" (31). It should be stressed that White's position in an out-of-the-way corner of England was not substantially different from that of many of Pennant's colonial informants. Pennant never visited Selborne, because he did not need to.

White's letters on Selborne constitute an ambiguous form of *autoethnography*: they seek to impart an insider's knowledge of a place on the periphery to metropolitan scientists, while displaying a clear knowledge of the difference between his methods and understanding of the place he describes and theirs.[8] White was already an astute observer of nature, experienced in keeping natural history records, when he met Pennant. Correspondence, however, required him to learn not only how to *see* nature but also how to *write* about it. It thus played a crucial role in encouraging him to consider natural history as a textual activity, particularly the process of translating his localized seeing into something that could be shared with others elsewhere. Matters of communication and translation were thus at the heart of the project. Pennant was not looking for a competitor, nor was he interested in White's writing beyond the information that it conveyed about the zoology of Selborne, so White grew increasingly uneasy about his subordinate position in this exchange and with the manner in which Pennant's idea of natural history disembedded or abstracted creatures from their living contexts, uprooting them from the natural and, he would argue, social communities in which they were found. By 1770, the correspondence was mostly at an end, mainly because White had found a more productive and satisfying correspondent in the figure of the lawyer/naturalist Daines Barrington, who had earlier published the *Naturalist's Journal*. "From many such journals kept in different parts of the kingdom," Barrington wrote in his preface, "perhaps the very best and accurate materials for a General Natural History of Great Britain may in time be expected, as well as many profitable improvements and discoveries in agriculture" (cited in *Selborne* xv).

At the same time, White was also corresponding with his brother John, who was chaplain at the naval garrison at Gibraltar and, like many colonial sojourners, was hoping to write a book on the natural history of that region. White mentored John, discussing the market for natural history books, what books he should consult, what topics to write about, and the importance of developing a systematic understanding of nature: "There is nothing to be done in the wide boundless field of Nat: hist: without system." "Stillingfleet gives you two very good models for Floras," he writes, and Hasselquist "will show You the manner in which the disciples of Lin[naeus]: investigate the nat: curiosities of a Country" (Foster, "Gibraltar Correspondence" 231). In advising his brother how best to write a colonial natural history, White eventually came to consider the possibility of writing his own "nat: Hist: of Selborne in the form of a journal for 1769" (*Selborne* 324). It also helped him to formulate for the first time his own unique understanding of what natural history should be and do. "Learn as much as possible the manners of animals," he writes to John. "They are worth a ream of mere descriptions. The best foreign faunists are very barren in the manners of y^{ir} animals" (323).

The "life & manners of animals are the best part of Nat: history" (325), he stresses in another letter.

Whereas most naturalists at the end of the eighteenth century understood natural history as a science of classification, dealing with the collection and identification of specimens, White is distinctive, as Ted Dadswell observes, for his focus on the detailed observation of animal behavior (*Selborne Pioneer* 60, 153–56). White's is, indeed, a living nature, and his effort to understand birds as social animals changed the study of natural history. His use of the phrase "life and manners," however, provides a clue that the framework structuring his observations was not so much scientific as moral literature. Indeed, many of his descriptions can be seen as minibiographies or moral portraits, emphasizing a specific psychological characteristic of an animal, as in his account of the dominant passion of gluttony in the blue titmouse: "The blue titmouse, or nun, is a great frequenter of houses, and a general devourer. Besides insects, it is very fond of flesh; for it frequently picks bones on dunghills: it is a vast admirer of suet, and haunts butchers' shops. When a boy, I have known twenty in a morning caught with snap mousetraps, baited with tallow or suet. It will also pick holes in apples left on the ground and be well entertained with the seeds on the head of a sunflower" (*Selborne* 92). A focus on "life and manners" was, indeed, a common feature of travel narratives, not in regard to animals, but instead in ethnographic accounts of the customs and ways of life of other peoples.[9] White's originality as a naturalist can be said, then, to lie in his application of the ethnographic techniques of these texts to the fauna of Selborne, as he looked on the animals of his parish in the same manner as a traveling naturalist might describe the "life and manners" of the people that he encountered in his travels. Much of the success of *Selborne* lay in White's ability to describe the common animals of England with the same close attention to physical and cultural differences which was customarily reserved for exotic peoples. His designation of birds in *Selborne* as a "feathered nation" may be conventional, but it nevertheless allows for semantic slippage between birds and indigenous peoples (202). His favorite birds, the hirundines or swallows, he denominates "a most inoffensive, harmless, entertaining, social, and useful *tribe* of birds" (134), while he similarly describes migratory birds as a "restless tribe" (45). *Selborne* is thus paradoxically both a familiar parochial history, like Crabbe's *The Borough*, in which White describes the life of the parish in the same manner as a parson might describe his parishioners, and a colonial "life and manners" ethnography, in which the region of Selborne is treated as a new territory composed of animal tribes that needed to be properly described.

White's own peripheral position as a "local" or "native" informant writing to a metropolitan scientific audience helped him to frame his descriptions. This parallel between isolated English counties and colonial ethnography can be seen in an anonymous contemporary review of Rev. John Collinson's *History and Antiquities of the County of Somerset* (1791), which suggests that "the history of a county is, in miniature, the history of a kingdom.... The little spot which before was confused in the aggregate, being magnified by the microscopic power of the observer, expands itself into an ample chart; and is found to boast its aborigines, its ancient monuments, laws, and customs, its eminent characters, and remarkable events; together with its appropriate blessings and curiosities of nature, in as interesting a degree as the most extensive regions" (60). At once an insider's narrative written for outsiders and a text that sees the local through the distancing lens of colonial ethnography, the aesthetic power of *Selborne* lies in its narrator's capacity to be inside and outside at the same time. Its author, we learn from "The Invitation to Selborne," is a "partial bard": both in his admiration of "his native spot;/ Smit with its beauties, loved, as yet a child" and in his willingness to restrict his subject to this place. The book represents an invitation to the reader—"my stranger"—to visit this "wildly majestic" place, the embodiment of "nature's rude magnificence" (*Selborne* 1).[10] Indeed, when Abraham and Felix Driver visited Selborne in 1794, its lack of agricultural improvement reminded them of a colonial nature: "We are sorry to observe such immense tracts of open heath, and uncultivated land, which strongly indicate the want of means, or inclination, to improve it, and often reminds the traveller of uncivilised nations, where nature pursues her own course, without the assistance of human art" (cited in Mabey, *Gilbert White* 31). White's decision to publish his natural history in epistolary form was a piece of genius, for it situated his nature within a context of exchange. On one hand, it allowed him to adopt the position of a local informant writing to metropolitan naturalists, and yet these "letters from the periphery" also came to be seen as "letters from England" (or as Ruskin recalled them, *Letters from Selborne*) (*Love's Meinie*, in Ruskin, *Works* 25:19), a correspondence in which the "life and manners" of the various animal tribes that make up England's village nature appeared for the first time. White defamiliarized English nature in order to see it clearly for the first time. Lucy B. Maddox is thus no doubt correct in arguing that the book successfully appropriates "the general assumptions and strategies of the travel account to its own local interests" ("Gilbert White" 47).

Although it was White's animal ethnographies that set him apart from many other eighteenth-century naturalists, what is most striking about them is the emphasis that he places on treating natural history as essentially a genre of storytelling focused on comprehending the language of these animal tribes. "Always procure what anecdotes you can of the life & conversation of animals," he writes to John. "They are the soul of Nat: History!" (Foster, "Gibraltar Correspondence" 320). White first used this phrasing a month earlier in a letter to Pennant in which he was critical of the localist naturalist Giovanni Scopoli for not being attentive enough to "the life and conversation of his birds" (*Selborne* 74), and he would return to it when he introduced his work to the Royal Society, stating that he was seeking to "promote a more minute inquiry into natural history; into the life and conversation of animals" (140).[11] The importance of this linguistic turn can hardly be overstated. First, it calls into question the conventional human-centered assumption that animals, like machines, can make sounds but cannot speak. Descartes stands at the head of a Western tradition that has denied animals language, writing that we must not assume "that animals speak, although we do not understand their language. For if that were true, they would be able to make themselves understood by us as well as by members of their species" (*Discourse on Method* 75). Although naturalists before White had recognized that every species of bird had its own unique song, they did not interpret these vocalizations as a mode of communication, but instead saw songs as being fixed mechanical or instinctual phenomena, innate to a species, and thus the spontaneous outflow of their natures. Birdsongs were useful in identifying birds, but not intrinsically different from the fixed external markings of their plumage. In emphasizing the "conversation of animals," White was radically calling this stereotype of the animal world into question. Secondly, he sought to make the study and interpretation of the languages and forms of communication with which animals and birds daily interact with each other a central element of natural history. This priority underpins his commitment to fieldwork and his belief in the superiority of local natural histories over those produced by metropolitan naturalists. "Faunists," he writes to Barrington, "acquiesce in bare descriptions, and a few synonyms . . . because all that may be done at home in a man's study, but the investigation of the life and conversation of animals, is a concern of much more trouble and difficulty, and is not to be attained but by the active and inquisitive, and by those that reside much in the country" (*Selborne* 125–26).[12] To John, he remarks, "true naturalists will thank you more for the life & conversation of a few animals well studied & investigated; than for a long barren list of half the Fauna of the globe" (Foster, "Gibraltar Correspondence" 492). Unlike other naturalists, who focused on the external characteristics, such as feathers and markings, that allowed animals to be identified and classified, White sought to capture their inner psychology and social life.

White's methods can be seen as both an extension and critique of John Locke, who famously opened the third book of *An Essay Concerning Human Understanding* by insisting that language was the basis of society: "God, having designed man for a sociable creature, made him not only with an inclination, and under a necessity to have fellowship with those of his own kind; but furnished him also with language, which was to be *the great Instrument, and common tie of society*" (2:3; my emphasis). Human beings, Locke argues, were created to be social, so God gave them "organs so fashioned, as to be fit to frame articulate sounds, which we call words." Making sounds, however, was not the same as using words, because language requires that sounds be used as the external, material signifiers of ideas. The use of language, which points to the existence of ideas, is thus a central means by which Locke distinguishes between human beings and animals. He writes, "Parrots, and several other birds, will be taught to make articulate Sounds distinct enough, which yet, by no means, are capable of Language" (2:3). Later in the *Essay*, he returns to the figure of the parrot as a creature that mimics speech rather than using it for expressive purposes. "Before a man makes any proposition," he writes, "he is supposed to understand the terms he uses in it, or else he talks like a parrot, only making a noise by Imitation, and framing certain Sounds which he has learnt of others; but not as a rational creature, using them for signs of ideas which he has in his Mind" (2:298). The force of Locke's arguments was partly weakened by his own recognition that often human beings were no more aware of what they were actually saying than parrots, but even so, Locke's rationalism denied that birds, or any other animals for that matter, had the capacity to use language. Lacking the capacity to form ideas, they also consequently lacked the capacity to enter into social interactions with each other. Language—or the ability to converse with one another—is thus for Locke restricted to human beings and becomes one of the key ways in which he denies that animals can have reason or form societies.

White's emphasis on the "life and conversation" of animals radically undercuts the Cartesian and Lockean epistemologies that forever excluded them from participating in a social world, however rudimentary. Since birdsongs were often thought to be innately impressed on a species, White delighted in noting that not all birds of a species sing in the same key (thus suggesting that they might have learned rather than come by their songs innately), and apparently even some of them suffered from tin ears:

> Neither owls nor cuckoos keep to one note. A friend remarks that many (most) of his owls hoot in B flat; but that one went almost half a note below A. The pipe he tried their notes by was a common half-crown pitch-pipe. . . . A neighbor of mine, who is said to have a nice ear, remarks that the owls about this village hoot in three different keys, in G flat or F sharp, in B flat and A flat. He heard two hooting to each other, the one in A flat, and the other in B flat. *Query*: Do these different notes proceed from different species, or only from various individuals? The same person finds upon trial that the note of the cuckoo (of which we have but one species) varies in different individuals; for, about Selborne wood, he found they were mostly in D: he heard two sing together, the one in D, the other in D sharp, who made a disagreeable concert. (*Selborne* 124)

In studying the social dimensions of animals, White is unique in that instead of focusing primarily on songs, he attended to the chirps, peeps, squeaks, croaks, hisses, and shrieks that constitute their everyday communications. In a letter to Barrington on their "notes and language," he writes that "many of the winged tribes" have "various sounds and voices adapted to express their various passions, wants, and feelings; such as anger, fear, love, hatred, hunger, and the like" (201). Here White is adopting the sentimentalist philosophy of writers such as David Hume and Adam Smith, for rather than seeing language as a form of reason or a mode of representation, he understands it as a means of expressing animal feeling. White has often been criticized for anthropomorphizing animals, but this criticism rests on the same exclusionary epistemology that White was seeking to dismantle, for it assumes that only human beings can have or communicate "passions, wants, and feelings." For White, human beings and animals exist on a social continuum, linked by feeling and language, and one of the great successes of *Selborne* is the manner in which it sees animals as being part of a shared community. The less reed-sparrow, he writes, is "a sweet polyglot, but hurrying: it has the notes of many birds" (102). Owls, he writes, have "very expressive notes" and can "hoot in a fine vocal sound" in "complacency or rivalry" with one another, or they can convey menace by a quick call, a "horrible scream," a snore, or a hiss. Adopting the popular distinction between "ancient" and "modern" languages, White comments that "the language of birds is very ancient, and, like other ancient modes of speech, very elliptical: little is said, but much is meant and understood." Because they only have a rudimentary language at their disposal, a few sounds must convey many meanings. Not all birds are "equally eloquent; some are copious and fluent as it were in their utterance, while others are confined to a few important sounds" (201). Ravens, for instance, are mostly restricted to croaks, or to a "solemn note that makes the woods to echo." Nevertheless, during the excitement of the breeding season, even they feel compelled to test the limits of their language and seek to break into song—unfortunately, "with no great success" (202). Every raven, we learn, in his heart of hearts wants to be a crooner. Domestic birds are no less rich in their linguistic capacities. No bird, he writes, possesses "such a variety of expression and so copious a language" as chickens. From the "twitterings of complacency" of four-day-old chicks on being offered flies, to the "easy soft note" of the pullet about to lay, or her "clamorous joy" when she has "disburdened herself," or the "new language" she adopts when her chicks are hatched, chickens display a "considerable vocabulary" (201–3).

An important outgrowth of Lockean epistemology was that it made language the key to understanding the ideas and psychology of others. Words, Locke argued, were the "sensible signs" by which the ideas, which each man holds "within his own breast, invisible and hidden from others," are made visible to others (*Essay* 2:4). Working within this epistemology, eighteenth-century explorers and naturalists focused on documenting indigenous languages not only to communicate with native peoples but also to gain access to their psychology, culture, and ideas. For instance, the published account of Cook's second voyage included a forty-page, 1,054-word "Vocabulary of the Society Islands" compiled by the surgeon's mate, William Anderson, who was soon to be the designated naturalist on the third voyage (Cook, *Voyage Towards the South Pole* 2:330). Traveling naturalists also inquired diligently into the indigenous names of the plants and animals that they collected, because the use of native informants was essential to their success. White's interest in the languages of birds is very much a reflection of these concerns. In his letter on "notes and language," White tells the story of a vizier who recited a conversation that he overheard between two owls in order to reclaim a sultan from the destructive path of war and destruction. Although White does not pretend to the same powers in his understanding of avian languages, many of the most memorable passages in *Selborne* ask the reader to join him in listening in on the conversation of birds. Linking their language to their psychology and behavior, White functions as a guide and translator, who allows the reader to see the world through their eyes. Birds speak in *Selborne* by virtue of the powers of this naturalist translator, and since all translations reflect as much the values of the cultures for which they are translated as the cultures from which they originate, one can hear in the exotic vocalizations of White's birds the traces of an English accent. *Selborne* is a natural history that reflects the sentimentalism of the age. White does not turn birds into objects of sentimental reflection, nor does he credit himself with a heightened sensibility, but he does seek to forge a bond of sociality between human beings and birds by enlisting his readers' sympathetic and imaginative identification with their psychology and inner emotional life. The resulting community of people and animals is based on this sharing of feeling, as White translates animal behavior and language—the external signs of feeling—into something with which the reader can identify. He is not always successful, and occasionally he slips into forms of sentimental excess, but instances of this sort are rare. More commonly, as Richard Mabey remarks, White portrays "the daily business of lesser creatures as a source not just of interest, but of delight and inspiration" (*Gilbert White* 2).

White's letters on the hirundines, which were read before and then published by the Royal Society, are classics in this art of translation and thus anticipate the emphasis on animal psychology which would later characterize the genre of the animal story, produced by Canadian writers such as Charles G. D. Roberts and Ernest Thompson Seton. For instance, in the letter on the swift, an "amusive" bird that he watched with "no small attention," White brings its language to life in order to bridge the distance between the reader and a bird that seems primarily to inhabit the skies. After giving information about when the swift usually appears in spring and about its nesting habits, White indicates that, despite his efforts, he has never seen one collecting nesting materials. This leads him to suspect that swifts usurp the nests of sparrows. We soon learn, however, that his evidence is entirely circumstantial, not so much based on personal observation as on avian hearsay: "I have seen them squabbling together at the entrance of their holes; and the sparrows up in arms, and much-disconcerted at these intruders"

(*Selborne* 154). One of the striking qualities of *Selborne* is the degree to which White demonstrates the "wonderful spirit of sociality in the brute creation" (165), even in their quarreling. Interpreting a bird whose vocalizations are essentially restricted to "squeaking" and a "harsh screaming note" (155), he nevertheless adopts the *tour de force* proposition that swifts copulate on the wing: "If any person would watch these birds of a fine morning in May, as they are sailing round at a great height from the ground, he would see, every now and then, one drop on the back of another, and both of them sink down together for many fathoms with a loud piercing shriek. This I take to be the juncture when the business of generation is carrying on" (155). In this spectacular image of swifts tumbling fathoms in their joyful embrace, White asks the reader to comprehend the meaning of their aerial shrieks of ecstasy. "How do you know but ev'ry Bird that cuts the airy way,/Is an immense world of delight, clos'd by your senses five?" asks the devil in a similar vein in William Blake's "Marriage of Heaven and Hell" (plate 7: 3–4). In his unassuming manner, White provides his own answer. Later, in describing the hens nesting in church walls and eaves, he comments that the males, "getting together into little parties, dash round the steeples and churches, squeaking as they go in a very clamorous manner." These, he says, are the sounds of the males "serenading their sitting hens," for they "seldom squeak till they come close to the walls or eaves." In turn, the females quickly respond with a "little inward note of complacency" (*Selborne* 156).

In moments such as these, White overcomes the division that separates human beings from animals. *Selborne* seeks to recover that Edenic time when human beings still conversed with animals, and they with each other. Milton was mistaken, White thought, in having Adam claim that "Much less can bird with beast, or fish with fowl,/So well converse, nor with the ox the ape" (166).[13] The best moments in his book are those when White, following in the footsteps of Adam, seems to recover this universal language, now existing only in fragments. These moments break the silence that normally envelopes our relationship with animals, and we are given a glimpse into this lost order of nature. For White, these moments of communicated feeling have almost the status of epiphanies, as when, "at a certain signal given," a mother swallow and her newly fledged nestling fly toward each other, "meeting at an angle," to share a meal, "the young one all the while uttering such a little quick note of gratitude and complacency." "A person must have paid very little regard to the wonders of Nature," White adds, with sly understatement, "that has not often remarked this feat" (145–46).

White's success in creating a community of shared feeling between readers and animals, which blurs the separation between human beings and animals, is certainly one of his great achievements as a writer, since—as I have been seeking to suggest—animals communicate with us in this book because White's translational activity allows them to do so. His book suggests that human beings have denied animals the capacity for language, not because animals lack signs for communication, but because human beings have lacked adequate translations. This achievement is all the more impressive inasmuch as it occurred at a time when most Europeans believed that animals were little more than automata. Milton's *Paradise Lost* was certainly in White's mind as he wrote a natural history whose mythic quality lies in its capacity to make its readers feel that at times they are inhabiting a world in which animals are no longer alien to us. This is not to say, however, that White's sentimentalist translation of the conversation of animals is not without its limitations. Because he values the animals that he knows best, and he knows best the ones that he can observe on a daily basis, his writing inevitably gives primacy to those creatures that have adapted themselves to living with human beings. He thus emphasizes a nature that is domestic and familiar, as is evident in the justly famous letters that he wrote on the "British hirundines" (*Selborne* 143). In three of these letters, those dealing with the swift, the martin, and the house-swallow, White represents these birds as "his feathered fellow-townsfolk" (*My Study Windows* 1), to quote James Russell Lowell, and he clearly enjoys writing in detail about their industry, sociality, and family values. The house-swallow, for instance, provides "a most instructive pattern of unwearied industry and affection" as she supports her family "from morning to night" (*Selborne* 146). Martins are "industrious artificers" (137), he writes, who may even have taught human beings how to build mud walls: "Thus careful workmen when they build mud-walls (informed at first perhaps by this little bird) raise but a moderate layer at a time, and then desist" (136).

In contrast, the monograph on the sand-martin (or bank-martin), though certainly more substantial than comparable efforts by Pennant or Bewick, proved difficult for White to write. He opens the letter with an apology: "it is scarce possible for any observer to be so full and exact as he could wish in reciting the circumstances attending the life and conversation of this little bird, since it is a *fera natura*, at least in this part of the kingdom, disclaiming all domestic attachments, and haunting wild heaths and commons where there are large lakes." With the sand-martin, White was dealing with a bird that did not wish to be a neighbor to human beings, one that did not nest in human dwellings. He notes that "several colonies of these birds" existed in the treeless Woolmer forest, but they "are never seen in the village; nor do they at all frequent the cottages that are scattered about in that wild district." This behavior strongly contrasts with the other swallows, which "are remarkably gentle and domesticated, and never seem to think themselves safe but under the protection of man" (*Selborne* 150). White also mistakenly assumes that the bird is relatively scarce in England. Whereas almost every village had its house-martins, swifts, and house-swallows, sand-martins, "scattered here and there, live a sequestered life among some abrupt sand-hills, and in the banks of some few rivers" (153). The other swallows are celebrated for their "architectonic skill" in nest building, while the burrowing sand-martin is seen as being "rude," less developed, less civilized, less human: "At the inner end of this burrow does this bird deposit, in a good degree of safety, her *rude* nest, consisting of fine grasses and feathers, usually goose-feathers, *very inartificially* laid together" (151). In contrast to his usually confident observations, this letter is filled with gaps and uncertainties: "I have never been able to discover" (151); "we have never yet been able to determine; nor do we know whether" (153); "Notwithstanding what has been advanced above, some few sand-martins, I see, haunt the skirts of London" (153).

White normally takes great pleasure in using the language and manners of an animal to register its inner life, but his comments about the language of the sand-martin are perfunctory and deal only with the bird's response to aggression: "these hirundines are no songsters, but rather mute, making only a little harsh noise when a person approaches their nests." Since none of the swallows can be said to be songsters, White's suggestion that the sand-martin is "rather mute," that it lacks language, further reaffirms the association, which he otherwise undercuts, between language and civilization. He seems not to question whether these birds, whose vocal skills are hardly less developed than other hirundines, seem mute because they do not converse, or because he has not been able to observe them in situations where he might have heard their conversation. This is one of those rare instances in the text when White does not portray a species of bird as being part of *his* world, but instead portrays himself as an intruder in *its* world. Although he has already noted that these birds are social, for they nest in "colonies," White nevertheless concludes that "they seem not to be of a sociable turn" (153). Clearly "sociability" does not refer to how these birds relate to each other, but to their willingness to adapt themselves to life with human beings. Thus, even Gilbert White's nature apparently has its colonial Others. Pointing to the sentimentalist limits of *Selborne*, White mistakenly concludes that the sand-martin "is much the rarest species" in England, not because of the scarcity of limestone formations in the Selborne area but because of its unsociable nature.

White's goal was certainly to capture in detail the rich polylingualism inherent in "the life and conversation of animals," but the creatures that he values the most and that receive his greatest attention are those that are most fully integrated into his own social world. In this sense, his translations do not seek to capture what is foreign in the subjectivity and culture of birds, but instead what is familiar. He domesticates the nature that he translates. As Henry Beston reminds us, we should not need to liken animals to ourselves in order to respect them: "The animal shall not be measured by man. In a world older and more complete than ours they move finished and complete, gifted with extensions of the senses we have lost or never attained, living by voices we shall never hear. They are not brethren, they are not underlings; they are other nations, caught with ourselves in the net of life and time" (*Outermost House* 25). White's sentimentalist translations of English nature inevitably marginalize those creatures who "live by voices we shall never hear," or those who only passed through the village on their way to somewhere else. The hinge to any critical assessment of its limitations thus lies in recognizing its silences. Lynn Festa has argued that sentimentalism should be understood not simply as an inclusive rhetorical practice, building human communities through an identification with others, but also as a differentiating activity, based on "the selective recognition of the humanity of other populations.... It is difficult to feel for peoples whose customs and manners are alien to our own," she writes. "Unfamiliarity breeds contempt" (*Sentimental Figures of Empire* 2–4).[14]

The selective character of White's sentimentalism is nowhere more apparent than in his letter to Barrington on Romani. "We have two gangs or hordes of gypsies which infest the south and west of England, and come round in their circuit two or three times in the year," he writes (*Selborne* 166). Here White adopts the "life and manners" stance of the naturalist, but rather than extending humanity to this group, the letter excludes them from the community of Selborne. Two "gangs," "hordes," or "tribes" (as he later describes them) "infest" southern England, like vermin. White was fascinated by bird migration, but in this letter there is no celebration of the periodic return of these nomadic outsiders to the village because these people, in his view, have no claim to a place in Selborne. Adopting the classificatory methods of the naturalist, he points out that although the two "tribes" are differently named, one has appropriated "the noble name of Stanley," while the other tribe, "Curleople," gives clear evidence of the true foreign origin of "these vagrants." Although learning their language might have made it possible to understand these people better, he rejects it as being "harsh gibberish," "jargon," "cant," a language that is so "corrupted" that only the "mutilated remains of their native language" can possibly be discerned. Whatever their original language might have been, these "vagabonds" have lost it through their "rovings." Mobility is here closely allied with degeneration. Since he has devoted so much attention to the nest-building activities of birds, White's turn, near the end of his letter, to the Romanis' modes of habitation and childbearing customs seems a cruel parody of his earlier monographs on the swallows. "While other beggars lodge in barns, stables, and cow-houses, these sturdy savages seem to pride themselves in braving the severities of the winter, and in living *sub dio* [in the open air] the whole year round" (167).

At this point, White communicates what would seem to be a conventional anecdote drawn from the arsenal of sentimentalist literature, the story of a young girl who gave birth to a child in his garden the previous autumn: "Last September was as wet a month as ever was known; and yet during those deluges did a young gypsy-girl lie-in in the midst of one of our hop-gardens, on the cold ground, with nothing over her but a piece of blanket extended on a few hazel-rods bent hoop fashion, and stuck into the earth at each end, in circumstances too trying for a cow in the same condition: yet within this garden there was a large hop-kiln, into the chambers of which she might have retired, had she thought shelter an object worthy her attention" (*Selborne* 167). White's description of the "nesting habits" of this Romani, his attention to the rudeness and poverty of the young girl's construction of the place where she intended to give birth to her child, is not intended to evoke pathos, but instead elicits a naturalist's comparison of the sociability of the Romani as a "tribe" with the state of improvement of the hirundines. Like the sand-martins, the young girl seems to prefer to weather "circumstances too trying for a cow in the same condition," rather than attempting to occupy the "large hop-kiln" that was nearby (as if, indeed, this were an option). Despite the obvious demand for assistance that a young woman in such circumstances should have placed on any observer, let alone a clergyman, the sympathetic identification that White accords to most of the animals of the village is absent from this description.[15] In this regard, *Selborne* conveys the limits not only of a domestic English nature but also of the society that promoted it. White observes animal communication in the same way as he studies Romani language: to postulate their psychology, but not to communicate with the beings that he observes.

Cracks in the Perfect Design

Despite these limits, *Selborne* has long been recognized as a landmark text in the history of ecological thought. As Donald Worster remarks, "White grasped the complex unity in diversity that made of the Selborne environs an ecological whole" (*Nature's Economy* 7). The complex, balanced interdependency of species, their being part of a single society, is, indeed, one of the fundamental arguments of the book. Naturalists before White had produced local and regional floras, yet these were primarily arid lists of species. What White attempted to do in *Selborne* was to document the natural world of Selborne in all its living complexity. "A good monography of worms would afford much entertainment and information," he writes, noting that "the most insignificant insects and reptiles are of much more consequence, and have much more influence in the economy of Nature, than the incurious are aware of; and are mighty in their effect.... Earth-worms, though in appearance a small and despicable link in the chain of Nature, yet, if lost, would make a lamentable chasm," not only because they serve as the food of half the birds, but because they augment the soil: "the earth without worms would soon become cold, hard-bound, and void of fermentation; and consequently steril [sic]" (*Selborne* 182–83).

White's commitment to the idea of a balance of nature reflected eighteenth-century conservative religious and political values. In John Ray's *The Wisdom of God Manifested in the Works of the Creation* (1691), William Derham's *Physico-Theology* (1713), William Paley's *Natural Theology* (1802), and the *Bridgewater Treatises* (1833–36), ecological thought is deeply bound up with the idea of a designed and unchanging creation. From the early 1700s onward, natural history was a popular pastime of the British clergy, who believed that the study of nature provided an ongoing demonstration of the wisdom of God.[16] Look "thro' Nature up to Nature's God," Pope famously declared in *Essay on Man* (*Poems* 3:332). This wisdom was displayed in the perfection of nature, the integration of all its parts to the achievement of a self-sustaining whole. Derham remarks that "the Balance of the Animal

World is, throughout all Ages, kept even, and by a curious Harmony and just Proportion between the increase of all Animals, and the length of their Lives, the World is through all Ages well, but not overstored" (Derham, *Physico-Theology* 171). Linnaeus's concept of an "economy of nature" expresses a similar idea, but he shifted the focus from theology to science and economy, from the appreciation of God through Nature to understanding the complex interactions of species within a system.[17] In the "Oeconomia Naturae," which White would have read in a translation by Stillingfleet, Isaac Bibery, who was supervised by Linnaeus, writes that "to perpetuate the established course of nature in a continued series, the divine wisdom has thought fit, that all living creatures should constantly be employed in producing individuals, that all natural things should contribute and lend a helping hand towards preserving every species, and lastly that the death and destruction of one thing should always be subservient to the restitution of another" ("Oeconomia Naturae" 40). White displays a similar understanding of the balance of nature, recognizing that all creatures sustain each other, even in death. In the journal entry for 13 November 1776, White writes, "Magpies sometimes, I see, perch on the backs of sheep, & pick the lice & ticks out of their wool; nay, mount on their very heads; while those meek quadrupeds seem pleased, & stand perfectly still, little aware that their eyes are in no small danger; & that their assiduous friends would be glad of an opportunity of picking their bones" (*Journals* 133). Nature for all these writers is a deeply conservative economy, for it assumes that all species express in their form and behavior the specific place or "station" that they have been assigned to occupy since the beginning of time.

White's achievement in assigning an ecological place to all creatures in *Selborne* is thus inseparably bound up with his conservative understanding of nature as a harmonious, hierarchically balanced, unchanging system. The creatures of Selborne belong to it in every sense, in their very nature, form, and behavior, since they were made to belong and occupy their station there. The success of this natural history in place making thus lies in its fusion of social belonging, registered in language and the capacity to identify with other species, with biological identity. This view of nature was clearly running into difficulties by the end of the eighteenth century, however, because it did not easily account for changing environments, for the movement of species into new environments, or for the obvious ecological imbalances and instabilities that were being visibly produced by this movement. If species had been created to be perfectly adapted to their respective environments, how and why did they move? Further, what would it mean, in a perfectly designed creation, for one species to displace another? How could a species become extinct if it was by nature fitted for an environment? How was one to understand those situations, particularly in the colonial world, where the balance of nature seemed to be under severe strain? Like Linnaeus, White was not opposed to the human improvement of physical environments. He believed that nothing would benefit a "northerly and grazing kingdom" more than the "study of grasses"; "to raise a thick turf on a naked soil would be worth volumes of systematic knowledge" (*Selborne* 196–97). He even entertained the idea of introducing canaries to England (36). White also echoed Linnaeus in seeing the differences in the global distribution of species as a valuable encouragement to commerce and the intercourse among nations: "As every climate has its peculiar produce, our natural wants bring on a mutual intercourse; so that by means of trade each distant part is supplied with the growth of every latitude. But, without the knowledge of plants and their culture, we must have been content with our hips and haws, without enjoying the delicate fruits of India and the salutiferous drugs of Peru" (195).

Given White's commitment to a stable order of nature, the point of greatest tension in *Selborne* has to do, perhaps not surprisingly, with mobility and migration. White was capable of grasping intellectually the idea that birds migrate, but he struggled emotionally with what this implied about the world that he inhabited. Throughout his life, he maintained against his own better judgement that at least some of the soft-billed birds did not migrate south in the winter. "What difficulties attend that supposition!" he writes, "that such feeble bad fliers (who the summer long never flit but from hedge to hedge) should be able to traverse vast seas and continents in order to enjoy milder seasons amidst the regions of Africa!" (*Selborne* 37). Coleridge could hardly contain himself when he read this passage. "Surely from Dover to Calais—and from Gibraltar (or even Toulon) to the Coast of Barbary cannot be called a traverse of *vast seas*" (*Marginalia* 4:147).

Most of all, White could not believe that his favorite birds, the swallows, actually left England every winter.[18] White could accept that most soft-billed birds, which feed on insects, needed to spend their winters elsewhere to survive, and his book celebrates the return of these expatriates to their native homes with emotions that could just as easily apply to human travelers. When it came to his beloved swallows, however, White, like his friend Daines Barrington, was "no great friend to migration," and he found it difficult to resign himself to the idea that they overwintered in Africa, even though his brother, stationed in Gibraltar, had seen "myriads" of them traversing "the Straits from north to south and from south to north, according to the season" (*Selborne* 121).[19] Perhaps their departure was only an illusion, that when faced with the choice of whether to stay or go, his favorite birds could not bring themselves to leave their homes in England and instead stayed close by, hibernating in some as-yet-undiscovered cells nearby. Late in the book, White was still attempting to solve the dilemma of the disappearing swallows. In October 1780, seeing an unusually large "last flight" of house-martins, he resolved to take particular notice of them. At sunset, he saw them all suddenly scud "in great haste towards the south-east," where they dove down into a thick covert of beechen shrubs. The same thing happened on two subsequent days, after which they no longer appeared. White was certain that he had found their "winter residence." He assures us that if he had owned the neighboring property, he would have "grubbed and carefully examined" the bushes to locate the birds' "secret dormitories," but he never carried out this plan, probably because he feared that he was likely to be proven wrong. The only place where White's swallows actually hid all winter was in his heart, which certainly warmed to the idea that "far from withdrawing into warmer climates," his martins "never depart three hundred yards from the village" (226–27).

Selborne was published on the eve of the French Revolution, yet as Donald Worster comments, "it would seem that the naturalist of Selborne could appreciate thoroughly the workings of the economy of nature, but almost nothing of the revolution going on in the politics and economy of man" (*Nature's Economy* 12).[20] White was certainly well aware of these events. In a letter to Rev. Ralph Churton on 4 December 1789, he expresses his amazement at the enormity and speed of revolutionary change in France: "Of all these strange commotions, the sudden overthrow of the French despotic monarchy is the most wonderful—a fabrick which has been now erecting for near two centuries, and whose foundations were laid so deep, that one would have supposed it might have lasted for ages to come: yet it is gone, as it were, in a moment!!" (Holt-White, *Life and Letters* 2:210–11). A journal entry of 21 January 1793 speaks volumes in its juxtaposition of the careful note-taking routine that White had practiced for almost three decades with registering the psychological shock of the execution of Louis XVI: "Thrush sings, the song-thrush: the missle-thrush has not been heard. On this day Louis 16th late king of France, was beheaded at Paris, & his body flung into a deep grave without any coffin, or funeral service performed" (*Journals* 422). In the huge gap between the continuities registered in the first sentence and the events happening in France, one senses the chasm that had opened between the order of nature as White understood it and the order of man.

With increasing frequency, in the latter part of *Selborne*, White's islanded security is disrupted by silences and by strange events that intimate change. Personally, he then often suffered from deafness, and in these periods he lost his primary link to the natural world. "When those fits are upon me, I lose all the pleasing notices and little intimations arising from rural sounds," he wrote, "and May is to me as silent and mute with respect to the notes of birds, &c. as August.... I am, at times, disabled: 'And Wisdom at one entrance quite shut out'" (*Selborne* 162). The allusion to Milton's blindness speaks volumes. Cut off from the conversation of animals, White was cut off from nature. For him, it was a fall as great as that suffered by Adam and Eve.

The weather also seemed to be changing, foreboding, as Stuart Peterfreund suggests, the "historical last days" ("Great Frosts" 102). First, there was the severe frost of January 1776, in which a prodigious snowfall made roads impassable and turned London into "a sort of Laplandian-scene, very wild and grotesque indeed" (*Selborne* 241). Most eerie was the absence of sound in the metropolis: "being bedded deep in snow, the pavement of the streets could not be touched by the wheels or the horses' feet, so that the carriages ran about without the least noise. Such an exemption from din and clatter was strange, but not pleasant; it seemed to convey an uncomfortable idea of desolation: ... ipsa silentia terrent [the very silence terrifies, Virgil, *Aeneid* 2]" (241–42). The hot and dry summer of 1781 was followed in 1783 by what came to be known as the "sand summer," as the ash from the massive Haki eruption in Iceland produced one of England's hottest summers and coldest winters. White described the summer as "an amazing and portentous one, and full of horrible phenomena; for, besides the alarming meteors and tremendous thunder-storms that affrighted and distressed the different counties of this kingdom, the peculiar haze, or smokey fog, that prevailed for many weeks in this island, and in every part of Europe, and even beyond its limits, was a most extraordinary appearance, unlike any thing known within the memory of man" (247–48).[21] Although White distinguished his own reactions from the "superstitious awe" of "country people," he nevertheless admitted that even "the most enlightened person" had reason "to be apprehensive" (248).

Cracks were also beginning to appear in the design of creation at Selborne. It was not just weather but nature itself that was changing. First, there were the local extinctions, the "gap[s] in the *Fauna Selborniensis*" that had appeared during his own lifetime (*Selborne* 21). The black grouse, once numerous in Woolmer forest, had long disappeared through overhunting, the last covey having been killed in the 1730s. With them went their habitat, the heathlands and bogs of Woolmer forest, which were now being replaced by plantations of economically valuable Scots pines. The red deer, which had once been common in the royal forest, were also now gone. Others had followed, among these the once numerous ravens and stone-curlews.[22] If most of these departures had occasioned sad reflections on the breakdown of a natural order, the imminent departure of one of these animals probably would have produced relief, under normal circumstances. By the 1750s, the common black rat, a creature famously connected with the Black Plague, which first arrived in England from Southeast Asia during Roman times, had become decidedly less common. Unfortunately, the decline of "the old genuine English house-rat" (R. Smith, *Universal Directory* 133) was not a sign that English village life was improving, however, for it was connected with the advent of a newcomer—the brown or Norway rat—which arrived in England in the 1720s and soon displaced its competitor. "One black rat was killed at Shalden some months ago, & esteemed a great curiosity," White notes in his journal in 1777. "The Norway rats destroy all the indigenous ones" (*Journals* 146). In his *British Zoology*, Pennant puzzled over whether he should provide the reader with a description of such a well-known animal as the common English rat, deciding to do so because it was likely that in the not-too-distant future a written account might be all that was left of the animal. Bigger, smarter, and more adaptable, the Norway rat, which originally came from the steppes of central Asia, quickly spread across England, reaching Scotland by 1754, France by 1750, and its ostensible home, Norway, by 1762, after which it headed across the Atlantic, where it arrived on the shores of North America in the 1770s (Yalden, *History of British Mammals* 183). White was surprised that even the owners of a secluded farmhouse "tho' so sequestered from all neighbourhood, & so far removed from all streams, & water, are much annoyed with Norway rats" (*Journals* 400). In a world in which animals were continually traveling and displacing other animals, naming them was also subject to confusion. Throughout the eighteenth century, in England this large brown rat was called either the

Norway rat or the Hanover rat, depending on one's political leanings or whether one believed that it had arrived on a ship from Norway or in the luggage of George I or one of his retainers.

Signs of change were also appearing in the insect world, where the balance of creation was being disrupted by sporadic periods of excess. White remembered a time in September 1741 when, rising before daybreak, he found "the whole face of the countryside" covered in cobwebs, as if "two or three setting-nets [were] drawn one over the other" (*Selborne* 163). By mid-morning, it seemed as if it were snowing gossamer: "a shower of cobwebs falling from very elevated regions, and continuing, without any interruption, till the close of the day . . . the flakes hung in the trees and hedges so thick, that a diligent person sent out might have gathered baskets full." This strange phenomenon, he discovered, was caused by innumerable tiny spiders that used their silk to fly long distances through the air. On 1 August 1785, at about three o'clock in the afternoon, it began to rain aphids:

> The people of this village were surprised by a shower of *aphides*, or smother-flies, which fell in these parts. Those that were walking in the street at that juncture found themselves covered with these insects, which settled also on the hedges and gardens, blacking all the vegetables where they alighted. My annuals were discoloured with them, and the stalks of a bed of onions were quite coated over for six days after. These armies were then, no doubt, in a state of emigration, and shifting their quarters; and might have come, as we know, from the great hop-plantations of Kent or Sussex. (223–24)

White's use of military metaphors to describe the sheer number of aphids that descended on the village conveys his sense of the threat posed by them. In 1785, he was giving a good deal of thought to traveling insects, particularly the "unaccountable manner" in which they "are often conveyed from one country to another" (223), for early in April he had discovered that a grapevine under his study window "swarmed" with an insect "which I little expected to have found in this kingdom" (221). The insect had apparently been infesting the plant for the previous four years, but it was not until that spring that White noticed it. Consulting Linnaeus, he learned that it was the wax scale insect (*Pulvinaria vitis*), "an horrid and loathsome pest" that affected grape vines in southern Europe. How the bug had found its way into his backyard was puzzling. Since his brother John had had a similar problem, White concluded that the insect had arrived in some boxes of plants and birds that his brother had sent him from Gibraltar in the early 1770s.

His fellow Oxonian John Lightfoot disagreed, drawing attention to the fact that these insects had also appeared on a vine at Weymouth in Dorset, having arrived there "by shipping" from France. White never found an answer.

At his home in London, Joseph Banks's vision of employing natural history for a global management of natures was also facing its own challenges. Having been instrumental in the transfer of innumerable species of plants and animals across the globe, he was learning to his dismay that plants and animals were just as capable of traveling on their own. In April of 1789, there appeared a report by the Privy Council Committee on Trade, which brought together submissions relating to the activities of the Hessian fly, a gall midge with a nasty scientific name, *Mayetiola destructor*, which was at that time ravaging the wheat crops in America. Entitled *Proceedings of His Majesty's Most Honourable Privy Council, and Information Received, Respecting an Insect, Supposed to Infest the Wheat of the Territories of the United States of America*, the report detailed the government's concern about an insect that throughout the 1780s had been spreading across American farmland at an alarming rate and which seemed poised to invade England.[23] England may have ruled the waves and been in the process of bringing over 150 million people under its rule, but this does not mean that it was not afraid of a fly.[24] The insect first appeared in the wheat crops of Long Island in 1777 during the War of Independence. What made its appearance singular was that it had never been seen before. Nothing was known about its origins, its life cycle, or how to combat it. Since the insect was first noticed where British troops had been stationed, the Americans concluded that it had been brought to America, perhaps intentionally, by George III's Hessian troops, so they named it accordingly. The third edition of the *Encyclopaedia Britannica* presents this interpretation as a matter of fact, noting that the fly "landed on Long Island with the force of Sir William Howe at an early period of the late war" and was first introduced into America "by means of some straw made use of in package, or otherwise."[25] The grubs of the insect, which eat through wheat stalks, caused enormous losses, often of entire crops. By the late 1780s, huge clouds of the fly were seen across most of the Northeast, from Canada to Pennsylvania. The printer Matthew Carey, writing in the inaugural issue of the magazine *American Museum*, concluded in 1787 that if something were not done, "the whole continent will be over-run—a calamity more to be lamented than the ravages of war" ("On the Hessian Fly" 133). The analogy between a biological and military invasion came easily to hand, not just because of the events that were thought to have given rise to the introduction of the insect, but because war, famine, and plague were seen as being apocalyptically intertwined with each other.

The Foreign Office in Britain first learned of the Hessian fly in May 1788 through its consul in Philadelphia, Phineas Bond, who painted a dark picture of the situation in America: "The growth of wheat, for several years past, in most of the middle states, has been greatly injured by an insect called the Hessian Fly, whose ravages have been progressive, and in some instances ruinous. Many farms have had the crops so completely cut off as to be left without bread, corn, or even seed" (Bond to Lord Carmarthen, 22 Apr. 1788, A. Young, "Proceedings" 406). Bond urged a prohibition of all wheat imports, arguing that "the Introduction of American Wheat may be the means of communicating the Insect to other Grain afterwards used as Seed, and the Consequences to the Agriculture of the Kingdom may be as fatal as they have proved to many Farmers in the Middle States of America" (407). The Privy Council turned to Banks for authoritative advice. When he received Lord Carmarthen's 30 May letter, Banks "was utterly ignorant what insect it was that the americans meant by the Hessian Fly," but he immediately recognized the need to investigate "a subject which appeared to him so pregnant with danger to this Country" ("Preface," Joseph Banks Manuscript Collection). Without adequate time to research the subject, and based on his reading of an article by Landon Carter in the *Transactions of the American Philosophical Society*, he initially concluded incorrectly that Bond's "Hessian Fly" was, in fact, the "Flying Wevil" of Virginia (L. Carter, "Observations Concerning the Fly-Weevil"). Because this moth fed on wheat grains and could thus be transported in grain shipments, he recommended that a ban be placed on all wheat imports from America, and the order was passed by the Privy Council on 25 June to prohibit wheat because it "may be infected with an Insect, the spreading of which would be injurious to the Grain of this Kingdom" (Joseph Banks Manuscript Collection 17). During the Romantic period, for the first time in history, an act was passed to prevent the spread of a biological pest. A week later, Banks realized that he had made a mistake when he read Samuel Mitchell's article on the Hessian fly in the *American Magazine*. Nevertheless, he continued to support the ban, arguing that the evidence that "two" species of insect were "at this time spreading rapidly in that country ... cannot but amply justify a temporary order to prevent the admission of any species of merchandize supposed capable of bringing with it the seeds of so dreadful a calamity"

(A. Young, "Proceedings" 425). The introduction of a new pest into England, he argues, would be "a calamity of much more extensive and fatal consequences than the admission of the plague." Whereas the plague might lead to "the extinction of a certain proportion of the human Species, which may be, and generally is replaced in the next generation," the Hessian fly would cause "a real diminution of population," by significantly diminishing England's ability to produce wheat (Banks, "General Report of Sir Joseph Banks" 444–45).

There was no love lost between America and England in the years immediately following the American Revolution, so Banks's recommendations were clearly partly motivated by a desire to punish the new republic economically, and he willingly accepted the Duke of Grafton's view that the Hessian fly was a "Scourge of Heaven . . . upon such ungratefull colonies and rebellious people" (cited in Pauly, "Fighting the Hessian Fly" 46). The Americans, for their part, were incensed. Jefferson, who was quick to defend America's nature against all forms of European aspersion, was less concerned about the prohibition on wheat exports to England than with the fact that England had published "a libel on our wheat sanctioned with the name of parliament, and which can have no object but to do us injury by spreading a groundless alarm in those countries of Europe where our wheat is constantly and kindly received. It is a mere assassination" (Jefferson to Benjamin Vaughan, 17 May 1789, *Papers* 15:133–34). Some Americans were even willing to entertain the idea of an appropriate counter-response. The farmer Col. George Morgan, writing to John Temple, the British consul in New York, in August 1788, could not resist the opportunity of reminding the British that "were a single straw containing this insect in the egg, or aurelia to be carried and safely deposited in the center of Norfolk in England, it wou'd multiply in a few years so as to destroy all the Wheat and Barley Crops of the whole Kingdom." Were the Fly "to reach Britain," he writes, "it would become the greatest scourge that island ever experienced, as it multiplies from heat and moisture, and the most intense frosts have no effect on the egg or aurelia" (Morgan to John Temple, 26 Aug. 1788, A. Young, "Proceedings" 467). Banks was appalled that an American would even consider the possibility of biological terrorism. He wrote that "obloquy" would fall on *all* Americans if "any one of them should wilfully bring over the Fly" (Banks to Vaughan, 10 Apr. 1789, Joseph Banks Manuscript Collection). As it turned out, the ban on wheat imports was short-lived. Faced with a severe corn shortage, the British government had to weigh the danger of working-class unrest against its fear of the fly, so it opted to lift the ban in December 1789. Although the Hessian fly would remain a serious pest in America throughout the nineteenth century, it was never successful in colonizing England, though throughout the 1790s authorities remained on high alert concerning this insect threat. In 1791, for instance, Arthur Young published in his *Annals of Agriculture* Banks's comments about a fly that Young had sent him whose similarity "to the Hessian fly was alarming." Banks assured him that in this case his fears were unjustified, for the fly was actually a pest that affected rye in Sweden. Young would use the article to thank Banks publically for protecting a now vulnerable England from foreign biological threats. In doing so, he found an appropriate metaphor for the naturalist as a counterinsurgent: "The public are indebted to the active spirit of Sir Joseph Banks, ever on the wing to disseminate the knowledge which may be beneficial to mankind" ("On the Musca Pumilionis" 176).

England may have successfully dodged the Hessian fly, but at the same time, a new and equally dangerous insect had appeared in metropolitan London, at a Sloane Street nursery where it had set about destroying thousands of apple trees. Banks first learned of the woolly aphid (*Schizoneura lanigera*), or, as it came to be known, the "American Blight," in 1790, and for the next eight years he closely observed its rapid spread across England. William Salisbury would speak of this aphid as "the most destructive insect we know ... which has found means, within a few years of extending itself all over the kingdom, and is every season gaining ground" (*Hints* 36–37). By 1816, William Kirby and William Spence had written that from London the aphid "now found its way into other parts of the kingdom, particularly into the cyder counties; and in 1810 so many perished from it in Gloucestershire, that, if some mode of destroying it were not discovered, it was feared the making of cyder must be abandoned" (*Introduction to Entomology* 1:201). The discovery at this time that a mixture of spirit of tar provided an effective means of control diminished the threat of this insect, so that when Banks's paper "Notes relative to the First Appearance of the Aphis Lanigera, or the Apple-Tree Insect, in this Country" was read at the Horticultural Society of London in 1815, the crisis had already been somewhat averted. However, the impact of the woolly aphid can still be seen in Thomas Andrew Knight's famous collection of aquatint engravings entitled *Pomona Herefordiensis*, published in 1811, at the height of the anxieties about the woolly aphid; many of the illustrations in the series portray apples that are infected with this pest.

The Black-Bobs of Selborne

White's interest in bird migration and in the invasive travels of insects suggests that, despite his insistence on the primacy of the local in natural history, he fully recognized the limits of his domestic naturalism and that to properly understand the life histories of many of the creatures that he loved so well, he had to look further afield, to grasp the interaction between the local and the global. White may have been a stationary naturalist, but he knew that the nature around him was beginning to move. In the latter pages of *Selborne*, though centered on the familiar life going on around him, White frequently looks outward into the silence and the unknown. One of these unknowns, with which he developed a fairly close intimacy in his last years, goes by the name of *Blatta orientalis*, otherwise known as the Asian cockroach, or "black-bob," as White called it. Although this insect had appeared in some of the major port cities of England by the late sixteenth century, most of the country was free of it until relatively late in the eighteenth century. Once these insects reached Holland and England, as L. C. Miall and Alfred Denny have indicated, they "gradually spread thence to every part of the world" (Miall and Denny, *Structure and Life-History of the Cockroach* 17).

Prior to their wholesale participation in European colonial expansion, cockroaches were primarily associated with the New World (particularly the Caribbean). John Smith, in *A Generall History of Virginia* (1624), is among the first to mention "a certaine India Bug, called by the Spaniards a Cacarootch, the which creeping into Chests they eat and defile with their ill-sented dung" (5:171). Three decades later, Richard Ligon, in *A True and Exact History of the Island of Barbados* (1657), speaks of the "cock-roche" as a pest that bites people in their sleep and destroys interior draperies. Eighteenth-century colonial accounts of the insect repeatedly comment on its disgusting behavior, its ravenous appetite, and its troublesomeness at night. "Like harpies, they defile whate'er they touch," writes James Grainger in *The Sugar-Cane* (1.339). Having returned sick and despondent from a harrowing journey up the Cottica River in eastern Surinam, the mercenary John Gabriel Stedman opened his stored trunks to find that he had "enemies at home as well as abroad; since most of my shirts, books, &c. were gnawed to dust by the blatta or cockroach, called *cakreluce* in Surinam: nay, even my shoes were destroyed, of which I had brought with me twelve pairs new from Europe, as they were extremely dear and bad in this country."

A description of the insect's habits follows: "By getting through the locks of chests or boxes, it not only deposits its eggs there, but commits its ravages on linen, cloth, silk, or any thing that comes in its way; by getting also into the victuals and drink of every kind, it renders them extremely loathsome, for it leaves the most nauseous smell, worse indeed than that of a bug" (*Narrative of a Five Years' Expedition* 1:194–95). Given what appears to be almost a universal consensus that the cockroach is a nasty, abject creature, Edward Long stands out for his minority-of-one celebration of its domestic virtues. Unfortunately, whatever points we might accord his cockroach panegyric are lost because it underpins a misogynistic diatribe against sloppy housewives. The insect, he writes, is "exceedingly industrious in gleaning up such filth and nastiness, as the slovenly neglect of bad housewives has left in holes and corners undisturbed by the broom." Natural history thus serves the purposes of moral fable: "Go to the *cockroach*, thou *slut*; consider her ways, and be *cleanly*" (*History of Jamaica* 3:887).[26]

Although most Britons during the eighteenth century would have had little experience with this newcomer, in one area they were particularly affected: by the latter half of the eighteenth century, commercial and naval ships had become floating insectaries. On his third voyage, Cook was particularly troubled by two species. According to William Anderson, the surgeon/naturalist on board the *Resolution*, the first of these, the Asian cockroach, "had been carried home in the ship" on the second voyage and had "withstood the severity of the hard winter in 1776, though she was in dock all the time" (Cook and King, *Voyage to the Pacific Ocean* 2:99). Cook learned of the other species, the German cockroach (*Blattella germanica*), shortly after he left New Zealand in February. Soon they were everywhere. The dietary situation on board was unimaginably nasty. It was bad enough that "if food of any kind was exposed, only for a few minutes, it was covered with them; and they soon pierced it full of holes, resembling a honeycomb," but the insects also left telltale signs of their presence, particularly on the bread, "which was so bespattered with their excrement that it would have been badly relished by delicate feeders" (2:99).[27] Opening the ship's sails was also a creepy experience, for "when a sail was loosened, thousands of them fell upon the decks." Sleeping was almost impossible, for at night, they would appear "in infinite numbers," making "every thing in the cabins seem as if in motion, from the particular noise in crawling about" (99). Although cockroaches have a reputation for being quite intelligent, Cook's were a particularly anti-intellectual mob that did everything in their power to subvert the higher scientific goals of the expedition. "They were particularly destructive to birds, which had been stuffed and preserved as curiosities," Cook writes. It might be said in their defense that they displayed a strong taste for scientific writing, but that was strictly from a gustatory standpoint. They "were uncommonly fond of ink; so that the writing on the labels, fastened to different articles, was quite eaten out; and the only thing that preserved books from them, was the closeness of the binding, which prevented these devourers getting between the leaves" (99).[28]

By the time the crew reached the Society Islands, desperate measures were necessary. Cook had stopped there in order to bring the islander Omai home. Brought to England in 1774 on the *Adventure*, Omai had been an instant celebrity, an embodiment both of the ideal of the "Noble Savage" and of a cosmopolitan. The visit to Huahine seemed the perfect occasion to get as many cockroaches onshore as possible. "While we lay in this harbour," Cook writes, "we carried ashore the bread, remaining in the breadroom, to clear it of vermin. The number of cock-roaches that infested the ship, at this time, is incredible" (*Voyage to the Pacific Ocean* 2:98). William Bayly, the astronomer on the *Discovery*, provides more details: "The 31st Capt Cook sent most of the chest's & other movables on Shore in order to destroy & get as many of the Cockroaches out of the Ship & smoaked the Ship with Gunpowder by which means Immence Numbers were kill'd & set on shore. These Insects were so Innumerable on board the Resolution that they run in every [part] of her so thick you would think the Ship Alive, even the closest box or trink [sic] were All Alive with them & they eat & destroy every thing they have" (Cook, *Journals* 3:238–39n). William Bligh did the same thing when he arrived in Tahiti on his first voyage of the *Bounty*, writing, "This morning I ordered all the chests to be taken on shore, and the inside of the ship to be washed with boiling water to kill the cockroaches. We were constantly obliged to be at great pains to keep the ship clear of vermin on account of the plants" (*Voyage to the South Sea* 121).

Given the indisputably low habits of cockroaches (Patrick Browne in 1756 called them "the most loathsome insects in America"), it is perfectly understandable why historians have preferred to focus their attention on the global travels of Omai rather than the *Blatta* tribe (P. Browne, *Civil and Natural History of Jamaica* 433). It is easy to ignore the historical importance of the landing of *germanica* and *orientalis* on the island of Huahine. There is also some evidence that cockroaches may have already arrived on the island, for among the words included in Anderson's "Vocabulary of the Society

Islands" is the Polynesian word for "cockroach": "potte potte" (Cook, *Voyage Towards the South Pole* 2:330). Such evidence is questionable, however, since Omai was probably the source for most of these words, so it is more likely that the insect that Omai named was not one he discovered on the island, but one of the many that were ready to hand on board the *Resolution*. Their very presence in the "Vocabulary," therefore, is probably a testimony less to Anderson's thoroughness as a field linguist than to the crew's understandable preoccupation with these troublesome stowaways. There can be little doubt that Cook must have seen the cockroaches as relatively annoying but unimportant actors in the history he was writing—and historians of the Pacific have generally agreed with him. None of these historians being cockroaches, it is perhaps understandable that they do not write colonial history from the standpoint of this insect, for which gaining a foothold in Huahine was not an unimportant event. The colonial period was a golden age for cockroaches, and long after the British and Europeans relinquished their colonies, the insect seems to have prospered. We are likely to see it as an abuse of terms to speak of cockroaches as cosmopolitans. Instead, we reserve the term for its almost mythic embodiment in the figure of Captain James Cook and, perhaps, for Omai, who stands as his complimentary cultural imitation, his "mimic man." Unlike his arthropodan messmates, Cook at least knew where he was headed and took notes and drew maps. Cockroaches, on the other hand, were historical tagalongs—at best, a traveling pest that was inconveniencing traveling Europeans. Mysteriously appearing on ships, and often disappearing just as mysteriously, these insects briefly scuttle into travel writing, eliciting momentary disgust, after which we pay them no more attention.

Nevertheless, the confidence with which the Enlightenment denied cosmopolitan status to almost all species except human beings, identifying cosmopolitanism with the perspectives of a traveling European elite, has been called into question in recent years by postcolonial critics who have developed new conceptions of cosmopolitanism which include traveling local cultures, ethnicities, diasporas, and demotic groups. Cosmopolitanism now includes, as Bruce Robbins remarks, "a new cast of characters. In the past the term has been applied, often venomously, 'to Christians, aristocrats, merchants, Jews, homosexuals, and intellectuals.' Now it is attributed, more charitably, to North Atlantic merchant sailors, Caribbean au pairs in the United States, Egyptian guest workers in Iraq, Japanese women who take *gaijin* lovers" ("Actually Existing Cosmopolitanisms" 1). Robbins does not include in his list of new cosmopolitan subjects the huge number of companion animals that resettled the globe along with Europeans, yet they too merit consideration as we seek to rethink what "cosmopolitanism" might mean. Certainly, biologists nowadays commonly use the term "cosmopolitan" in referring to cockroaches, because many species of this insect have a worldwide distribution and their biogeographical origins are lost in time. Over much of its recent history, the roach has been far more successful at establishing lasting colonies than were either the British or the Romans. Even at the height of its power, Britain could only paint about one-fifth of the landmass of the globe pink. Roaches would require a less rosy color, and they would probably want to eat the map along with the territory, but their present territorial claims would be almost universal. Although some of them live in the wild, most of these insects, like White's swallows, thrive in close association with human beings.

Since Linnaeus and most of his followers assumed that all species were created for the places in which they were found, he often named species after the geographical location in which they were collected. This was an unfortunate procedure when it came to cockroaches, which by this time were becoming established world travelers. *Blatta orientalis* was named on the basis of a specimen sent to Linnaeus from China. Unfortunately, it did not originate there, but instead probably came from Africa. By the time Linnaeus came upon the insect, it had already spread through transatlantic shipping first to the West Indies and America and then to the East. Similarly, the other cockroach introduced into England during the seventeenth century—the German cockroach (*Blattella germanica*)—was no more German than the Hessian fly was English. When Linnaeus named this species in 1767, the insect had probably been living in Europe for about a thousand years, having probably originated in Sudan or Ethiopia, from whence it traveled with cargo to the Mediterranean coast, spreading first to Asia Minor and then through colonial trade to Europe and the rest of the globe. Linnaeus was nevertheless not averse to using scientific names to satisfy nationalistic prejudices, and in this he was no different than the Germans, who normally referred to the insect as the French, Russian, or Swabian cockroach. The great Swedish naturalist was hardly more successful in naming many other cockroaches. Like the Australian cockroach (*Periplaneta australasiae*), the American cockroach (*Periplaneta americana*) appears to have come from Africa and probably arrived in America sometime around 1625 on slave boats. American naturalists, for their part, sought to publicly disown all connection with this bug by repeatedly describing it as "non indigenous." If the association of geography and naming was confused during the eighteenth century by the traveling habits of cockroaches, it has only gotten worse since. In 1771, in his *Catalogue of the Animals of North America*, Forster listed four species of cockroach as inhabiting North America: *B. orientalis*, *B. germanica*, *P. americana*, and *B. livida*. Now there are at least fifty-five (around thirty-five in Texas alone), the most recent being the Asian cockroach (*B. asahinai*), discovered in Florida in 1986. In nineteenth-century America, alongside the nationalistic and taxonomic issues of this bug naming, there was also an additional moral dilemma. As B. D. Walsh noted in the 1830s, "'Cock-roaches' in the United States are always called 'roaches' by the fair sex, for the sake of euphony" (*Comedies of Aristophanes* 1:89n).

In England during the eighteenth century, most meetings with cockroaches were first encounters. In his 1770 *Catalogue of British Insects*, Forster includes both *orientalis* and *germanica*. In 1816, Kirby and Spence speak of the cockroach as "one of the evils which commerce has imported." After becoming a permanent resident in metropolitan dockyards and warehouses, it soon became a regular denizen of bakery shops and cellars. "In the London houses, especially on the ground-floor, they are most abundant, and consume every thing they can find, flour, bread, meat, clothes, and even shoes" (*Introduction to Entomology* 1:241). Known by the name black-beetle or mill-beetle, the insect, as Ebenezer Sibly remarked, "seems to have made great progress of late years in extending itself throughout the kingdom" (*Magazine of Natural History* 13:173) and was "now completely naturalized to this climate" (170).

Although it took these insects some time, it appears that *orientalis* finally arrived in the village of Selborne by 1790, for on 28 March White was surprised to learn that one of his neighbors was "over-run" with a kind of "black beetle" or "black-bob" that "swarmed in her kitchen" when she got "up in a morning before daybreak." To his utter astonishment, he soon discovered that he was no better off, and that the only real difference between him and his neighbor was that he was a later sleeper. Consulting the works of Linnaeus and of the Renaissance entomologist Thomas Mouffet (of "Little Miss Muffet" fame), White soon identified the bug as *Blatta orientalis* (Mouffet's *Blatta molendinaria*). He was fully aware of the global aspirations of this insect and the role that colonial commerce had played in its spread. "These insects," he writes, "belonged originally to the warmer parts of America, &

were conveyed from thence by shipping to the East Indies; & by means of commerce begin to prevail in the more northern parts of Europe, as Russia, Sweden, &c. How long they have abounded in England I cannot say; but have never observed them in my house till lately" (*Journals* 355). By the fall of that year, White's kitchen was beginning to look a lot like the cabin of Cook's *Resolution*, even if, to look on the bright side, the cockroaches and crickets seemed to be getting along with each other: "After the servants are gone to bed, the kitchen-hearth swarms with young crickets, & young *Blattae molendinariae* of all sizes from the most minute growth to their full proportions. They sem [sic] to live in a friendly manner together, & not prey the one on the other" (370). Although White seems to have been reasonably comfortable with his new roommates, his live-in sister-in-law, Mrs. John White, was not, perhaps because she had read Edward Long or because she was simply an inveterate enemy of insect intruders. For the next two years she engaged in an all-out war of extermination against their new guests. By June 1792, after destroying "thousands" and killing "some hundred every night," she seemed on the verge of complete victory having "almost subdued" them (406). Two months later, however, even she had to concede defeat, because "fresh detachments" of *orientalis* would fly in every night through the open casement windows "from the neighbouring houses which swarm with them" (410). White was philosophical: "These, like many insects, when they find their present abodes overstocked, have powers of migrating to fresh quarters." In the final years of his life, White was still spending quite a bit of time thinking about migration. Now, however, instead of celebrating the joy of observing flocks of expatriates returning each spring, confirming by so doing that, despite their mobility, Selborne continued to be their "native" home, White now struggled with a nature that had come from elsewhere, another kind of winged nature that was even more intimately involved in his household, a foreign nature that was soon to become as local and native as his beloved swallows. Teeming with crickets and cockroaches, Gilbert White's domestic world had come to mirror the larger world around him.

CHAPTER FIVE

William Bartram's *Travels* and the Contested Natures of Southeast America

Of the natural history and travel writings that Coleridge and Wordsworth read in Somerset, no book had a greater impact on their understanding of the natural world and how to write about it than William Bartram's *Travels through North and South Carolina, Georgia, East and West Florida, the Cherokee Country, the Extensive Territories of the Muscogulges, or Creek Confederacy, and the Country of the Chactaws* (1791), a book that William Hedges judges to be "the most astounding verbal artifact of the early republic" ("Toward a National Literature" 190). Explicit in its ungainly title was Bartram's recognition that he was moving through contested lands, where European settlers and the southeastern American Indians, living in close proximity, had been engaged in a constant, protracted, and violent struggle for territory. Where Gilbert White could imagine that he was living in a comfortably domestic island nature, reflective of a homogeneous and stable "English" order, Bartram confronted a world in which the struggle among peoples extended to nature itself and was everywhere reflected in the landscapes that he visited. Like White, Bartram was not writing about nature in general, but instead about the natural world of the southeastern American Indians, a region that was undergoing enormous political, social, and ecological changes. In a style that combines the language of American pastoral with religious spiritualism and scientific modes of description, he sought not simply to catalog the natural history of "the Indian countries" ("Report to Dr. Fothergill" 430), but to provide testimony of its ecological importance. As a trained botanist, Bartram could appreciate, in ways that his compatriots often could not, the unique beauty and ecological richness of the natural world that the Native Americans had produced and sustained for centuries and that was about to be lost forever. He could see what had changed and what was changing, and thus he could also see into the future.

When in the 1780s Bartram wrote the *Travels*, he must still have hoped that an appreciation of nature might serve as the basis for reenvisioning society. As Wordsworth phrased it, a love of nature might lead to a love of mankind. Bartram also must have believed that by celebrating the natural history of the "Indian countries," he might persuade the newly emergent republic of the United States to ensure that the needs and rights of the people who had originally welcomed Europeans to their land would be respected. "Who has a stronger claim to this Country than the Indians?" he asked ("Some Hints & Observations" 197). The fundamental precepts of Christianity, he writes, "teach us that all the nations & Tribes of Men, are Brethren & the offspring of one Family: They instruct us that every nation is equally entitled to those bountiful blessings of Providence, the furniture and produce of the Earth, & the benign influence of the Planets" (195–96). By this time too, however, he also must have recognized that the idea of a beneficent nature, which he sought to promote in the *Travels*, was being contradicted by the realities of the frontier. In this sense, as others have suggested, Bartram anticipates Darwin. Even as he sought to demonstrate that the natural world could provide the new republic with a model of a diversity of peoples, animals, and plants living in peaceful coexistence, this commitment was continually being undercut by violence. Consequently, Bartram continued to hold onto the slim hope that the frontier nature that he portrayed in the *Travels*—to a degree a fiction—might provide a blueprint for a future republic even as he probably knew that the way of life and nature that he was describing would soon only exist in memory, in stories, and in the written testimony provided by his book.

Bartram's *Travels* can be said to be structured by the double consciousness that is often associated with Wordsworth, for in this narrative there are two Bartrams: Bartram the young man who explored the colonial American Southeast during the 1760s and 1770s, and the older post-Republic writer, who in the 1780s reflected on that experience and sought to convey its meaning in the *Travels*. Against the young man's enthusiasms and ecstasies, the *Travels* strikes an elegiac chord; it is essentially a memory book, an older man's reflection on past experiences that he had written about more than a decade earlier. Although critics have often argued that Bartram wholeheartedly adopted the myth of the American Adam, outside society, confronting the wilderness, and forging a new self-identity, Bartram knew that the landscapes he visited were not a primordial Eden, but a nature that reflected the values and agency of the Native Americans who had inhabited the Southeast for centuries.[1] When in the 1780s he began to prepare the *Travels* for publication, much had changed since he undertook his two journeys to Florida, first with his father, the preeminent American botanist and plant collector John Bartram, in 1765–66, and then again on his own in 1773–76. His father was now dead, and the botanical correspondence network that he and Peter Collinson had developed, and which had enlisted William as a collector and illustrator, was a thing of the past. His patron, John Fothergill, died in 1780, followed two years later by Daniel Solander, who had been charged with classifying the plants that Bartram had collected.[2] Consequently, most of Bartram's botanical discoveries would go unacknowledged. At the same time, the new republic of the United States had come into being, and Bartram would use his book to reflect on the future of this new nation. Most importantly, the natural world that he had encountered in Florida, a place whose "Face & constitution" he believed were expressive of its people ("the country is Indian wild"), was disappearing with the steady encroachment of white settlers on Native American lands and with every new land settlement treaty (William to John Bartram, 27 Mar. 1775, *Correspondence of John Bartram* 770).

In *Travels* Bartram straddles two worlds, trying to establish a middle ground between the expansionary ideology of white settlers and the southeastern American Indians. And the places that he visited, whether controlled by the English or indigenous peoples, were also mixed and layered. Even the food that he consumed, as Kathryn E. Holland Braund has shown, "reveals products of a landscape already changed by imported plants and animals— and by new settlers and slaves" ("William Bartram's Gustatory Tour" 33). By the early 1850s, Bartram's vision of the frontier as a space that still held the possibility of human beings living together in benevolence and mutual respect seemed, in Carlyle's view, "immeasureably *old*. All American Libraries ought to provide themselves with that kind of Book; and keep them as a kind of future *biblical* article" (Carlyle to Emerson, 8 July 1851, *Collected Letters of Thomas and Jane Welsh Carlyle* 26:105). Carlyle powerfully captures the unique manner in which the *Travels* portrays the struggles and contradictions that shaped the emergence of the United States of America. In the 1820s, Coleridge was recommending the *Travels* as "the latest book of travels I know, written in the spirit of the old travellers" (12 March 1827, *Table Talk, Collected Works* 14, pt. 2, 57). Coleridge was, perhaps, too humble to recommend his own poetic imitation of such travels—of an ancient mariner who seeks to forge social renewal from a tale about violence and how it taught him to love nature and others.

Colonial Botanical Networks

At the beginning of the *Travels*, Bartram provides a conventional Enlightenment justification for his journey, writing that, "continually impelled by a restless spirit of curiosity in pursuit of new productions of natures, my chief happiness consisted in tracing and admiring the infinite power, majesty and perfection of the great Almighty Creator, and in the contemplation, that through divine aid and permission, I might be instrumental in discovering, and introducing into my native country, some original productions of nature, which might become useful to society" (*Travels* 48).[3] In this "restless . . . pursuit" of new plants, in his belief that the study of nature leads to an appreciation of the divine wisdom of God, in tasking himself to archive the flora and fauna of the Southeast, and in his desire to introduce useful plants into society, Bartram adopts ready-to-hand ideas about the social, religious, scientific, and economic importance of natural history. Indeed, it was these same beliefs that for forty years had governed the insatiable collecting practices of his father. In 1728, John established on the outskirts of Philadelphia the first botanic garden in the Americas, and this garden, as William remarks, "became the Seminary of American vegetables, from whence they were distributed to Europe, and other regions of the civilized world" ("Preface to a Catalogue" 587). It was in the autumn of 1734 that John began to sell and exchange seeds, cuttings, and roots with Peter Collinson, and over the next forty years this exchange fostered an enormous influx of North American plants into England.[4] For Collinson, this correspondence was all about acquiring new plants. "There is no end of the Wonders in Nature," he commented, "the More I see the more I covet to see" (Collinson to John Bartram, 25 July 1762, *Correspondence of John Bartram* 565). In another letter, Collinson urges his American friend to "look out sharp for" the broad-leaved kalmia and mountain laurel, "for one can Never have too Many" (520).

John's last major botanical expedition, in 1765–66, at the age of sixty-six, was to survey the territories of East and West Florida newly ceded to Britain by France and Spain as part of the 1763 Treaty of Paris. John knew a good collecting opportunity when he saw one, so he had written to Collinson in October 1764, asking him to use his influence "to enable me to travail A year or two through our kings new acquisitions to make A thoro natural & vegitable search either by publick authority or private subscription" (Bartram to Collinson, 15 Oct. 1764, *Correspondence of John Bartram* 641). Six months later, Collinson informed him that he had been appointed the king's botanist to the Floridas, with a £50 annuity. "Lord Bute & ye Earl of Northumberland declared that it was nessesary that ye floridas should be searched & that I was ye Properest person to do it" (John to William Bartram, 7 June 1765, *Correspondence of John Bartram* 651), he boasted to his son. William was twenty-six at the time, and though his real love was natural history illustration, his father had unfortunately set him up as a merchant in Cape Fear because he did not believe that William could make a "resonable liveing" as an artist (Bartram to Collinson, 27 Apr. 1755, *Correspondence of John Bartram* 384). William, however, was miserable as a merchant, so he was happy to settle up his accounts and join his father on the expedition as his artist and assistant. The experience of Florida was transformative, and at the end of the journey, he decided to settle in East Florida as a rice planter. The project ended badly, however, and, after a brief period as a draftsman on the royal survey of East Florida, he was soon back in Philadelphia, first working as an agricultural laborer and then once again, sadly, as a merchant.

Around this time Bartram began to be recognized for his natural history drawings. For more than a decade he had been sending Collinson illustrations of American plants and insects ("Billys Elegant drawings are admired by all that see them.... His Butterflies, Locust are Nature itself his yellow Fly is admirable") and was at the time receiving commissions from wealthy patrons such as the Duchess of Portland and Dr. John Fothergill (Collinson to John Bartram, 28 May 1766, *Correspondence of John Bartram* 667).[5] "When art is arrived to such perfection to copy close after nature," wrote Collinson, "who can describe the pleasure, but them that feel it, to see the moving pencil display a sort of paper creation, which may endure for ages, and transfer a name with applause to posterity!" (Collinson to William Bartram, 28 July 1767, W. Darlington, *Memorials* 290). When in 1772 he once again ran into financial difficulties, he turned to Fothergill with the proposition that he be privately engaged to undertake a natural history excursion into Florida. Fothergill, who possessed one of the largest botanic gardens in England at Upton, second only to Kew Gardens, welcomed the chance to obtain new plants and additional drawings, so he agreed to pay Bartram £50 per annum plus expenses, hoping that the young man might "get into a good livelihood by sending boxes of plants and seeds to Europe from those less frequented parts of America" (Fothergill to William Bartram, 1772, *Correspondence of John Bartram* 753–54). Bartram was also encouraged to follow his father's example by keeping "a little journal, marking the soil, situation, plants in general[,] remarkable animals, where found, and the several particulars relative to them as they cast up" (Fothergill to William Bartram, Corner and Booth, *Chain of Friendship* 402).

Commissioned to serve as a professional plant collector for an English gentleman, Bartram showed every indication that he would become, as Thomas P. Slaughter suggests, "the same explorer for empire that his father became" (*Natures* 188), yet despite the fact that his *Travels* emerged out of the protocols and conventions of Enlightenment natural history, the book ultimately became something quite different.[6] During the early stages of the expedition, William literally followed John's footsteps, retracing much of the previous journey of 1765–66. Repetition was thus an important structuring element in both his life and his *Travels*: Bartram the writer repeats and reworks in narrative form a journey that he had taken a decade earlier, which was itself based on an even earlier journey undertaken with his father. Repetition allowed him to adopt the profession of his father while symbolically differentiating himself from it. Whereas John traveled in order to exploit nature—to discover, collect, and exchange plants—William was traveling in order to better understand himself and the world around him; his travels allowed him to represent relationships—of creatures to one another and of human beings to one another and to the natural world.

Much of the complexity of *Travels* derives from its often confusing mixture of the generic aspects of a natural history travel narrative with a young man's quest to find his identity and place in the world. Objective, impersonal botanical descriptions and seemingly endless lists of plants with their Latin names are combined with a deeply personal, religious, and aesthetic veneration of nature. Nature as a scientific business is set against nature as inspiration. What began as a form of appropriative translation, of possessing another peoples' nature by exploring, collecting, mapping, and naming it, became something much larger and more complex—a reflection on the deep relationship between the natural history of a place and the history of its people. A detailed descriptive survey of the flora and fauna of the American Southeast became a narrative about one man's religious and deeply ecological experience of a nature that, like the people who inhabited it, was new to him. Moments of intense aesthetic pleasure in the landscapes that he was seeing punctuate this narrative, as in the account of the Alachua Savanna, where he writes that the attention is "wholly engaged in the contemplation of the unlimited, varied, and truly astonishing native wild scenes of landscape and perspective, there exhibited: how is the mind agitated and bewildered, at being thus, as it were, placed on the borders of a new world!" (*Travels* 120). Ironically, it was also in the same place, in the nearby town of Cuscowilla, that young Bartram found his true self, not as "William Bartram" alone, but also as Puc Puggy or the "Flower Hunter," a name given to him by the Seminole chief Ahaye, whom the British knew as Cowkeeper. In this regard, the *Travels* supports Tim Fulford's argument that the Romantics transformed "Indians into fictional figures, some uncanny, some alien, [and] redefined British identity in ways that could not be adequately formulated by the existing social and political discourses." By so doing, they "sketched out new, half-realised forms of self, in which foreign and familiar met—'strange meetings' which seemed, and still seem, prophetic of the possibility and the difficulty of living at one with what we like to think we are not" (Fulford, *Romantic Indians* 33). Years later, as an older Bartram reflected on his journey and revised the journal that he had written as a young man for John Fothergill, he sought to explain how he had been positively translated by experiencing the nature that he had come to survey. The resulting text voices the contradictory hope that somehow the new Republic would ensure that there would be space enough for a true diversity of people and natures, and for similar translations.

Bartram's Natural Republic

Richard Poirier's claim that "the most interesting American books are an image of the creation of America itself" certainly applies to Bartram's *Travels* (*World Elsewhere* 6). One of the central conundrums of Bartram criticism has had to do, in Francis Harper's words, with how it was possible for him to write "an account of travels extending all the way from Pennsylvania to Florida and the Mississippi, and including nearly two years of the Revolution, without once referring to that momentous conflict" ("William Bartram and the American Revolution" 574). For Harper, the answer lay in Bartram's proto-Romantic disposition, which looked to nature as an escape and refuge from the social and political turmoil of his times. In this regard, the *Travels* can be read as an example of what Jerome McGann has called "romantic ideology," that is, a belief that an imagined relationship to nature might stand as a replacement for the disappointments of history. In recent years, however, a more complex understanding of the social and political dimensions of this text has emerged, which sees it not so much as an escape from history and politics into nature as an attempt to use natural history to reflect critically on the future of the new republic, particularly on the rights of Native Americans.[7] "Nature does not speak with the immediacy of one of Benjamin Franklin's political tracts," comments Douglas Anderson, "but to a surprising degree the wilderness enables Bartram to place his own eventful times in a context wide enough to provide a basis for measured skepticism as a corrective for patriotic fervor" ("Bartram's *Travels* and the Politics of Nature" 5). Edward J. Cashin also concludes that Bartram's tendency not to write directly about "the world in which he lived" was less an expression of a political omission than of his desire to use his book to further an idea of "America as he hoped it would be, purged of wars and ugliness, filled with wonderful plants and animals for the betterment of mankind.... He dwelt upon the good in the hope of advancing the good" ("Real World of Bartram's *Travels*" 12). Charles H. Adams argues that "the great events of the 1770s may not appear in the text, but the South that Bartram constructs at his writing table through the turbulent 1780s offers, in its rendering of both the history of nature and the nature of history, a powerful imaginative response to the Revolutionary era" ("William Bartram's *Travels*" 114). For Bartram, nature and history were not alternatives, but were integrally bound up with each other. As a discipline that involved the study of both, natural history provided a framework for comprehending the radical changes that were taking place in the social and natural landscape of the Southeast and for representing his own vision of what a truly enlightened social polity might be, suggesting at the same time that the new American republic was not living up to its ideals.

It is not difficult to see why Lane Cooper considered *Travels* to be "a typical specimen of romantic nature worship," for the book frequently represents nature as an open temple in which all creatures seem to enjoy their place within the Creation ("Bartram redivivus?" 152). "This world, a glorious apartment of the boundless palace of the sovereign Creator," writes Bartram, "is furnished with an infinite variety of animated scenes, inexpressibly beautiful and pleasing, equally free to the inspection and enjoyment of all his creatures" (*Travels* li). It is hardly unusual for an eighteenth-century writer to speak of nature as being, in Daniel Boorstin's words, "a complete and perfected work of divine artifice" (*Lost World of Thomas Jefferson* 54), but what makes Bartram unique, as Bruce Silver has observed, is that he does not use this idea primarily for theocentric purposes (for as a Quaker he takes the existence and wisdom of a benevolent God as a given), but instead to encourage his readers to appreciate nature on its own terms ("Eighteenth-Century Accounts of Nature"). In many ways unique for his time, Bartram did not see the world in anthropocentric terms as having been created to serve only human needs and perspectives, but instead he believed that it was created to be appreciated by all the creatures that compose it. Theology, aesthetics, and politics are thus brought together in the idea that nature in all its diversity, in its "infinite variety of animated scenes," has been freely provided for the pleasure and the intellectual and moral benefit of all its creatures, and that all animals have been accorded the rational capacity to "inspect" and to "enjoy" it.

Without the hierarchies normally brought to bear within the politics of sympathy, Bartram was ahead of his time in seeing animals as moral and intellectual beings.[8] In a letter to Benjamin Smith Barton, written within a year of the publication of the *Travels*, he asks, "Can any Man of sense & candour, who has the use of his Eyes, Rational Faculty, doubt that Animals are rational creatures?" (Bartram to Benjamin Smith Barton, 29 Dec. 1792, Hallock and Hoffman, *Search for Nature's Design* 168). Like Gilbert White, Bartram recognized that "every Animal hath a Language, Both by words or sounds, perfectly articulate & By actions which is perfectly understood apparently without error or mistake by every Individual, both old & Young of the same Tribe or Race," and from this fact he concluded that "if Animals have a vocal Language, it is self evident that they have Intelligence, they have Ideas & Understanding" ("[Dignity of Human Nature]" 355). "If we bestow but a very little attention to the economy of the animal creation," he writes in *Travels*, "we shall find manifest examples of premeditation, perseverance, resolution, and consummate artifice, in order to effect their purpose" (*Travels* lvii). Reason was, for the Quaker Bartram, the fundamental expression of the Inner Light; it was the "Divine Intelligence" that "penetrates & animates the Universe . . . the Immortal Soul of Nature, of Living moving beings" and "of Vegetables" ("[Dignity of Human Nature]" 353). For this reason, he was unwilling to accept the idea that creatures were bereft of the Inner Light, and he was just as unwilling to accept the notion that nature was created only for the benefit of human beings. For Bartram, animals were not excluded from enjoying the natural world of which they were a part and, thus, of knowing and loving the God who had created it (Walters, "'Peaceable Disposition' of Animals" 168). "I cannot, believe I cannot, be so impious," he writes. "Nay my soul revolts, is destroyed, by such conjectures as to desire or imagen, that Man who is guilty of more mischief & Wickedness than all the other Animal[s] together in this World, should be exclusively endued with the knowledge of the Creator" ("[Dignity of Human Nature]" 353).

When Bartram draws on the traditional "argument from design," by comparing nature to a watch or piece of brocade, he is less interested in the complexity of the Creation as a physical system than in a corresponding intellectual diversity: "If then the visible, the mechanical part of the animal creation, the mere material part is so admirably beautiful, harmonious and incomprehensible, what must be the intellectual system? that inexpressibly more essential principle, which secretly operates within? that which animates the inimitable machines, which gives them motion, impowers them to act, speak, and perform, this must be divine and immortal" (*Travels* lvi). Bartram emphasizes the intellectual and moral integrity of animals, their role as living agents within the Creation, sharing reason ("divine and immortal") and engaged in the pursuit of their own forms of happiness. Bartram is a deeply ecological writer, even before the term "ecology" had been invented, for he understands nature as a system of interrelationships among creatures that think, feel, and seek as much as they can to get along with one another. It is not composed of objects or machines, but of a diversity of subjectivities, each created "for a certain & indispensable purpose in this Vast System of Creation, as instruments, Members or Organical beings design'd & created to form a part in the Whole & Act & perform a certain part" ("[Dignity of Human Nature]" 348). Even plants contribute to and share in the pleasures of the Creation, having "some sensible faculties or attributes, similar to those that dignify animal nature; they are organical, living and self-moving bodies" (*Travels* liv). Wordsworth, after reading the American naturalist, began to make similar claims, such as his statement of faith, in the April 1798 "Lines Written in Early Spring," "that every flower / Enjoys the air it breathes."

Like Gilbert White, Bartram was deeply committed to the sociality of nature, but in contrast to White, who maintained that animals act on the basis of unreflective instinct, Bartram countered that all animals, including human beings, were given instinct as a form of divine guidance, and that human beings "act most Rationally & virtuously when our Actions seem to operate from simple instinct, or approach nearest to the manners of the *Animal creation*" ("[Dignity of Human Nature]" 353). Human beings, however, then and now, often disregard these dictates, which is why Bartram believed that natural history provided a better model for how human beings should act than human history. Bartram adopts a primitivist philosophy, arguing that since the southeastern American Indians are closer to nature, they provide better models for human society than Europeans. As he argues in *Travels*, "the primitive state of man," like that of animals, was "peaceable, contented, and sociable" (71). Eighteenth-century primitivism has been readily dismissed by those who believe that the human resides in what separates it from nature. Such views, however, were part and parcel of the wholesale devaluation of non-European and indigenous societies, so we should be careful not to dismiss such views too quickly.

While White maintains the separation between himself and the living creation that he observes, Bartram is fundamentally ecocentric, that is, he understands nature as a living and diverse community to which he belongs. Some of the most powerful passages in the *Travels* are those in which Bartram observes nature not to separate himself from it, but in order to share in its pleasures. Describing the flight of "the loud, sonorous, watchful savanna cranes," he writes that with "musical clangor," "they form the line with wide extended wings, tip to tip, they all rise and fall together as one bird; now they mount aloft, gradually wheeling about; each squadron performs its evolution, encircling the expansive plains, observing each one its own orbit; then lowering sail, descend" (93). Here Bartram is emphasizing the beauty, harmony, and order of these birds as they fly together as a single unit, "as one bird," a "squadron," communicating with each other. Bartram enjoys what he sees, but he is not alone, for, clearly, so do the birds. They contribute to the order and beauty of the Creation in synchronizing their flight. Likening it to planetary motion, Bartram clearly recognizes that to achieve this goal, each of these "watchful" birds must watch the other, "observing each one its own orbit," wing tip to wing tip. Bartram's cranes, in other words, are observers too, and the order that they create is an expression not only of their *being seen* but also of their *seeing* in the world. Elsewhere, on the Alachua Savannah, Bartram describes how the cranes bring a typical day to a close: "The sonorous savanna cranes, in well disciplined squadrons, now rising from the earth, mounted aloft in spiral circles, far above the dense atmosphere of the humid plain; they again viewed the glorious sun, and the light of day still gleaming on their polished feathers, they sung their evening hymn, then in a straight line majestically descended, and alighted on the towering Palms or lofty Pines, their secure and peaceful lodging places. All around being still and silent, we repaired to rest" (121). Here in this naturalized version of an evensong, two forms of consciousness and two kinds of seeing are brought together: those of the observing naturalist and those of the cranes, who rise together to enjoy their last view of a "glorious sun" that "still gleam[s] on their polished feathers." Touched by the sight of the sun and touched by its rays reflecting off their "polished" feathers, they sing "their evening hymn" to a Creation that they both enjoy and create, before retiring for the night. Security and peace reign, and Bartram, like these diurnal animals, follows their example and repairs to his own rest. Bartram's world is not one in which human beings enjoy a creation made exclusively for themselves, but a vast interrelated system in which all creatures seek to be happy and contribute to the pleasures of creation. *Travels* is filled with Edenic resonances, but underlying such recoveries is a naturalist who believes that he is communicating with a nature that looks back at him.

Bartram's sense of participating in the nature that he observes is also emphasized at sunrise when the birds, roused by the "reanimating appearance of the rising sun," awaken him and he responds to their "chearful summons" with his own song. "Ye vigilant and faithful servants of the Most High!" he affirms: "ye who worship the Creator, morning, noon and eve, in simplicity of heart; I haste to join the universal anthem. My heart and voice unite with yours, in sincere homage to the great Creator, the universal sovereign" (*Travels* 65). There is no doubting the deeply religious character of this passage, but its real emotional impact comes from Bartram's participation in a "universal anthem" of praise, "the universal shouts of homage, from thy creatures"—"My heart and voice unite with yours." And Bartram's prayer, which the reader is allowed to overhear, asks human beings to adopt an ethics of stewardship toward nature:

> O sovereign Lord! since it has pleased thee to endue man with power, and pre-eminence, here on earth, and establish his dominion over all creatures, may we look up to thee, that our understanding may be so illuminated with wisdom and our heart warmed and animated, with a due sense of charity, that we may be enabled to do thy will, and perform our duty towards those submitted to our service, and protection, and be merciful to them even as we hope for mercy. Thus may we be worthy of the dignity, and superiority of the high, and distinguished station, in which thou hast placed us here on earth.

This may sound as if Bartram is adopting the conventional idea that human beings are better than the animals over whom they have power, but what he is not saying here is that though "Man . . . is the first order of Beings in this World . . . this does not prove because he is the most powerful that he is the most divine" ("[Dignity of Human Nature]" 352–53). Power is not the same as goodness. Human beings must learn from the Creation, and it is their duty to protect the creatures that together contribute to its beauty and complexity.

Here one can see the affinities between Bartram's ecology and the core principle of indigenous knowledge that "all Creation is important; all must be respected" (McGregor, "Coming Full Circle" 388). It also anticipates the ecological ethics of Aldo Leopold, whose "land ethic" affirms the right of soils, waters, plants, and animals "to continued existence, and, at least in spots, their continued existence in a natural state. In short, a land ethic changes the role of Homo sapiens from conqueror of the land-community to plain member and citizen of it. It implies respect for his fellow-members, and also respect for the community as such" (*Sand County Almanac* 204). In a passage in the manuscript of *Travels*, having described the beautiful Vale of Cowe, near the site of what was then the abandoned lower Cherokee town of Keowee, and having been "wellcomed, by communities of the splendid Turkey, the capricious Roe buck, and all the happy tribes which inhabit those prolific fields" to a feast of wild strawberries in its verdant fields, Bartram takes his leave of "my sylvan friends," promising that "I would never disturb them in a hostile way or in any maner dispute with them their equal Right to the enjoyment of those prolific plains and happy peaceable abodes" ("Genetic Text" bk. 3, 4).

Other examples of Bartram's effort to show that nature is a community of intellectually diverse creatures who respond rationally and aesthetically to the order that they help to create can be found throughout the *Travels*. Describing the subterranean journeys taken by fish in Florida's underground artesian springs, Bartram suggests that during these travels they behold "new and unthought of scenes of pleasure and disgust," and that when they return to "the surface of the world, [they] emerge again from the dreary vaults, and appear exulting in gladness, and sporting in the transparent waters of some far distant lake" (*Travels* 111). No wonder that Coleridge, upon reading Bartram, became a convert to the idea of the "One Life," and in "This Lime-Tree Bower My Prison," a poem that refers directly to Bartram, he imagines a similar journey, this time taken by human beings, from a subterranean cavern to the surface of the earth: "Now, my Friends emerge / Beneath the wide wide Heaven—and view again / The many-steepled track magnificent / Of hilly fields and meadows" (*Poetical Works* 1.1:349–54, lines 20–23). Where Coleridge reserves aesthetic experience for his friends, Bartram imagines that it is an everyday event, even for fish. Describing the myriads of mayflies that, after having spent an entire year as nymphs in the cold gravel of a streambed, rise into the air for the short evening that constitutes their adult lives, taking "for a few moments . . . a transient view of the glory of the Creator's works" (*Travels* 53), Bartram imagines their insect joys: "How awful the procession! innumerable millions of winged beings, voluntarily verging on to destruction, to the brink of the grave, where they behold bands of their enemies with wide open jaws, ready to receive them. But as if in-sensible of their danger, gay and tranquil each meets his beloved mate in the still air, inimitably bedecked in their nuptial robes. . . . With what peace, love, and joy do they end the last moments of their existence." There are many passages in *Travels* where the natural world appears to be a peaceful community, so it is easy to forget—though Bartram as a meticulous observer of the natural world could not—that these usually appear as moments of respite in what otherwise seems to be an ongoing state of war. The ephemera may embody "peace, love, and joy," but they do so "behold[ing] bands of their enemies with wide open jaws, ready to receive them" (53).

Although Bartram saw himself as "an advocate or vindicator of the benevolent and peaceable disposition of animal creation in general, not only towards mankind, whom they seem to venerate, but also towards one another, except where hunger or the rational and necessary provocations of the sensual appetites interfere" (*Travels* 168), it is clear that violence and destruction are the norm in the *Travels*. Bartram's idea of the inherent peace and reason of nature was thus at odds with the reality of nature on the frontier. Much of the violence he encountered was caused by human beings. In the introduction, Bartram provides a graphic account of the needless destruction of a mother bear (ostensibly for her skin and for oil) followed by the "cruel murder" (lvii) of her cub, which stood beside its dead mother and "appearing in agony, fell to weeping and looking upwards, then towards us, and cried out like a child" (lvii). Hunger and sexual desire also lead animals to destroy one another. Immediately following this description, Bartram recounts in gory detail a spider's predatory attack on a bumblebee, "with the circumspection and perseverance of a Siminole, when hunting a deer." Having described how the bee eventually succumbs to "the repeated wounds of the butcher" who all the while kept "a sharp eye upon me," Bartram suggests that the killing will not stop here, for soon enough the predator will become the prey of another, "the delicious evening repast of a bird or lizard" (lix). At the edge of a stream, he witnesses a "war [that] seemed to be continual" (28), taking place between crayfish and goldfish. At Battle Lagoon, the naturalist described the thunderous roar of the alligators and their violent competition for mates. Having barely escaped from their attacks, he watches in horror as "hundreds of thousands" of fish journeying upstream through the river narrows are devoured by these reptiles: "the horrid noise of their closing jaws, their plunging amidst the broken banks of fish, and rising with their prey some feet upright above the water, the floods of water and blood rushing out of their mouths, and the clouds of vapour issuing from their wide nostrils, were truly frightful" (79). In Bartram's world, even plants display avid carnivorous impulses, most notably the Venus flytrap, which Bartram was the first person to illustrate. In one of his drawings of the insectivorous pitcher plant (*Sarracenia*), Bartram provides an emblem of a world in which all creatures eat and are eaten.

Insects crawl across the foreground, little knowing that they are about to become the prey either of the surrounding pitcher plants, of the lizard stalking them in the left-hand corner, or of the horned toad concealed in the corner on the right. Fortunately—at least for the would-be prey—the frog, which is being eaten by a northern scarlet snake, has been removed from the equation. Despite their being part of this nightmare of violence and predation, Bartram recognized that it was required of life. Even "the Ravens & Vultures must be fed," he commented ("Genetic Text" bk. 2, 277). Violence and destruction are so prevalent in nature that the moments of peace and respite seem all the more impressive. At Lake George, while Bartram lay sleeping, a wolf stole the fish that he had strung above his head, and the naturalist was astonished that the animal did not choose the easier and safer option of killing him in his sleep: "I say would not this have been a wiser step, than to have made protracted and circular approaches, and then after, by chance, espying the fish over my head, with the greatest caution and silence rear up, and take them off the snags one by one" (101). Similarly, after almost stepping on a rattlesnake, he wondered why "the generous, I may say magnanimous creature lay as still and motionless as if inanimate" (167), when it could just as easily have bit him.

Bartram's *Travels* marks the transition from an Enlightenment conception of a stable nature, expressing the design of a Creator, to one in which struggle, violence, and change seem to lie at the very heart of things. I would

William Bartram, *Sarracenia Flava or Pitcher, insectivorous plant with snail, snake eating a frog,* sepia drawing, 1754, British Museum, London, Great Britain / The Art Archive at Art Resource, NY, image reference AA327279

argue that his experience of the colonial frontier fostered this shift. What drives his natural history forward is his attempt to understand the conditions under which human beings and animals might coexist in spite of the continuing presence of violence and predation. This preoccupation is brought into focus in a passage that had an enormous impact on Coleridge, particularly in his composition of "Kubla Khan."[9] In the transparent waters of "an inchanting and amazing crystal fountain" that Bartram discovered near Lake George, he saw "innumerable bands of fish in the most brilliant colours," swimming near "the voracious crocodile," "the devouring garfish," and "inimical trout," as if they belonged to one society: "all in their separate bands and communities, with free and unsuspicious intercourse performing their evolutions: there are no signs of enmity, no attempt to devour each other; the different bands seem peaceably and complaisantly to move a little aside, as it were to make room for others to pass by" (*Travels*

105). "Though real," he informs us, the scene had all the appearance of "a piece of excellent painting." This passage provides an allegory for Bartram's art, for although the waters seem to be an idyllic embodiment of a prelapsarian nature, a "paradise of fish" exhibiting "a just representation of the peaceable and happy state of nature which existed before the fall," he nevertheless was fully aware that "in reality it... [was] a mere representation; for the nature of the fish is the same as if they were in Lake George or the river." The fish had not changed their natures, but only the conditions under which they interacted with each other. "The water or element in which they live and move, is so perfectly clear and transparent," Bartram declares, that "it places them all on an equality with regard to their ability to injure or escape from one another; (as all river fish of prey, or such as feed upon each other, as well as the unwieldy crocodile, take their prey by surprise; secreting themselves under covert or in ambush, until an opportunity offers, when they rush suddenly upon them:) but here is no covert, no ambush, here the trout freely passes by the very nose of the alligator and laughs in his face, and the bream by the trout." Douglas Anderson has insightfully suggested that this passage be read as "a kind of dark parody of Madison's argument in Federalist 10 that republican 'representation' functions as a restraint upon the bitterness of faction" ("Bartram's *Travels* and the Politics of Nature" 9).

In more general terms, Bartram is reflecting on the need for transparency as a condition of peace within diverse political and social communities. The establishment of open relationships with others is a necessary condition for living together, especially in societies where every animal tribe sees things differently. Countering the tendency in many studies to see natural history as a straightforward imposition of European ways of seeing on the world, Ian Marshall has suggested that Bartram extended his ecological commitment to the richness and diversity of the natural world to his understanding of intercultural relationships in the emerging republic (*Story Line* 35–50). He hoped that the representation of nature provided in the *Travels* might provide a pattern for the new republic. Indeed, Bartram's *Travels* powerfully captures the richness and vitality of a frontier nature where one could still imagine the continuance and peaceful coexistence of the diversity of animal and human tribes in one polity. The book is hardly an expression of Bartram's retirement from politics, therefore, but instead an attempt to use natural history to conceive a "natural republic," a society that would recognize different subjectivities and where the violence of faction would be restrained by sharing space and mutual transparency.

Drifting on the waters of the Suwannee River near Manatee Springs in Florida, Bartram saw his ideas about the relationship between space, transparency, and peace confirmed. Looking into the waters of the main channel, which seemed to him "as transparent as the air we breathe," he beheld "the watery nations, in numerous bands roving to and fro amidst each other, here they seem all at peace; though incredible to relate, but a few yards off, near the verge of the green mantled shore there is eternal war, or rather slaughter." As an image of a peace constituted in the midst of a world of ongoing violence, the river is an emblem of Bartram's art and its politics. In the main channel, Bartram writes, "there is nothing done in secret," even though, "incredible to relate," on the flowery green verges of the stream, there is constant destruction (*Travels* 145). Bartram's natural history does not imagine a social polity in which human beings and animals have changed their natures, but one in which hostilities have been suspended by the maintenance of relationships of openness and transparency and, crucially, the provision of adequate space for everyone. "Give place to one another," he wrote in the manuscript of the *Travels*; the "Gifts of Providence . . . are abundantly sufficient to satisfy the necessities and even the conveniencies of every creature, without a necessity of contention; & these are equally attainable, as well as the right of every one" ("Genetic Text" bk. 1, 280).

Bartram admired the southeastern American Indians, and one of the primary goals of his narrative is to defend them "as moral equals" from the negative stereotypes promoted by US expansionists (Waselkov and Braund, *William Bartram on the Southeastern Indians* 204). Rather than blaming indigenous peoples, therefore, for the violence that was an ongoing aspect of life on the frontier, he attributed it to the continual encroachment on their lands by European settlers. He writes, "Our injustice & avarice, in pressing upon their Borders & dispossessing them of their Lands, together with the outrage committed against their Persons & encroachments made on their hunting Grounds by the Frontiers, provoke them to retaliation" ("Some Hints & Observations" 194).

The young naturalist's meeting in the forest with a Seminole tribe member, who had just recently suffered mistreatment at a nearby trading post and had left there vowing to murder the first white person that he met, provides an allegory of Bartram's commitment to a politics grounded on openness in our dealings with others. Alone and unarmed in the wilderness, Bartram recounts that "on a sudden, an Indian appeared crossing the path, at a considerable distance before me. On perceiving that he was armed with a rifle, the first sight of him startled me, and I endeavoured to elude his sight . . . but he espied me, and turning short about, set spurs to his horse, and came up on full gallop." Discovered, Bartram feels certain that he is about to be killed: "I was in his power. . . . The intrepid Siminole stopped suddenly, three or four yards before me, and silently viewed me, his countenance angry and fierce, shifting his rifle from shoulder to shoulder, and looking about instantly on all sides." Rather than showing fear, Bartram "advanced towards him, and with an air of confidence offered him my hand, hailing him, brother." Instead of conflict, the encounter produces mutual recognition and respect: "In fine, we shook hands, and parted in a friendly manner, in the midst of a dreary wilderness; and he informed me of the course and distance to the trading-house, where I found he had been extremely ill-treated the day before" (*Travels* 15). This encounter provides a striking alternative to the cycle of violence that was shaping the southeastern frontier, and perhaps it is for this reason that Wordsworth, also thinking about the ways in which war was devastating British society, probably drew upon this episode in composing the "Discharged Soldier" episode of *The Prelude*.

From the start, Bartram was aware that he was exploring contested space, for it had been necessary, for his own safety, to delay his excursion into the American Indian countries until the conclusion of the Second Treaty of Augusta (1773), which ceded more than two million acres of Native American territory to the Georgia colony (Bellin, *Demon of the Continent* 123–30). When he finally explored the territory, it was as part of a company of surveyors delegated to establish the new boundaries: "we joined the caravan, consisting of surveyors, astronomers, artisans, chain-carriers, markers, guides and hunters, besides a very respectable number of gentlemen, who joined us, in order to speculate in the lands" (*Travels* 23). Bartram may have recognized that the historical and political events that had made this region available to him as a botanist were also those that were opening it up to settlement. Nevertheless, believing that "every nation is equally entitled to those bountiful blessings of Providence, the furniture and produce of the Earth, & the benign influence of the Planets" ("Some Hints & Observations" 195–96), he hoped that "a judicious plan" might be developed that would make it possible for them to have a "union with us" (*Travels* lxi). The *Travels* remain committed to the idea of a republic in which the Native American lands are preserved, and the narrative provides a powerful testimony to the natural world produced and sustained by the southeastern American Indians.

"Traces of a settlement which yet remain": Bartram and the Historical Ecology of the Frontier

Bartram did not see the natures of the Southeast as being stable, homogeneous, or organic wholes. Instead, he understood them as reflecting the territorial struggles that had been taking place there for centuries, first, among a succession of different tribes that had migrated to the region from the West, displacing the original First Nations inhabitants, and then, more recently, between Native Americans and European settlers. Bartram recognized that the nature that he was encountering was not a pristine wilderness, but instead one that was infused with history, where the traces of indigenous culture were powerfully registered in the landscape, in ancient fields, in huge monumental earth mounds, and in the shell middens that were a distinctive aspect of southeastern American Indian agriculture. Importantly, he recognized that the history of indigenous peoples and of European settlement could be seen in the landscapes that each of these peoples had produced, so one of the distinctive concerns of the *Travels* was to use natural history, combined with anthropology and archaeology, to discover the historical evidence of the places where the southeastern American Indians had settled and the impact that they had had on the landscape. Moving through landscapes shaped by mobility, conflict, and displacement, Bartram sought to write a history of the land and the changes that it had undergone, seeing nature less as a symbol than as a lens for reading human history. Bartram saw the frontier as a rich historical tapestry documenting the impact of human beings on nature, and particularly of the role that the southeastern American Indians had played in creating an extraordinary ecology, which embodied their respect and love for the natural world.

A characteristic example of the manner in which Bartram used natural history to recover the ecological history of the regions he visited can be seen in his description of the failed settlement of Charlotia, in Florida. When he first visited the village, with his father in December 1765 and January 1766, it had just been founded by Denys Rolle, Member of Parliament for Barnstaple, who had brought approximately fifty indentured laborers from England to work his eighty-thousand-acre plantation, stretching for twenty-three miles along the east side of the St. Johns River.[10] Soon after, these laborers would be supplemented by black slaves, so by 1770, despite steady losses as laborers disappeared or died by disease, the village reached a population of two hundred people. When Bartram returned in 1774, it had become a ghost town. With the exception of the mansion house and the blacksmith's shop, "the remaining old habitations" were "mouldering to earth" (*Travels* 61). We do not often associate Bartram with the poetics of ruins, but the abandoned settlements of both Europeans and Native Americans appear regularly in the *Travels*. Significantly, Charlotia was not the first village built on this spot, for Bartram recognized that the high bluff on which it was situated, "fifteen or twenty feet perpendicular to the river," was on ancient American Indian shell midden:

> The upper stratum of the earth consists entirely of several species of fresh water Cochlae, as Cochelix, Coch. Labyrinthus, and Coch. Voluta; the second, of marine shells, as Concha mytulus, Concostrea, Conc. peeton, Haliotis auris marina, Hal. patella, &c. mixed with sea sand; and the third, or lower stratum, which was a little above the common level of the river, was horizontal masses of a pretty hard rock, composed almost entirely of the above shell, generally whole, and lying in every direction, petrified or cemented together, with fine white sand; and these rocks were bedded in a stratum of clay. I saw many fragments of the earthen ware of the ancient inhabitants, and bones of animals, amongst the shells, and mixed with the earth, to a great depth. (60)

Although Bartram's careful listing of the different kinds of freshwater and marine shells that composed the shell midden, like the lengthy lists of plants elsewhere in *Travels*, might seem excessive, it should not prevent one from recognizing the depth of historical vision that it embodies, combining geology and archaeology, as he demonstrates that Rolle's plantation town was not built in a pristine wilderness, but instead on the remains of an ancient Native American town. His attention to the shells indicates that he realized that the very ground on which the village stood had actually been built up by the southeastern American Indians. The sheer depth of the bluff, which was fifteen to twenty feet above the river, combined with the recognition that at some point their immediate food supply changed from saltwater to freshwater shellfish, suggests the enormous length of time that the village had occupied this site.

Extending his gaze beyond the shell midden bluff, Bartram also discerned the remains of an "Indian old field" and the ridged terraces, also composed of earth, sand, and shells, where plantings of corn, beans, squash, and tobacco were intermixed with fruit and nut trees and other useful plants:

This high shelly bank continues, by gentle parallel ridges, near a quarter of a mile back from the river, gradually diminishing to the level of the sandy plains, which widen before and on each side eastward, to a seemingly unlimited distance, and appear green and delightful, being covered with grass and the Corypha repens, and thinly planted with trees of the long leaved, or Broom Pine, and decorated with clumps, or coppices of floriferous, evergreen, and aromatic shrubs, and enameled with patches of the beautiful little Kalmea ciliata. These shelly ridges have a vegetable surface of loose black mould, very fertile, and naturally produce Orange groves, Live Oak, Laurus Borbonia, Palma elata, Carica papaya, Sapindus, Liquid-amber, Fraxinus exelsior, Morus rubra, Ulmns, Tilia, Sambucus, Ptelea, Tallow-nut, or Wild Lime, and many others. (*Travels* 60)

Bartram was not seeing a wilderness, but the remains of a Native American mixed crop field. His description emphasizes the aesthetic dimensions of the field, noting how the grassland is set off by "clumps, or coppices of floriferous, evergreen, and aromatic shrubs," culminating in the "enamel[ing]" produced by *Kalmia ciliate* (calico bush or spoonwood), a plant with a beautiful white flower which Native Americans also used to make spoons. The list of the plants that Bartram noted growing in the field is thus an important example of early field archaeology and ethnobotany, as he recognizes and appreciates the fact that the plants that he was seeing did not just happen to be there, but had been chosen by the southeastern American Indians because they found them useful, nutritive, aromatic, or pleasing to the eye: live oak, royal palm, oranges, papaya, elderberry, soapberry (for soap), sweet gum (for chewing gum, tea, and salve), white ash (for tools, furniture, and medicine), red mulberry (a popular medicinal), and basswood (for rope, mats, fish nets, and baskets).

Bartram was not just seeing a present-day landscape; he was also using his knowledge to recover the traces of another ghostly landscape that for centuries had occupied this site. The passage is thus essentially an ecological history of a place, read as much through *the nature left behind* as through its man-made artifacts. In what is an extraordinary example of understatement, Bartram displays his deep appreciation of the natural legacy provided by the southeastern American Indians: "The aborigines of America, had a very great town in this place, as appears from the great tumuli, and conical mounts of earth and shells, and other traces of a settlement which yet remain. There grew in the old fields on these heights great quantities of

Callicarpa [beautyberry] and of the beautiful shrub Annona [pawpaw]: the flowers of the latter are large, white and sweet scented" (61). Much of Bartram's appreciation of the southeastern indigenous culture comes from his recognition that the beauty of the natural world that he was experiencing, down to the very soil in which their plants grew, was the result of their long-term interaction with the environment around them, as they selected and cultivated the plants that they most valued and provided a rich soil in which they could thrive. In the shelly old fields on Drayton's Island, Florida, Bartram came upon three spectacular flowers: "a beautiful species of Lantana" (*Lantana camara*), which grew "in coppices in old fields" (67); the gingerbush (*Pavonia spinifex*), which he identified as a yellow hibiscus; and the morning glory (*Operculina dissecta*). In the *Travels*, Bartram repeatedly affirms that he and the southeastern American Indians shared a deep intellectual bond, centered on their mutual appreciation of natural beauty.

It can hardly be an accident that Bartram's description of the failed settlement of Charlotia is immediately preceded by a description of a contemporary Native American village that he passed on the river. There he saw children fishing and playing in the water, women "hoeing corn," and elderly men and women relaxing under "the cool shade of spreading Oaks and Palms, that were ranged in front of their houses" (*Travels* 59). In addition to a well-maintained orange grove, he noted that there were several hundred acres of cleared land, a considerable portion of which was planted with corn, sweet potatoes, beans, pumpkins, squashes, melons, and tobacco, "abundantly sufficient for the inhabitants of the village." The people, he observed, "were civil, and appeared happy in their situation." Not only does this account provide a radical contrast to what had been the conditions suffered by the indentured laborers on Rolle's plantation, but it also indicates the manner in which he attempted to connect contemporary Native American culture with an important part of its past.

For Bartram, botany was a crucial aspect of historical understanding, for it allowed him not only to discern the traces of previous settlements in the plants that had been left behind but also to assess the relationship between past cultures and the natural world. When he visited Fort Frederica, the first English settlement in Georgia, on St. Simons Island, the fortress, plantations, and town had long since been abandoned. Bartram noted that "a very large part of this island had formerly been cleared and planted by the English, as appeared evidently to me, by vestiges of plantations, ruins of costly buildings, highways, &c. but it is now overgrown with forests"

(*Travels* 40). Bartram saw plenty of evidence of the previous presence of the English on the island in the moldering architectural ruins that were now buried by forests, but what is distinctive about his approach to this landscape is his interest in the nature that the British left behind: "The ruins of a town only remain; peach trees, figs, pomegranates, and other shrubs, grow out of the ruinous walls of former spacious expensive buildings" (40). Pomegranates growing out of ruined walls: nothing speaks more powerfully of Bartram's ability to use nature to see into the past. Natures come and go, just like people, and what grows in a place often grows there because someone planted it sometime in the past.

Bartram discovered abandoned southeastern American Indian fields in the same way. Although most of the "old Indian settlements, [were] now deserted and overgrown with forests," the "antient cultivated fields" were marked by the presence of those trees and shrubs that they liked to cultivate: "1. Diospyros [persimmon], 2 Gleditsia triacanthos [honey locust], 3 Prunus Chicasaw [Chickasaw plum], 4. Callicarpa [French mulberry], 5. Morus rubra [red mulberry], 6. Juglans exaltata [shell-barked hickory], 7. Juglans nigra [walnut], which inform us, that these trees were cultivated by the ancients, on account of their fruit, as being wholesome and nourishing food" (*Travels* 25). Bartram had never seen a Chickasaw plum growing anywhere else than "in old deserted Indian plantations," so he even speculated that the plant was not native to the Southeast, but had been "brought from the S.W. beyond the Mississippi, by the Chicasaws" (25).[11] In an old meadow beside the Altamaha River, in Georgia, he noted that the plants that had been cultivated were not just food plants but also aromatic shrubs: "Myrica cerifera [wax myrtle], Magnolia glauca [white bay], Laurus benzoin [spicebush], Laur. Borbonia [sweet bay], Rhamnus frangula [buckthorn], Prunus Chicasaw, Prun. Lauro cerasa [laurel cherry], and others" (32). Once again, what a people valued could be seen in the plants that they cultivated.

Among the most impressive and enigmatic features of the Native American landscapes that Bartram encountered in his travels were the huge pre-Columbian earthen mounds. Though no one at the time knew who had built them or what purpose they had served, Bartram was among the first to study them closely. On the east bank of the Altamaha River, he writes, "lie the famous Oakmulge fields, where are yet conspicuous very wonderful remains of the power and grandeur of the ancients of this part of America, in the ruins of a capital town and settlement, as vast artificial hills, terraces, &c. already particularly mentioned in my tour through the lower districts of

Georgia" (*Travels* 241). Bartram correctly recognized the antiquity of these earthworks: "Many very magnificent monuments of the power and industry of the ancient inhabitants of these lands are visible . . . the work of a powerful nation, whose period of grandeur perhaps long preceded the discovery of this continent" (25). He also recognized that these were "public works" requiring "united labour and attention" (206), so they were expressions of a highly developed society. As Waselkov and Braund comment, "Bartram's litany of massive mounds, avenues, and artificial lakes, numerous abandoned fields and townsites, and Indian myths of lost tribes and ancient battles endowed the American landscape with an unsuspected antiquity" (*William Bartram on the Southeastern Indians* 209–10). When he questioned Native Americans about these mounds, however, no one appeared to have any knowledge of their origin or purpose, so Bartram quite reasonably concluded that they had been constructed by an ancient tribe that had long since disappeared from the region. Thus, in a manner similar to William Stukeley or Sir William Jones, Bartram postulated a sophisticated ancient indigenous culture that he simultaneously isolated from their contemporary descendants.[12] When he imagined the Southeast's pre-Columbian past, he did not see a wilderness inhabited by savages at war with one another, but a peaceful, sophisticated agricultural society that integrated itself with its surroundings and had learned to balance its needs with a deep respect for the natural world.

One consequence of this line of thought was that it led Bartram to the somewhat extraordinary conclusion that none of the tribes that he encountered on his travels were "strictly consider'd, Aborigines of these Countries" ("Some Hints & Observations" 193). All the people who lived in the Southeast, he believed, Native Americans and Europeans alike, had come from somewhere else; all were colonizers who had settled there by forcefully displacing those who had arrived there before them. Most of the tribes, he concluded, had come from New Mexico, fleeing the Spanish invaders and arriving in successive waves at about the same time as the Europeans were establishing colonies in New England, Virginia, Carolina, and Florida.[13] Although American expansionists would later develop a mythology that claimed that the "mound builders" were not actually Native Americans and by so doing would justify the expulsion of the southeastern American Indians from their traditional lands, Bartram never questioned that Native Americans built these monuments, and he insisted on the contemporary tribes' right to their land: "they have possess'd the Soil, for some Ages past, and the Bones, and other sacred reliques of their Ancestors lie buried in the dust about their Towns" (193). Instead of adopting the interpretive models of American imperial historiography, Bartram saw in the disappearance of these people a warning about the forgetting that promotes empire. At the core of the *Travels* is a reflection on the violence and destructiveness that shape empire building, whether by Europeans or their native counterparts. In support of the view that the colonial Southeast had been the site of constant resettlement, Bartram referred to a tradition that the Cherokee held in "common with the other nations of Indians, that they found them [the mounds] in much the same condition as they now appear, when their forefathers arrived from the West and possessed themselves of the country, after vanquishing the nations of red men who then inhabited it, who themselves found these mounts when they took possession of the country, the former possessors delivering the same story concerning them" (*Travels* 232). It was in the ceremonial and burial mounds of the Ocmulgee fields that the Creeks "sat down (as they term it) or established themselves, after their immigration from the west, beyond the Mississippi, their original country" (35). This was the place where they made their last stand, after suffering tremendous losses in battles with other tribes, "being to the last degree persecuted and weakened by their surrounding foes." It was here also that they saw their fortune turn as they entered into an alliance with the English colonists of Carolina. From then on, "they never ceased war against the numerous and potent bands of Indians, who then surrounded and cramped the English plantations, as the Savannas, Ogeeches, Wapoos, Santees, Yamasees, Utinas, Icosans, Paticas, and others, until they had extirpated them" (35).

The parallel between the imperial dreams of the Creek Confederacy and those of European colonialists clearly shapes Bartram's understanding of the history of the Southeast. All these nations were so intent on "contending for empire & the honor & glory of their tribes, they in part forgot or disregarded their ancient lineage & affinity" ("Observations on the Creek and Cherokee Indians" 143). Empire produces violence and forgetting, as the recognition of their ancient brotherhood was lost in the rivalry for empire. Bartram's history of the Southeast is thus not the story of the conquest of an indigenous people, rooted in place, by invading newcomers, but instead one of competing imperial visions. He believed that its landscapes preserved the traces of these ongoing conflicts as a succession of migrating peoples struggled to gain land already occupied by other immigrants. The effects of European colonial expansion, in forced migrations and the consequent struggle for territory, were thus also being played out among the southeastern tribes themselves.

Bartram's recognition of the lengthy history of tribal conflict that had also been taking place in the Southeast makes the *Travels* distinctive, for he not only was interested in documenting the changes introduced by Europeans but also sought to write an ecological history of the settlement of the Southeast by Native Americans. This history had not been written down, only existing in oral traditions and in the changes that were registered in the land. Although Bartram was traveling through territory now possessed by Europeans and contemporary Native American tribes, he knew that these lands had once been occupied by tribes that had long since vanished from the region, "the Savannas, Ogeeches, Wapoos, Santees, Yamasees, Utinas, Icosans, Paticas, and others" who left their abandoned towns, overgrown fields, and imposing earthwork temples, tumuli, and fortresses scattered across the landscape as testimony to a sophisticated civilization that had been destroyed by waves of colonial migration. With colonialism came the wilderness, which now stood as a living screen covering the remnants of these sophisticated agricultural communities. When Bartram visited Alachua, the ancient city of the Potanos, he passed through "a great extent of ancient Indian fields" where thousands of people had once assembled for "ball play and other juvenile diversions and exercises" (*Travels* 175). Now these fields were "grown over with forests of stately trees, Orange groves and luxuriant herbage," but there was no question in Bartram's mind that he was walking "over those, then, happy fields and green plains . . . as almost every step we take over those fertile heights, discovers remains and traces of ancient human habitations and cultivation" (175). Traveling through the Southeast, Bartram was traveling through the past. On Amelia Island, Georgia, he visited some large mounds, the "Ogeeche mounts," as they were called, which contained the last remains of the Ogeeche nation, which had taken "shelter here, after being driven from their native settlements on the main near Ogeeche river. Here they were constantly harrassed [sic] by the Carolinians and Creeks, and at length slain by their conquerors, and their bones intombed in these heaps of earth and shells" (43). On the St. Johns River, near St. Francis, while in search of firewood for the night, Bartram discovered another "ancient burying ground," covering two or three acres, the "tumuli of the Yamasees, who were here slain by the Creeks in the last decisive battle, the Creeks having driven them into this point, between the doubling of the river, where few of them escaped the fury of the conquerors"

(88). Establishing a precedent that Wordsworth would follow in "The Thorn," Bartram measured the mass graves—"they were oblong, twenty feet in length, ten or twelve feet in width and three or four feet high"—and, in typical fashion, he notes that the place was "now overgrown with Orange trees, Live Oaks, Laurel Magnolias, Red bays and other trees and shrubs, composing dark and solemn shades" (88). No wonder that even as nature often produces moments of rhapsody in the *Travels*, it also evokes more somber feelings as Bartram reflects on the long and painful colonial history that lay concealed in the wilderness, a history that continued to cast a dark shadow on the present-day Native Americans whom he met on his journeys.

Bartram probably would have agreed with the idea that the quality of a civilization is expressed in the quality of the nature that it leaves behind. Certainly, there is much to suggest that he saw English settlement in Florida as a tragedy. Along with Charlotia, Bartram visited three other deserted plantations on the St. Johns River, all having lasted for little more than two years. Typical of most of the Florida plantations, they were built on land that had previously been cleared by American Indians. About the abandoned plantation of William Stork, who had published John Bartram's *Journal* as part of his *Description of East-Florida* (1769), Bartram had little to say. However, in the old fields and orange groves that still remained, he observed "many lovely shrubs and plants," most notably "several species of Convolvulus [moon flower] and Ipomea [morning glory], the former having very large, white, sweet scented flowers" (*Travels* 159). Implicit in this botanist's recognition of the beauty of these plants is appreciation for the people who had brought them to this place and cultivated them there. At Rawdon Hall, owned by the absentee Earl of Moira, Bartram found "a very spacious frame building . . . settling to the ground and mouldering to earth." The nature of the place, however, had been devastated. Except for "some scattered remains of the ancient Orange groves," the "extensive old fields" had been put under the plow to make way for cotton and indigo (160). Given the ecological destruction and wastefulness introduced by English planters, it is no wonder that Bartram could hardly contain himself: "I have often been affected with extreme regret, at beholding the destruction and devastation which has been committed, or indiscreetly exercised on those extensive, fruitful Orange groves, on the banks of the St Juan, by the new planters under the British government, some hundred acres of which, at a single plantation, has been entirely destroyed to make room for Indigo, Cotton, Corn, Batatas, &c. or as they say, to extirpate the musquitoes, alledging that groves near their dwellings are haunts and shelters for those persecuting insects" (160). In Bartram's criticism of the southern plantation system and his defense of southeastern American Indian mixed crop plantings, one can recognize a thoroughly modern concern about the destructive aspects of monocultural farming practices. As early as the 1760s, John Bartram was already communicating anxiety about the loss of species through forest clearances and the introduction of European livestock. In a letter to Collinson, he reflected on the extraordinary fact that in thirty years of collecting plants, he had never found "one single species in all ye after times that I did not observe in my first Journey through ye same province but many times I found that plant ye first that neither I nor any person could find after which plants I suppose was distroyed by ye cattle" (John Bartram to Peter Collinson, 30 Sept. 1763, *Correspondence of John Bartram* 608).

The last abandoned plantation Bartram visited was that of John James Perceval, the second Earl of Egmont, at Mount Royal, where he came upon "a magnificent Indian mount" that provided "a most enchanting prospect of the great Lake George, through a grand avenue" (*Travels* 64). Built by the Timucua people, who lived there from about AD 1200 to 1600, it was composed of a large sand mound that stood at the head of a long causeway, fifty yards wide and bordered by banks, which ran in a straight line for three-quarters of a mile through an orange grove and then "an awful forest, of Live Oaks ... terminated by Palms and Laurel Magnolias," which stood on the verge of an artificial pond that lay like a "brilliant diamond, on the bosom of the illumined savanna, bordered with various flowery shrubs and plants." Around the mound, which was twenty feet in height, was an "Orange grove, together with Palms and Live Oaks," and from whence there spread out "a considerable extent of old fields" (64). Bartram first visited the place on 25 January 1766, with his father, who was also similarly impressed by the order and grandeur of its design:

> It must be very ancient, as live-oaks are growing upon it three foot in diameter; what a prodigious multitude of Indians must have laboured to raise it? ... [D]irectly north from the tumulus is a fine straight avenue about 60 yards broad, all the surface of which has been taken off, and thrown on each side, which makes a bank of about a rood wide and a foot high more or less, as the unevenness of the ground required, for the avenue is as level as a floor from bank to bank, and continues so for about three quarters of a mile to a pond of about 100 yards broad and 150 long N. And S. (*Diary of a Journey* 45)

For William, Mount Royal was a testimony to a culture that built environments integrating the natural world with human life: "Neither nature nor art, could any where present a more striking contrast, as you approach this savanna . . . as we advance into the plain, the sight is agreeably relieved by a distant view of the forests, which partly environ the green expanse, on the left hand, whilst the imagination is still flattered and entertained by the fair distant misty points of the surrounding forests, which project into the plain, alternately appearing and disappearing, making a grand sweep round on the right, to the distant banks of the great lake" (*Travels* 64–65). Displaying a knowledge of plants which only a botanist could appreciate, the Timucua knew how to build a natural landscape in the flatlands of Florida which was aesthetically dynamic and harmonious, linking the tumulus prospect with the orchards and fields, stretching outward to encompass the oak groves, the pond, and, in the distance, Lake George.

In describing Mount Royal in the *Travels*, Bartram was drawing upon his memory, because when he visited Mount Royal again, in 1774, the place, which had "possessed an almost inexpressible air of grandeur . . . was now entirely changed" (*Travels* 65). In 1867, shortly after Bartram's first visit, the area was granted to Perceval for development. Over the next three years, the orange grove, palms, and oaks were cleared and the "grand highway" plowed under in order to plant indigo, corn, and cotton. The only reason why the mount was not also destroyed was that Governor James made it a condition of the grant that the mount not be disturbed (Schafer, *William Bartram* 51). By 1770, after all this destruction, with costs mounting, and with the death of Perceval, the land was abandoned. "That venerable grove is now no more. All has been cleared away and planted with Indigo, Corn, and Cotton, but since deserted: there was now scarcely five acres of ground under fence. It appeared like a desart, to a great extent, and terminated, on the land side, by frightful thickets, and open Pine forests" (*Travels* 65). A place that had spoken powerfully of human beings living in harmony with the natural world was now a ghastly ruin. All that remained in this empty solitude was a huge mound of earth.

CHAPTER SIX

"I see around me things which you cannot see"
William Wordsworth and the Historical Ecology of Human Passion

∽

> [I] stood alone
> Beneath the sky, as if I had been born
> On Indian Plains...
> (W. Wordsworth, 1805 *Prelude* 1:300–302)

In the late spring of 1832, after visiting the Great Plains of Illinois, William Cullen Bryant wrote the poem "The Prairies," which expresses the powerful impact that this landscape had on him. Bryant had never seen anything like it. The opening lines describe an encounter with a radically different nature, so unfamiliar, Bryant observes, that in "the speech of England" there was "no name" for it, so the Old French word "prairie" was adopted (*Poetical Works* 1:228). In its representation of a solitary individual's encounter with a vast nature, "The Prairies" has all the markings of a conventional Romantic nature poem. Alone in this expansive unpeopled space, this "encircling vastness" of "unshorn fields, boundless and beautiful" (228), it is easy for Bryant to conclude that he is seeing a pure and original nature, untouched by human hands since the beginning of time. "Man hath no power in all this glorious work" (229), he declares. No sooner does Bryant postulate that this nature is the direct expression of "the hand that built the firmament" (229), however, than he retracts this claim, asking himself whether, in riding across these empty plains, he is not being "sacrilegious," treading on the graves of the dead:

> I think of those
> Upon whose rest he [the horse] tramples. Are they here—
> The dead of other days?—and did the dust
> Of these fair solitudes once stir with life
> And burn with passion? (229)

With these questions a previously unacknowledged history of a people and a place floods into the poem, and what had seemed an originary landscape existing in an eternal present, separate from human passions, agency, and interests, is suddenly pervaded with a past. The absence of people in the landscape, in other words, does not mean an absence of ghosts. Bryant wonders whether these "fair solitudes" (229) are not, in fact, a vast unmarked graveyard. This contradiction between present and past natures constitutes the interpretive hinge of the poem. How does one recover the history of a place when the only document of that history is the land itself? What can one know of a people who produced no written records and whose primary historical documents were the landscapes that they have left behind? How important can a history be if nature itself seems intent on erasing it? And how is one to recover the human passions that once brought this dust to life?

For Kevis Goodman, in *Georgic Modernity and British Romanticism*, these are questions that prompt georgic poetry, which uses the furrows of verse not to hide the past, but to interrogate it. Bryant's prairies speak of violence and forgetting. Noting the mysterious large tumuli that everywhere lay scattered across the plains—"mighty mounds / That overlook the rivers, or that rise / In the dim forest crowded with old oaks"—Bryant constructs an imperial myth, imagining that the Great Plains were once inhabited by a "disciplined and populous race . . . that long has passed away" (*Poetical Works* 1:229). As I indicated in the previous chapter, the idea that America had once been inhabited by an ancient race of mound builders was a popular myth that emerged in the wake of Bartram's *Travels*, notably in the writings of Benjamin Smith Barton, Thomas Jefferson, and Caleb Atwater. Bryant imagines that at the same time as the Greeks were building the Parthenon, the prairies were being settled by a sophisticated people who cleared and cultivated the land, even going so far as to domesticate the bison: "here their herds were fed / When haply by their stalls the bison lowed, / And bowed his maned shoulder to the yoke." Bryant speculates that where he now stands, "lovers" once

> walked, and wooed
> In a forgotten language, and old tunes,
> From instruments of unremembered form,
> Gave the soft winds a voice. (230)

"The Prairies" can be said to be Bryant's "Grecian Urn" on a vast scale, but where Keats's urn stands separate from time, Bryant's landscape artifact has suffered change. Now "the gopher mines the ground, / Where stood their swarming cities" (230).

In a letter to his friend, the poet and critic Richard Dana, Bryant indicated that the poem had its origin in the moment when he realized (mistakenly) that the prairies looked like a vast abandoned field because that was, in fact, what they were: "These prairies, of a soft fertile garden soil, and a smooth undulating surface, on which you may put a horse to full speed, covered with high thinly growing grass, full of weeds and gaudy flowers, and destitute of bushes or trees, perpetually brought to my mind the idea of their having been once cultivated. They looked to me like the fields of a race which had passed away, whose enclosures and habitations had decayed, but on whose vast and rich plains smoothed and levelled by tillage the forest had not yet encroached" (*Letters* 1:360). It is hard to imagine a grander geographical idea than that of the Great Plains, stretching from Illinois to the Rockies and from Texas to Canada, being the ruined fields of a highly sophisticated race of agriculturalists who left no other historical record of their existence than the plains themselves and the mysterious tumuli that dotted them. The prehistoric Druidic earthworks of Stonehenge paled in comparison to this image of America's past. From such a perspective, the Great Plains did not represent an Edenic nature that had remained the same throughout time, but instead a massive anthropogenic landscape in ruins. As the historian and archaeologist Caleb Atwater communicated to the editor of the American Antiquarian Society, Ohio was "nothing but one vast cemetery of the beings of past ages" (Anon., *Archaeologia Americana* 5).

Both Bryant and Bartram approached American landscapes as the manifest expression of human history, but whereas Bartram recognized that Native Americans had been instrumental in selecting and preserving the characteristic wilderness life of Georgia and Florida, Bryant used the myth of the mound builders to deny First Nations a positive agency in the creation of the Great Plains environment and thus any claim to the land. Their different environmental histories reflect contrary views on the role of Native peoples in the making of the American landscape. Whereas Bartram stressed the need to respect Indian sovereignty and rights, Bryant had inaugurated his career as chief editor of the *New York Evening Post* by writing editorials, from January to May 1830, promoting Andrew Jackson's Indian Removal Bill (Galloway, "William Cullen Bryant's American Antiquities" 727). In "The Prairies," he adopts the popular expansionist view that the great earth mounds had not been built by early Native Americans, but instead by a nonnative settler race whose civilization was destroyed by the coming of the "red man":

> The red man came—
> The roaming hunter tribes, warlike and fierce,
> And the mound-builders vanished from the earth.
> The solitude of centuries untold
> Has settled where they dwelt. (*Poetical Works* 1:230)

Bryant thus represents Native American culture as a destructive force that "settled" the region with "solitude"; that is, they unsettled the land. When Bryant visited Illinois, Native Americans had already been expelled from their ancestral lands, though Chief Black Hawk had just led a band of Sauk and Fox across the Mississippi in an effort to recover them. Bryant portrays the forcible eviction of native people from the northern plains and the disappearance of the fauna that supported them, that is, the beaver and the huge herds of bison that once roamed there, as a voluntary decision on their part, as they "sought / A wider hunting ground." At the same time, imperial history finds its justification in the myth of the inevitable advance of civilization. "Thus change the forms of being," he writes:

> Thus arise
> Races of living things, glorious in strength,
> And perish, as the quickening breath of God
> Fills them or is withdrawn. The red man, too,
> Has left the blooming wilds he ranged so long. (231)

Bryant is here adopting the same model of history that Andrew Jackson had employed in his Second Inaugural Address to justify his Indian policy: "one by one have many powerful tribes disappeared from the earth. To follow to the tomb the last of his race and to tread on the graves of extinct nations excite melancholy reflection. But true philanthropy reconciles the mind to these vicissitudes as it does to the extinction of one generation to make room for another" (cited in Zanger, "Premature Elegy" 16).

In writing "The Prairies," Bryant certainly would have had in mind Robert Southey's epic *Madoc*, which recounts the legend of a Welsh nobleman who establishes a colony in the New World during the twelfth century. More importantly, however, Bryant is rewriting Wordsworth's "Michael" on an epic scale. Where Wordsworth used a pile of stones to tell the story of a way of life and of human passions which had created the characteristic landscapes of the Lake District, Bryant writes an imperial myth about the rebuilding and resettling of a continent. Trained as a lawyer, however, Bryant did not see the world through the eyes of either Michael or Wordsworth. Standing amid this "great solitude" (Bryant, *Poetical Works* 1:232), echoing the "utter solitude" (W. Wordsworth, *Poetical Works* 2:80–94, line 13) of "Michael," Bryant's narrative justifies the forced displacement of Native Americans from their native lands as the expression of both a divine plan and voluntary choice. The unpeopling of the prairies is thus both a precondition for his solitary musings and the space from which he summons up the vision of the reappearance of a new settler nation to which he belongs. After noting the many insects, flowers, birds, reptiles, and "gentle quadrupeds" that still inhabit the place, he singles out a newcomer: "The bee, / A more adventurous colonist than man, / With whom he came across the eastern deep." In the domestic hum of this fellow colonizer, Bryant hears the voice of prophecy: "the sound of that advancing multitude / Which soon shall fill these deserts" (Bryant, *Poetical Works* 1:232). This prophecy comes directly from Wordsworth's *Excursion*, where the Wanderer likens Britain's colonial settlers to bees and welcomes the idea of her "cast[ing] off / Her swarms and, in succession send[ing] them forth" (W. Wordsworth, *Poetical Works* 5, bk. 9, lines 377–78) to populate "the smallest habitable rock, / Beaten by lonely billows" (lines 387–88). Such places, he argues, will "hear the songs / Of humanised society; and bloom / With civil arts, that shall breathe forth their fragrance, / A grateful tribute to all-ruling Heaven" (lines 388–91). During the nineteenth century, even bees, it seems, were humming the songs of empire. In "Thanatopsis," a poem that indicates the enormous impact that the *Lyrical Ballads* had on a youthful Bryant, he argues that communion with Nature in "her visible forms" (Bryant, *Poetical Works* 1:17) can bring hope and consolation to human beings even in the face of human mortality. In "The Prairies," this traveling nature serves as a divine signal that the European colonization of the prairies is both a recovery and a repossession of a vast deserted garden that for centuries has been lost to weeds and savages.

Recognizably an American rewriting of "Michael," the poem is just as importantly, if less obviously, influenced by "The Ruined Cottage," which first appeared as book 1 of *The Excursion* in 1814. One can see that Bryant was already thinking of the poem when he wrote to Dana, for despite his having studied botany with Amos Eaton, the author of the *Manual of Botany for the Northern States* (1817), while he was at Williams College (Muller,

William Cullen Bryant 109), Bryant nevertheless refers to the Great Plains as being "full of weeds and gaudy flowers." Bryant's recasting of Wordsworth's narrative about how Margaret's garden is progressively taken over by weeds as the story of the ruination of a continent can thus be read as an example of the westward diffusion and remediation of British literature. In this chapter, however, I want to use the poem to consider an exchange that went in the opposite direction, from America back to England. The reason why Bryant so easily adapted Wordsworth's poetry to interpreting the ancient history of the American continent was that Wordsworth had already learned from Bartram how to read the life of human beings in the natures that they left behind. The reason why the straggling heap of stones in "Michael" was able to speak so powerfully about the disappearance of a way of life was that Wordsworth had learned how to make them speak through Bartram. In Bartram, in other words, Wordsworth recognized that nature was an artifact half created by human agency. Both writers focused on landscapes that held within them the story of their dispossession from the people who created them: for Bartram, it was the First Nations people of the American Southeast; for Wordsworth, it was the pastoral farmers of the Lake District.

Wordsworth and "the secret spirit of humanity"

When in 1797–98 Coleridge and Wordsworth first began to write nature poetry in earnest, they did so with Bartram's *Travels* in their hands.[1] For Coleridge, it was "a *delicious* book," and not "a Book of Travels, properly speaking; but a series of poems, chiefly descriptive, *occasioned* by the Objects, which the Traveller observed" (*Marginalia* 1:227). In it, Bartram had shown that nature was a vital presence that could bring spiritual healing and wholeness to a divided life. "Deep joy," Coleridge called it, in "This Lime-Tree Bower My Prison," a poem that not only refers explicitly to the *Travels* but also indicates the manner in which Coleridge patterned his nature writing on this American model. Living in a deeply divided England, the poet identified with the ecstatic moments in the *Travels*, where the American naturalist seemed to be immersed in nature in all its spiritual plenitude. Coleridge read the book as proof that the recovery of unity between the self and nature could serve as the basis for a renewed social brotherhood. By overcoming our alienation from nature, he believed that human beings could establish the foundations of social and personal renewal. Wordsworth followed Coleridge's lead, declaring that one could be "tuned by Nature / to sympathy with Man" ("*The Ruined Cottage*" and "*The Pedlar*"

MS. B, 5ᵛ, 149). "Why is it we feel / So little for each other but for this," he wrote in the Alfoxden Notebook: "That we with nature have no sympathy / Or with such things as have no power to hold / Articulate language" (*"Ruined Cottage"* 121).

For Wordsworth, the capacity to relate to the nonhuman world was a precondition of social feeling, and "in all things," he writes, "I saw one life, and felt that it was joy; / One song they sang, and it was audible—" (1805 *Prelude* 2:429–31). Unlike Coleridge, however, he also learned from Bartram to see nature in historical terms. Alongside Bartram's joy in *nature present*, Wordsworth could also discern the lingering ghosts and spirits of a *nature past*, one that appeared in remnants and whose silences spoke of "bond[s] / Of brotherhood" that were "broken" and of absences and losses that were permanent and enduring (*"Ruined Cottage"* MS. D 84–85). A core aspect of Bartram's *Travels* is its reflection on natures abandoned (often forced) by the people who brought them into being. Wordsworth learned to look on these dispossessed natures as the primary elements in the history of human passion, seeing nature as a historical document that preserved "the spiritual presences of absent things" (*Excursion* 4.1234).[2] Where Coleridge sought to achieve a unity of mind and external forms, seeking in nature the communication of a divine spirit, the "One Life" animating all thoughts and things, Wordsworth's nature often speaks about what it means to be abandoned; it survives as ruins, haunted by change and loss, speaking of better times. And so do many of the characters who inhabit his poems. In lines originally intended for "Michael," Wordsworth writes that "I look into past times as prophets look / Into futurity" (*Lyrical Ballads*, DC MS. 30, *Michael* MS. 1, 4ʳ). To see the lingering presence of the past within the present, to see a nature whose loss was obscured by a *nature present*, required a heightened form of prophetic vision, which Wordsworth made his own. Bartram's *Travels* was thus the great precursor text of the *Lyrical Ballads*.[3] In reading this great American naturalist, both men came to see themselves as nature poets: Wordsworth, for a lifetime, and Coleridge, up until that moment when he finally concluded that the unity and community that he so desperately sought in nature could not be found there, but only within himself.

"I see around me here things which you cannot see," declares the Pedlar in "The Ruined Cottage," as he begins a lesson in how to "no longer read / The forms of things with an unworthy eye" (*"Ruined Cottage"* MS. D 67–68, 510–11). Here, the Pedlar claims that he possesses a kind of second sight, "an eye" that can look "deep into the shades of difference / As they lie hid in all exterior forms" (MS. B 94–96). "The Ruined Cottage" seeks to teach us how to see the human history that lies hidden in landscapes. Although critics often reduce the poem to "the story of Margaret," Wordsworth clearly intended it to be a narrative about the history of a place, one that would illuminate the integral relationship between the passions of a certain class of people and the physical landscapes that they had created. The landscape that the Pedlar describes is not a static object, but instead a living entity, which holds within it, when read properly, the story of Margaret, her passions, and her deep connection to this place. As William Faulkner remarked, "The past is never dead. It's not even past" (*Requiem for a Nun* 92).

Since a 1790s narrative about the abandonment of a young woman was hardly without precedent, the real experimentalism of "The Ruined Cottage" lies in its unique formal structure, which expresses Wordsworth's commitment to the idea that the mind and nature are mutually fitted to each other. This concept would later be encapsulated in the 1800 poem "Home at Grasmere," where the poet argues that his distinctive task is to demonstrate "How exquisitely the individual Mind . . . to the external World / Is fitted—and how exquisitely too . . . The external World is fitted to the Mind" (*Home at Grasmere* MS. B 1006–11).[4] Although Blake, in his annotations to *The Excursion*, famously dismissed this idea, commenting that "You shall not bring me down to believe such fitting & fitted I know better" (*Poetry and Prose* 656), this commitment to both a material understanding of the imagination and a creative, cultural understanding of nature constitutes the cornerstone of his poetry. Wordsworth is, as Jonathan Bate argues in *Romantic Ecology*, an ecological poet, and Andrew Hazucha rightly considers him "the first true English ecocritic" ("Neither Deep nor Shallow but National" 62).[5] In 1801, Coleridge would draw out a similar analogy between mind and nature in Wordsworth. When, in reading Bartram, he came upon a description of the soil and vegetation around Wrightsborough, Georgia, he applied it "by a fantastic analogue & similitude to Wordsworth's Mind": "The soil is a deep, rich, dark Mould on a deep Stratum of tenacious Clay, and that on a foundation of Rocks, which often break through both Strata, lifting their back above the Surface. The Trees, which chiefly grow here, are the gigantic Black Oak, Magnolia, Fraxinus excelsior, Platane, & a few stately Tulip Trees" (*Biographia Literaria, Collected Works* 7.2:155 and note). Fifteen years later, in the *Biographia Literaria*, he would return to this passage, declaring that it provided "a sort of allegory, or connected simile and metaphor of Wordsworth's intellect and genius."

Wordsworth must have been flattered by the notion that his mind was like nature itself, that it required the language of natural history, of geology and botany as much as psychology, to delineate its characteristic qualities. Coleridge's metaphor also captures Wordsworth's understanding of the mind and nature as being temporal structures, built up layer upon layer like a geological formation. What it misses, however, is Wordsworth's understanding of this mutual fittedness in historical terms. Nowhere has Wordsworth been more misread than in the assumption that he believed, as Gillian Beer encapsulates the view, that there is "a pre-existing harmony between the external and internal world" (*Darwin's Plots* 69). Wordsworth did not believe that there was a preestablished or fixed relationship between the mind and nature, but instead that both were the products of a dynamic interaction that changed over time. That is why he speaks of the "progressive powers" not only of the individual mind but also "perhaps no less / Of the whole species (MS. B 1007–8). In this regard, Wordsworth can best be understood as anticipating the methods and concerns of historical ecologists, who seek to trace "the ongoing dialectical relations between human acts and acts of nature, made manifest in the *landscape*. Practices are maintained or modified, decisions are made, and ideas are given shape; a landscape retains the physical evidence of these mental activities." And both Wordsworth and contemporary historical ecologists believe that "past and present human use of the earth must be understood in order to frame effective environmental policies of the future" (Crumley, "Historical Ecology" 9). Dave Egan and Evelyn Howell encapsulate the complexity of such a task when they write that "by doing historical ecology research you are entering into a dialogue with a place, its people, plants and animals, and its past. Out of this process of uncovering, analyzing, and interpreting cultural and biological data will come a narrative because everything we touch involves us in a history with it, and provides us with the opportunity to create a referential narrative history; if we keep or discover the proper records" (*Historical Ecology Handbook* xxiii). Tim Ingold makes a similar point, arguing that a "landscape is constituted as an enduring record of—and testimony to—the lives and works of past generations who have dwelt within it, and in so doing, have left there something of themselves." Ingold further adds that "the landscape tells—or rather is—a story. It enfolds the lives and times of predecessors who, over the generations, have moved around in it and played their part in its formation" ("Temporality of the Landscape" 152).

In Wordsworth, we see a similar commitment not just to ecology but to reading landscapes as a primary historical medium and register of human passions. In the Fenwick note to "The Thorn," for instance, he stresses that "the Reader cannot be too often reminded that Poetry is passion; it is the history or science of feelings" (*Poetical Works* 2:513). Thus, in commenting on the Pedlar, Wordsworth writes that

> To him was given an ear which deeply felt
> The voice of Nature in the obscure wind,
> The sounding mountain and the running stream.
> To every natural form, rock, fruit, and flower,
> Even the loose stones that cover the highway,
> He gave a moral life; he saw them feel
> Or linked them to some feeling. (*"Ruined Cottage"* MS. B 77–83)

In this universe shaped by action and feeling, neither landscapes nor the passions that they give rise to and which in turn shape them exist apart from time and change. "We die, my Friend," the Pedlar declares,

> Nor we alone, but that which each man loved
> And prized in his peculiar nook of earth
> Dies with him or is changed, and very soon
> Even of the good is no memorial left. (MS. D 68–72)

Here the Pedlar affirms that the nature that human beings "half create" ("Tintern Abbey," *Poetical Works* 2:259–63, line 106) and that expresses their individual and collective passions is as historical as any artifact or event. That is why the earth has so many stories to tell. In "The Ruined Cottage," Wordsworth set out to write a history of the passions that had brought a special place into being. His goal was to show not simply the productive interaction between human passions and a place but also how, as those passions changed, so too did the garden they had produced.

"The Ruined Cottage" is, as David Simpson remarks, "a ghost story" (*Wordsworth, Commodification and Social Concern* 44). What makes it extraordinary in this regard is that the Pedlar is haunted as much by things as by people. The discovery of a broken bowl by a water's edge, of "a well / Half-choked [with willow flowers and weeds]" (MS. D 62–63), of weeds and spear grass growing out of a ruined wall, of a "gaudy" sunflower "Look[ing] out upon the road" (MS. D 114–15)—all these things speak to the Pedlar of a woman and her family, and of the deep emotional connection that she once had to a place that, like her, now exists as an afterlife. In his story, the Pedlar awakens modes of presence that, if they are not exactly "human life," come "very near / To human life" (MS. D 372–74). Human feeling and the forms of nature are so intimately bound up with each other that objects are the primary vehicles for the expression of passions, and things, like people, can "haunt the bodily sense" (MS. B, 4^v, 145). Long before she died, Margaret "lingered in unquiet widowhood / A wife and widow" (MS. D 447–48), a living ghost haunting a lost world; now dead, she and the nature that mattered to her haunt the Pedlar's seeing and his story. Yet they do so in a very particular manner. As Geoffrey Hartman suggests, "Wordsworth does not make absence present in the manner of gothic, ghostly, or surrealist fantasies, or animated specters. The absent one remains absent in his representations. What is depicted is, as it were, the legacy of this absence" (*Unremarkable Wordsworth* 29). The natures that live on in Wordsworth's poetry do so in this strange spectral manner, for these are worlds that appear as absences. Thus, Wordsworth's engagement with the past is not so much a recovery as an attempt to convey as much as possible what we have lost; in a language that looks as much to the future as to the past, he thus adopts a subjunctive prophetic gesture toward a recovery that is to come. "I would give," he writes,

> as far as words can give
> A substance and a life to what I feel
> I would enshrine the spirit of the past
> For future restoration. (1805 *Prelude* 11:339–42)

The Pedlar refuses to separate the history of Margaret from that of her garden and cottage, for each serves as a lens for understanding the other, as mind and nature, psychology and ecology, mutually interpenetrate each other. Unfortunately, this aspect of the poem has been much maligned by New Historicist critics, who have treated the natural imagery of the poem as if it were an elaborate form of cover-up that evades history rather than engaging with it. For instance, Jerome K. McGann argues that "to read Wordsworth's re-telling of this pitiful story is to be led further and further from a clear sense of the historical origins and circumstantial causes of Margaret's tragedy. The place of such thoughts and such concerns is usurped, overgrown." As if the poem were as weedy as Margaret's garden, McGann claims that "Armytage, poet, and reader all fix their attention on a gathering mass of sensory, chiefly vegetable, details" (*Romantic Ideology* 83). Annabel Patterson similarly claims that Wordsworth naturalizes history by meditating on "the *naturalness* of the forces that have brought the cottage to ruin, from the bad harvests that initiated the cycle of poverty to the beautiful weeds that encroached on Margaret's garden" (*Pastoral and Ideology* 277). In Alan Liu's study of the emergence of the Romantic image in "The Ruined Cottage," vegetation that *tropes* or *turns away* from history is so pervasive in the poem that the landscape has become tropical. In this "vegetable tropics forced by the sun" (*Sense of History* 318), botany and aesthetics represent an evasion of history: "Wordsworth must find a way to plant what I will call the vegetable 'tropics' of the revisionary eye around his blind spots" (315); "the Pedlar throws around the blind spots of the cottage and Margaret one luxuriant still life after another" (317); "Vegetation, in sum, substitutes for the human being and the human place" (317). The idea that Wordsworth's close attention to plants is a blind or escape from history is, as I hope I have suggested, fundamentally mistaken. Rather than recognizing that the history of colonialism was being read out of the very plants that composed a place, because the natural was one of the primary sites of transformation, conflict, and change during the period, New Historicism treats nature as something that is static and unchanging. For this criticism, botanical knowledge is not a form of historical knowledge, and looking closely at plants can only be a distraction from history.

A surprising aspect of Wordsworth's exploration of the passions that link human beings to the earth is that he does so by focusing on what it means *to lose* that connection. In a narrative poem about the marriage of mind and nature, Wordsworth tells the story of a marriage breakdown and its impact on a woman's connection to a place. Abandoned by her husband, Margaret abandons herself, her family, her house, and her garden, even as, tragically, she cannot move beyond them. Consequently, her life exists in an uneasy, troubled state of separation from all that matters to her, and she forsakes those creative relationships of sense, feeling, and passion which would otherwise link her to the world of things. In "this world of feeling & of life," we must not live in dreams, Wordsworth writes, but instead must use the faculties that we have been given—"apprehension, reason, will & thought / Affections, organs, passions"—to live fully in the world: "Thus disciplined / All things shall live in us & we shall live / In all things that surround us" ("*Ruined Cottage*" 269). The Pedlar speaks of this breakdown when he remarks that poets, in calling in their elegies upon "the hills and streams to mourn" the departed, do not do so "idly," but instead in recognition of the "creative power / Of human passion" (MS. D 75–79):

> a bond
> Of brotherhood is broken: time has been
> When every day the touch of human hand
> Disturbed their stillness, and they ministered
> To human comfort. (84–88)

Amid the ruins and the frozen silence of this breakdown, the Pedlar delineates the passions that have marked both this marriage and this place, feelings of love, anger, pain, loss, and hopeless despair. Loss too, for Margaret is lost to herself and to others: "what I seek I cannot find. / And so I waste my time: for I am changed; / And to myself... have done much wrong" (351–53). Alexander Regier draws upon the work of Walter Benjamin to read the poem as an exploration of a nature that mourns the breakdown of the language that once connected it to human life (*Fracture and Fragmentation* 30–51), and this view is nicely captured in Alan Liu's sense that the poem is "a sort of fossilized narrative or drama" (*Sense of History* 313). Such a form is particularly suited to the subject matter of the poem, for one of the major consequences of Margaret's loss of her husband is that she believes that her world has come to a standstill.[6] Denied a future, or, at least, believing that she cannot move forward without the return of Robert, she lives in a kind of temporal limbo, as a kind of living fossil or ghost. But unlike a ghost or fossil, she and her world are not immune to time and decay. Like a geologist reading the ruins of the past, the Pedlar tells the story of what it means to lose one's purchase on the present, to linger on in a world that is already past, even as he recounts the forces—war, economic depression, disease—that ultimately caused the destruction of this place and the family that inhabited it.

Set in a poverty-stricken region of southern England, "The Ruined Cottage" is structured as both a colonial and a working-class elegy, which explores not only the political and economic factors that were leading to the disappearance of an entire way of life but also how these changes were affecting the psychic and emotional well-being of the rural poor. The poem explores the contradictory hope and despair that accompany a troubled individual's struggle to hold on to her way of life in the face of abandonment, poverty, and misery, yet this narrative about the ruin of an English cottage garden was also intended to illuminate broader changes that were affecting the rural working class at the end of the eighteenth century. In essence, it is a narrative about the decline of the English rural working class. The idea for the poem emerged from Wordsworth's observation of the extreme poverty of cottagers in Dorset. In a 1795 letter to William Matthews, he writes that "the country people here are wretchedly poor" (*Letters: Early Years* 154), and a month later Dorothy expands on this theme, writing that "the peasants are miserably poor; their cottages are shapeless structures (I may almost say) of wood and clay—indeed they are not at all beyond what might be expected in savage life" (162). Seeing savagery not as a natural state of human being, but instead as a product of human degradation, Wordsworth's goal in the poem was not, as some have suggested, to naturalize Margaret's suffering, but instead to explore the ways in which a capitalist economy was impoverishing (in all senses of the word) the rural working class. Just as importantly, he wanted his readers to feel and critically understand what was being lost.

In using the disappearance of an English weaver's cottage and garden to symbolize a major change in English working-class life, Wordsworth was taking up a topic that would become central to the nineteenth-century understanding of the history of the working class and would culminate in the work of twentieth-century leftist historians, such as E. P. Thompson. Wordsworth was not the first, however, to draw this kind of parallel, for a little more than a year earlier, in September 1795, the political activist John Thelwall had done the same. In the essay "On the Causes of the Late Disturbances," published in the *Tribune* on 23 September 1795, Thelwall argued that the immediate cause of worker unrest sweeping across England in the 1790s was the distress and misery of the lower orders of society, who were the first to feel the effects of political corruption and economic mismanagement. Arguing that "it is the duty of every member of society to see that the laborious classes of mankind are enabled to maintain themselves, in comfort and abundance, by their labours," Thelwall observes that in the city of Norwich alone, which was once a "flourishing mart of trade and manufacture" ("On the Causes of the Late Disturbances" 316), twenty-five thousand people out of a population of forty thousand were at that time out of work. In order to convey the huge impact of these changes on working-class life, Thelwall drew upon his own experience as the son of a Spitalfields silk mercer, observing that whereas once a weaver enjoyed "a small summer house and a narrow slip of a garden, at the outskirts of the town, where he spent his *Monday*, either in flying his pidgeons, or raising his tulips ... [t]hose gardens are now fallen into decay. The little summer-house and the Monday's recreation are no more; and you will find the poor weavers and their families crowded together in vile, filthy and unwholesome chambers, destitute of the most common comforts, and even of the common necessaries of life" (317–18).

Poverty in Thelwall's Spitalfields has the same effect as poverty in Dorset. When Thelwall first heard "The Ruined Cottage," during his visit to Alfoxden from 17 to 27 July 1797, he would not have thought that the Pedlar's focus on the state of Margaret's garden was irrelevant to the major social questions of the day, but instead would have recognized that it reflected the terms by which the social, economic, and moral conditions of the working class were at that time being registered. As Robin Veder has remarked, in a paper on the significance of working-class gardening in the historical understanding of the working class, "the disappearance of the Spitalfields weavers' flowers became a key element in romantic narratives of loss. From the late eighteenth century, Thelwall and others employed the flowers as a potent symbol of lost leisure and independence, and a dividing marker between the mythic identities of the independent 'intelligent artisan' and the impoverished, 'ignorant labourer' in industrial Britain" ("Flowers in the Slums" 262). Situated on the dividing line between "work" (which implies specialist knowledge, technical skill, and creativity) and "labour" (which does not necessarily require them), the disappearance of the weaver's garden, as one of the most significant cultural expressions of artisan culture, was understood as marking, to cite E. P. Thompson, "the difficult and painful nature of the change in status from artisan to depressed outworker" (*Making of the English Working Class* 270).

Cottage gardens first appeared in England during the seventeenth century, with the wave of French Huguenot weavers who arrived in Spitalfields, Norwich, and other parts of England, fleeing from religious persecution. By the late eighteenth century, they came to be seen as a distinctively English style of gardening, reflecting the industry, needs, skills, and interests of the British working class. In *An Encyclopaedia of Gardening*, John Loudon remarks that "wherever the silk, linen, or cotton manufactures, are carried on by manual labor, the operators are found to possess a taste for, and to occupy part of their leisure time in the culture of flowers" (86). Whereas wealthy landowners used plantings to make landscapes look like European paintings, the cottage garden met the multiple needs—recreational, aesthetic, culinary, and economic—of the working class. In their dense plantings of perennials, flowering shrubs, culinary herbs, vegetables, fruit trees, and medical plants, these gardens created their own kind of aesthetic, one that combined informality and economy with a love of flowers. Of particular interest were the "florists' flowers," the showy hyacinths, ranunculi, narcissi, tulips, auriculas, anemone, pinks, and carnations that were the prized elements of a burgeoning nursery trade. As Loudon further remarks, "The taste for florists' flowers, in England, is generally supposed to have been brought over from Flanders with our worsted manufactures, during the persecutions of Philip II; and the cruelties of the Duke of Alva, in 1567, was the occasion of our receiving, through the Flemish weavers, gillyflowers, carnations, and provins roses" (84). The president of the Linnean Society, Sir James Edward Smith, makes a similar point, writing that "the taste for the cultivation of flowers was, probably, imported from Flanders along with our worsted manufacture" ("Biographical Memoirs" 295–96).

The working class was thus deeply associated during the eighteenth century with the love and culture of flowers. Concomitant with the rise of English cottage gardens was the establishment in many provincial towns of annual flower shows and feasts, where prizes were given for the best "florists' flowers," with auriculas being the primary focus in the spring and carnations in the summer.[7] As the gardener William Hanbury commented in 1770, gardening artisans "often exhibit the fairest flowers" and "a weaver, or the like, will often run away with the prize at those feasts" (*Complete Body of Planting and Gardening* 1:286). The cultivation and breeding of choice flowers required a care and attention that was rarely found among the hired help on the larger estates, so the best flowers were also most frequently found in cottage gardens (K. Thomas, *Man and the Natural World* 229).[8] Throughout the eighteenth and early nineteenth centuries, the weavers of Lancashire, Paisley, and Spitalfields were famous for their knowledge of botany and their breeding and cultivation of flowers. Many of the newest and fanciest flower varieties were developed by working-class gardeners, and as Collinson notes in a 25 December 1767 letter to John Bartram, they were also often willing to spend what was for them large sums of money to obtain special varieties, which could then be used for breeding newer varieties: "So great is the Itch, that a poor Raged Shoemaker a Weaver or a Baker will give half a Guinea or a whole one for a New flower—such is the infatuation" (J. Bartram, *Correspondence* 693).

The extraordinary care and attention to detail required in the selection, cultivation, and breeding of show flowers were often subject to caricature, as in Samuel Johnson's claim in *Rasselas* that it is not the "business of the poet" to "number the streaks of the tulip," or in George Crabbe's Letter on "Trades" in *The Borough*, where he describes the typical entomological and botanical pursuits of a weaver as being a frivolous, if harmless, pastime:

> There is my Friend the *Weaver*; strong desires
> Reign in his breast; 'tis Beauty he admires:
> . . .
> For him is blooming in its rich array,
> The glorious Flower which bore the palm away;
> In vain a Rival tried his utmost art,
> His was the Prize, and joy o'erflow'd his heart.
> "This, this! is Beauty; cast, I pray, your eyes
> On this my Glory! see the Grace! the Size!
> Was ever Stem so tall, so stout, so strong,
> Exact in breadth, in just proportion, long?
> These brilliant Hues are all distinct and clean,
> No kindred Tint, no blending Streaks between;
> This is no shaded, run-off, pin-ey'd thing;
> A King of Flowers, a Flower for England's King:
> I own my pride, and thank the favouring Star
> Which shed such beauty on my fair *Bizarre*." (*Complete Poetical Works*
>   1:433–39, lines 69–106)

Crabbe's use of the highly technical language of florists in referring to a "bizarre" carnation, which he defines as a variegated flower "with three or more colours irregularly and indeterminately" (1:436n), indicates that, as a serious botanist himself, he was not immune to such interests. "Beauty, in the eye of a florist," he adds, "contrary to the opinion of Thomson, and I believe of some other poets . . . depends upon the distinctness of their colours: the stronger the bounding line, and the less they break into the neighbouring tint, so much the richer and more valuable is the flower esteemed" (1:436n). Elsewhere, in *The Parish Register* (1807), he speaks more favorably of "the Cot! where thrives th' industrious Swain, / Source of his pride, his pleasure, and his gain," with its attached garden, where "on every foot of that improving ground" can be found chives, leeks, legumes, "Apples and Cherries grafted by his hand," and "cluster'd Nuts, for the neighbouring market stand" (*Complete Poetical Works* 1:212–80, bk. 1, lines 31–32, 134, 145–46). And close to the cottage, a reed fence encloses a "favourite spot"

> Where rich Carnations, Pinks with purple eyes,
> Proud Hyacinths, the least some Florist's prize,
> Tulips tall-stemm'd and pounc'd Auricula's rise. (lines 148–51)

Although Johnson was a proponent of the idea that great art portrays a generalized or representative nature, the floral taste of weavers for the beauty of the particular not only was reflected in their gardens but also had an enormous impact on the design of eighteenth-century textiles, especially in the Spitalfields figured silks. As Deborah Kraak remarks, florists' flowers "became the core of a tasteful vocabulary of motifs used in disproportionate numbers on English textiles and embroideries" ("Eighteenth-Century English Floral Silks" 846).

As many historians have argued, by the late eighteenth century the economic independence of weavers was severely compromised by new modes of production, and these pressures only increased during the first half of the nineteenth century. Thus, as E. P. Thompson comments, "the history of the weavers in the 19th century is haunted by the legend of better days" (*Making of the English Working Class* 269). In focusing on a weaver's family, cottage, and garden, Wordsworth was writing his own history of the flowering and collapse of a class of artisans which was famous for its commitment to natural history. In this regard, we should not overlook the fact that Margaret and Robert had a small number of books, some of which must have been on botany and gardening, and that in Margaret's decline these precious commodities end up "scattered here and there" (MS. D 84–85) across the cottage, "open or shut / As they had chanced to fall" (MS. D 408–9). Covering the ten-year period that saw the disappearance of Margaret and her world, the narrative seeks to show how one might write a history of the passions, specifically those of Margaret and Robert, through the plants that they left behind. "The Ruined Cottage" superficially has all the appearances of a conventional gothic narrative evoking the pleasurable melancholy that arises from seeing nature and time reclaiming the remains of a rural cottage. Of crucial significance, however, is the fact that the nature that Margaret loved is itself in ruin. It is lost and cannot be reclaimed. The Pedlar is the exemplar of this new way of reading the natural world, what Wordsworth, in an excised draft, called the "Colours & forms of a strange discipline" (MS. B 43ᵛ). Through the interactive process of seeing and telling, the Pedlar illuminates the complex interweaving of narrative and sight which underpins the Wordsworthian goal of reading the history of passions in the history of things.

"All things," he writes in an early draft, "shall speak of man and we shall read / Our duties in all forms" (MS. D Additions 68ʳ).

The poem begins with the poet's account of his having accidentally come upon the ruined cottage, when, in seeking relief from the heat and gnats, he noticed in the distance a clump of high elms, a sign of human habitation in what is otherwise a forsaken waste. There, he meets an old friend, the Pedlar, who directs him to a place where he can find refreshment. After climbing over a stone wall, the poet narrator discovers "a plot / Of garden-ground, now wild" (MS. D 54–55) and a "well / Half-choked" (MS. D 63–64) with weeds and willow flowers, the latter having fallen from the willows that once served as a hedge enclosing the garden. The overgrown willow hedge now shades the entire garden, producing "a cheerless spot . . . a damp cold nook" (MS. D 60, 62) where very little grows. Nevertheless, the poet can still discern the remnants of what once was a berry patch, with a few lanky gooseberries and currants producing scanty fruit on "their leafless stems" (MS. D 58). To a botanist, there was perhaps no clearer sign of this garden's having once belonged to a working-class family than the gooseberries, which were arguably the preeminent cultivar bred by them. As Loudon remarks, "in Lancashire, and some parts of the adjoining counties, almost every cottager who has a garden, cultivates the gooseberry, with a view to prizes given at what are called gooseberry-prize meetings; of these there is annually published an account, with the names and weight of the successful sorts, in what is called the *Manchester Gooseberry-Book*" (*Encyclopaedia of Gardening* 732). We learn that the husband Robert was an avid gardener, busy with his spade in the garden "till the day-light / Was gone and every leaf and every flower / Were lost in the dark hedges" (MS. D 179–82). If he did not win prizes for his gooseberries, he certainly won them for his "carnations once / Prized for surpassing beauty, and no less / For the peculiar pains they had required" (*Excursion* 1:724–25). Of the ornamental plants that had once graced the garden, there remains but one, a single "gaudy" sunflower looking "out upon the road" (MS. D 53–54), as if, having taken up the vacated position of Margaret, it is hopelessly seeking her (or Robert's) return. Everywhere he looks, the poet sees ruin and decay, weeds and debris. Where a cottage once stood clothed in an "outward garb of houshold flowers, / Of rose and sweet-briar" (105–6), there now stands four walls "tricked / With weeds and the rank spear-grass" (107–8). Where a gate and arbor led into the garden, now "Nettles rot and adders sun themselves" (109–10). A garden that once contributed to the emotional and economic well-being of a working-class family is now a pathogenic space.⁹

The Pedlar's role in the poem is to use narrative to convey to the poet the sense of a world that has been lost and which cannot, even in words, be recovered. We are never really given a description of how the garden actually once appeared; instead, we catch glimpses of bits and pieces of it as they arise in the narrative, often at the very moment when they are about to disappear forever. Since the garden was inherently the product of skill and industry, fitted to Margaret's and Robert's economic, intellectual, aesthetic, and recreational needs (their "minds" as Wordsworth might say), its disappearance indicates that they no longer are agential presences in the landscape. It thus embodies their displacement. When, at the end of summer, the Pedlar visits the cottage, he notices that the garden plants are no longer being pruned or kept in place, as "the honeysuckle crowded round the door" and "knots of worthless stone-crop started out / Along the window's edge, and grew like weeds" (MS. D 308–11). In the garden, a climbing rose trained to grow against a stone wall is described as being "bowed . . . down to the earth" by the "unprofitable bindweed." The Pedlar's use of the words "worthless" and "unprofitable" indicates the degree to which the cottage garden was, indeed, a product of working-class industry and economy. We learn that the daisies, thrift, chamomile, and thyme have left the borders of the garden and have "straggled out into the paths / Which they were used to deck" (MS. D 318–20). The Pedlar also relates that he noticed that sheep were now entering the garden, indicating that the hedges and fencing are no longer being maintained. By the following spring, when he again visits the place, the state of the garden clearly indicates "that poverty and grief / Were now come nearer" (MS. D 413–14) to Margaret. The ground is no longer being tilled, and "weeds" and "knots of withered grass" have replaced the "herbs and flowers," many of which have been "gnawed away / Or trampled on the earth" (MS. D 418–19) by animals. A young apple tree has had its bark girdled "by truant sheep" (MS. D 422), and an ongoing state of idleness born of hopelessness seems to have settled on Margaret and her cottage. The only thing that grows in this place is Margaret's desperate hope, "Fast rooted at her heart" (MS. D 490), that Robert will return. When she eventually dies, the cottage has become a "wretched spot" (MS. D 487), a complete expression of her wretchedness. And for those who know how to read this landscape, especially as it is informed by narrative, what remains continues to tell her story long after she is gone. Everywhere it speaks of poverty, pain, and suffering, and what it means to be abandoned.

"The Ruined Cottage" is a poem about the relationship between landscape and narrative. "'Tis a common tale ... a tale of silent suffering, hardly clothed / In bodily form" (MS. D 232–34), the Pedlar remarks, as he attaches a story to a place that speaks of a way of life and a nature that no longer exist. Responding to the Pedlar's story, the poet comments, "the things of which he spake / Seemed present" (MS. D 211–12). Having learned how to see the traces of humanity registered by landscapes, grieving as much for what has happened to the place as for Margaret, the poet turns back to these ruins, tracing

> with milder interest
> That secret spirit of humanity
> Which, 'mid the calm oblivious tendencies
> Of nature, 'mid her plants, and weeds, and flowers,
> And silent overgrowings, still survived. (MS. D 502–6)

Here Wordsworth shows as great an awareness of the forgettings of nature as any contemporary New Historicist critic, as the "calm" that he describes is expressive of nature's "oblivious tendencies," her "silent overgrowings." But Wordsworth is certainly not arguing for these destructive forgettings. The nature that has displaced Margaret's garden is not an adequate substitute for what has been lost, even if it does hold a tragic kind of beauty. Wordsworth recognized in "A Night on Salisbury Plain," as Anne Janowitz notes, that "a ruined nature is a sure symptom of national corruption" (*England's Ruins* 94), and he continued that critique in "The Ruined Cottage."

As many critics have indicated, Wordsworth struggled to find a way to conclude the poem. This was not because he did not fully appreciate the tragedy that was registered by this nature become ruin, nor was it because he wanted to minimize or, worse yet, market this loss as poetry. The real problem had to do with how to respond to that history. The poem can be said to enact what Joseph Roach has termed "surrogation," which seeks to compensate for the "cavities created by loss through death or other forms of departure" (*Cities of the Dead* 2–3). The poem ends with the poet and the Pedlar leaving "the shade" and entering a milder air "peopled" by the songs of the linnet and thrush. Wordsworth's poetry is structured by the contradictions of fully responding to the claims of the present and to those that the past makes on us. "Waking from the silence of my grief," and looking at the ruins around him, Wordsworth writes that

> to some eye within me, all appeared
> Colours & forms of a strange discipline
> The trouble which they sent into my thought
> Was sweet . . . (MS. B, 43ᵛ)

Wordsworth was troubled by an art whose knowledge was built on the deepest recognition of historical suffering mixed with the pleasure produced by form. "The Ruined Cottage" grapples with this "strange discipline" and clings to it at the same time. That is why the poem continues to trouble its readers with its strange and haunting pain and beauty. "It falls short as cure," Goodman remarks, "but not as historiography" (*Georgic Modernity* 118).

"Michael"

Although Wordsworth is usually read as being committed to the idea of an eternal and unchanging nature, his poems are most often about what it means to live with a nature that is passing out of existence. Much of his poetry is about the experience of losing one's nature, either because one has lost touch with it (the great Wordsworthian theme of aging and loss of vision) or by witnessing its disappearance or replacement by another. This is a conservative position, but at a time when everywhere across the globe indigenous people were seeing their lands being translated into something else under the banner of modernization or "improvement," Wordsworth's poetry contributes to a politics of resistance in stronger terms than his liberal and postcolonial detractors might care to admit. In writing about the Lake District, Wordsworth adopted a nativist position. "These Lakes and mountains," he writes, "are my native Country . . . among which I have passed the greatest part of my life" (Wordsworth to J. Pering, 2 Oct. 1808, *Letters: Middle Years* 1:271–72). Instead of blaming Wordsworth for having foisted his daffodils on the rest of the world, therefore, it would be fairer and more productive to see him as a nativist poet who sought to preserve the historical ecology of the Lake District in the face of ecological, economic, and social changes being introduced by outsiders. His arguments for preserving native places in *A Guide through the District of the Lakes* are worth stressing: "surely there is not a single spot that would not have, if well managed, sufficient dignity to support itself, unaided by the productions of other climates, or by elaborate decorations which might be becoming elsewhere" (*Prose Works* 2:218).

There is no question that Wordsworth often portrays nature as a mainstay against the changes and uncertainties of contemporary life, especially in the early poetry. In the preface to *Lyrical Ballads*, for instance, he indicates that he chose subjects drawn from "low and rustic life" because their passions "are incorporated with the beautiful and permanent forms of nature" (*Prose Works* 1:124). He goes on to claim that his poetry is built on two indestructible pillars, the human mind and nature, declaring that he has had "a deep impression of certain inherent and indestructible qualities of the human mind, and likewise of certain powers in the great and permanent objects that act upon it, which are equally inherent and indestructible" (1:151). *The Prelude* argues that a natural education produces lasting results because it intertwines the passions "that build up our human soul / Not with the mean and vulgar works of man, / But with high objects, with enduring things, / With life and Nature" (1:434–37), and this idea is repeated in book 7 in Wordsworth's reference to "the forms / Perennial of the ancient hills" (7:726–27). In "Tintern Abbey," nature is also represented as "the anchor" of his "purest thoughts" (*Poetical Works* 2:259–63, line 110), yet the poem itself tells a different story. Wordsworth might have initially hoped that nature would serve as a permanent ground of social feeling and community, but that was not the nature he got. In "The Ruined Cottage," he tried to convince himself that the beauty of "the weeds, and the high spear-grass" (MS. D 514) that had taken over Margaret's garden could provide some element of consolation for the tragedy manifested there, yet the remnants of her garden speak just as powerfully of change and loss, and the destruction of the garden is itself a microcosm of a broader ruination of nature explicit in a landscape that has become reduced to "a bare wide Common" (MS. D 20).

Having used the disappearance of a rural English garden to talk about the increased poverty and loss of independence of the working class, Wordsworth sought in 1800 to extend this idea to a "series of pastorals ... laid among the mountains of Cumberland and Westmoreland" (*Poetical Works* 2:467n). In these poems, he confronted the increasingly obvious fact that the way of life that had produced the unique landscapes of the Lake District was now a thing of the past and that the nature that he had known all his life was about to disappear. In "The Brothers" and "Michael," Wordsworth was less interested in refashioning a traditional literary genre than in using it to examine the historical ecology of the Lake District, and particularly the ways in which the distinctive beauty of this nature reflected the passions and land practices of a small "class of men" who, as he observed in a well-known letter to Charles James Fox, were "rapidly disappearing," the small landowners or "statesmen" who for centuries had combined traditional crop farming in the valleys with the open grazing of sheep and cattle on the fells (Apr. 1801, *Letters: Early Years* 315). In these poems, Wordsworth began to explore what would become a lifelong concern: it was no longer a matter of preserving nature in the abstract, but instead a unique historical ecology (that is, a nature that had come into being through specific modes of human agency), when the socioeconomic conditions that had originally given rise to it no longer existed.

Displacement is a core theme in both of the pastoral poems he wrote. "The Brothers" is about a failed homecoming, for Leonard Ewbank learns that the way of life and the home to which he had hoped to retire, after spending years in colonial trade, no longer exist. His family name, which links him to sheep and to the land, also alludes to the "Matron's Tale" of the 1805 *Prelude*, which Wordsworth had originally written as an episode for "Michael." The tale centers on the search for a lost sheep and a young boy's discovery of it by virtue of a unique characteristic of the Herdwick variety of sheep, bred in the Lake District; once these animals have been attached, that is, "hefted," to a specific piece of pasture on the hillside, they instinctively return to it for the remainder of their lives:

> Drive one of these poor creatures miles and miles,
> If he can crawl he will return again
> To his own hills, the spots where when a lamb
> He learnt to pasture at his mother's side. (1805 *Prelude* 8:254–57)

Nothing is more local than a hefted sheep, and in both "The Brothers" and "Michael" Wordsworth sets human mobility against "hefting," or the desire to stay in place. Ironically, this breed's homing instinct played a crucial role in the upland pastoral economy of the Lake District, because it allowed farmers to graze their flocks on the distant open fells without using fences or constant shepherding to keep them in place. Since the upland grazing of sheep is one of the primary factors in producing the distinctive landscapes of the region, Harry Griffin, a regular contributor to the *Guardian*, rightly observes that "the fell farmers and their sheep—particularly the hardy, indigenous Herdwicks—are really the architects of the scenery." When the sheep disappear, "the whole landscape will harshen and deteriorate. The Lake District will disappear" ("If they go").

In "The Brothers," both siblings suffer from forms of nostalgia, or homesickness, and lose their way. In a tragic irony, James, the sickly brother who stayed home, having "gone forth among the new-dropped lambs, / With two or three companions" (*Poetical Works* 2:1–13, lines 358–59) to heft the new lambs, is lost on the fells when, sleepwalking, he tumbles from a precipice called the Pillar. Leonard, whose "soul was knit to this his native soil" (line 298), was forced "to try his fortune on the seas" (line 306), when, with the death of his octogenarian grandfather, Walter Ewbank, the estate and house, "buffeted with bond, / Interest, and mortgages" (lines 214–15), were sold, along with the flock of sheep which, in the view of the Priest of Ennerdale "Had clothed the Ewbanks for a thousand years" (line 303). On his return, Leonard loses his path "through fields which once had been well known to him" (line 93), finding "Strange alteration wrought on every side / Among the woods and fields, and that the rocks / And the eternal hills, themselves were chang'd" (lines 97–99). Some of these changes undoubtedly reflect changes in Leonard, but he is not wrong in feeling that his native landscape has become alien and that the "eternal hills" are not as eternal as he had thought. The fact that the Priest does not even recognize Leonard makes it clear that little remains of the community that he would seek. At the end of the poem, Leonard relinquishes any hope of a homecoming and returns to his life as "A seaman, a grey-headed Mariner" (lines 435).

Where "The Brothers" explores the disappearance of pastoral life in a dialogue riddled with dramatic irony, "Michael" follows directly from "The Ruined Cottage" in seeking to recover a history preserved by a landscape, in those things that "you might pass by, / Might see and notice not" (*Poetical Works* 2:80–94, lines 15–16). Initially, Wordsworth himself seems not to have noticed the central symbol of the poem, the sheepfold. In what is likely the first draft of the introduction to the poem, he tells how he came upon the ruined sheepfold:

> For me
> When it has chanced that having wandered long
> Among the mountains I have waked at length
> From dream of motion in some spot like this
> Shut out from man some region one of those

> That hold by an inalienable right
> An independent life and seem the abode
> Of nature & of unrecorded time
> If looking round I have perchance perceivd
> Some vestiges of human hands some steps
> Of human passion they to me have been
> As light at day break or the sudden sound
> Of music to a blind man's ear (*Lyrical Ballads* 615).

Wordsworth's reference to having come upon "a spot" that seemed to him the very "abode / Of nature & of unrecorded time" reminds us of how much a radically new conception of geological time informs his poetry, and also of the ways in which he often uses landscapes to envision an earth that is as yet untouched by human agency, listening to sounds, he says in *The Prelude*, "that are / The ghostly language of the ancient earth, / Or make their dim abode in distant winds" (1805 *Prelude* 2:327–29). Events prove, however, that this belief is an illusion, for in "looking round" he sees "some vestiges of human hands some steps / Of human passion." Prior to that moment, he had been living in a dream, blind or in the dark, seeing but noticing not.

This failure in seeing constitutes the visionary underpinnings of "Michael," for in reflecting on the impact that seeing these stones had on him, Wordsworth writes, "They are as a creation in my heart" (*Poetical Works* 2:482n). Over the next two months, Wordsworth repeatedly visited these ruins, finding his own creative power in responding to the passion that they communicated, as he composed a poem that would convey in narrative the way of life marked by the unfinished sheepfold and the nature that had disappeared with his way of life. Dorothy first mentions the poem in her journal, writing that "we walked up Greenhead Gill in search of a sheepfold" (11 Oct. 1800, *Journals* 44). Four days later, she remarks that "Wm again composed at the sheep-fold after dinner" (45), and on 18 October her description of Wordsworth sounds a lot like a description of Michael himself: "William worked all the morning at the Sheep-fold but in vain" (46). Charles J. Rzepka sees Wordsworth as "our first poet of the *pre*-historical imagination, our first *archaeological* poet" ("Sacrificial Sites" 211). Rightly so, for the pile of unhewn stones that begin and end the poem hold a story of passion, the unlocking of which is the primary creative task of the narrative. The work of the poet in communicating a history of passion in words thus parallels the work of the shepherd in stones and on the land.

That Wordsworth recognized that the physical character and aesthetic beauty of the landscapes of the Lake District were the historical expression of the pastoral labor of people like Michael over many centuries is clear, for in the proem he explicitly declares that even before he first heard the tale of Michael, he "already loved" these shepherds, "not verily / For their own sakes, but for the fields and hills / Where was their occupation and abode" (lines 24–26). In his letter to Fox, he explains that for the statesmen land was not a thing, separable from themselves, but instead was the essential medium for the realization, expression, and communication of their passions. "Their little tract of land serves as a kind of permanent rallying point for their domestic feelings, as a tablet upon which they are written which makes them objects of memory in a thousand instances when they would otherwise be forgotten. It is a fountain fitted to the nature of social man" (Apr. 1801, *Letters: Early Years* 314–15). This semiotic understanding of the relationship between identity and landscape is further developed in "Michael," where the valleys, streams, rocks, fields, and hills are described as "a book" composed of the incidents "impressed . . . upon his mind / Of hardship, skill or courage, joy or fear" (*Poetical Works* 2:80–94, lines 67–69), even preserving "the memory / Of the dumb animals, whom he had saved, / Had fed or sheltered" (lines 70–72). In a phrase that captures how deeply Michael's life is fused with the fields and hills around him, Wordsworth writes that they "were his living Being, even more / Than his own blood" (*Poetical Works* 2:82n), constituting "The pleasure which there is in life itself" (line 77). In the proem to "Michael," Wordsworth extends this understanding of the social dimensions of landscape to himself, for he argues that "the power / Of Nature," expressed through "the gentle agency / Of Natural objects," led him "to feel / For passions that were not my own" (lines 28–31). Like an anthropologist, Wordsworth differentiates himself from Michael's class and culture; he does not belong to their time. Yet what they share is a landscape and the stories that were born from it. In these mountain landscapes, Wordsworth read the strength, independence, and creativity of the people who had created them.

When Michael is called upon to pay a debt incurred by a distant nephew, he is forced to make what for him is an impossible choice between his son and the land. His choice proves tragic for his family, but also for the land that had been in his family for centuries. Luke disappears, never to return, and the pastoral labor that had previously been an unceasing aspect of the place comes to an end. In this regard, the unfinished sheepfold, described in an overflow line as "a dying form" (*Poetical Works* 2:482n), is expressive of a broader social breakdown affecting all elements of the landscape. The historical environment of the main narrative, which everywhere registers the active presence of Michael, is replaced by one in which he and his family and ancestors have all but disappeared from the landscape:

> No habitation can be seen; but they
> Who journey thither find themselves alone
> With a few sheep, with rocks and stones, and kites
> That overhead are sailing in the sky.
> It is in truth an utter solitude;
> Nor should I have made mention of this Dell
> But for one object which you might pass by,
> Might see and notice not. Beside the brook
> Appears a straggling heap of unhewn stones! (lines 9–17)

Except for a nameless heap of stones, the place shows little sign of human presence and is inhabited only by "a few sheep" and by the scavenging red kites that feed on them. The "hidden valley" that the vale opens onto is little different, for we learn at the end of the poem that almost nothing is left of the farm: the cottage, once named "The Evening Star," "Is gone, the ploughshare has been through the ground / On which it stood" (lines 486–87), while the "Clipping Tree," which stood beside it, is now just an old oak standing alone in a field—"[G]reat changes have been wrought / In all the neighbourhood" (lines 478–79).

A Guide through the District of the Lakes

Although Wordsworth was reluctant to add his name to the long list of writers who, from the late 1760s, had produced a flood of guide books on the Lake District, financial necessity seems to have induced him in 1809 to write the introduction to Joseph Wilkinson's folio edition of engravings *Select Views in Cumberland, Westmoreland, and Lancashire* (1810). Revised as part of the *River Duddon* collection (1820) and then published separately in 1822, it was one of Wordsworth's most successful works, with a fifth edition appearing in 1835 under its received title *A Guide through the District of the Lakes*. Its popularity was well deserved, for the *Guide* is a groundbreaking attempt, well ahead of its time, to write an ecological history of the Lake District. This history had its beginnings, first, in "The Ruined Cottage," and then in "Michael," which examined the moral and intellectual values of a pastoral way of life which shaped and were shaped by the mountainous landscapes in which the old shepherd lived. In the *Guide* Wordsworth expanded his focus to provide a richer material account of the historical processes that first brought the physical landscapes of the Lake District into being and why, as a result of changes in the economic and cultural relationship of the people to the land, these landscapes were on the point of disappearing forever.

Dorothy was proud of what Wordsworth had achieved, declaring in an 1809 letter to Catherine Clarkson that the introduction to *Select Views* "is the only regular and I may say *scientific* account of the present and past state and appearance of the country that has yet appeared" (*Letters: Middle Years* 1:372). The historical geographer Ian Whyte confirms this assessment, noting that Wordsworth's "holistic approach to the study of the Lake District landscape deserves to be better known in relation to the development of nineteenth-century geography" as a "masterly synthesis in which he developed generalizations about the Lake District landscape from personal observation and experience" ("William Wordsworth's 'Guide to the Lakes'" 101–2). Whyte singles out the "discussion of the evolution of Lake District woodlands" as "an early classic of writing in biogeography ... not paralleled in other guides" (101–2, 104). The core of the *Guide*, captioned as a "Description of the Scenery of the Lakes," registers Wordsworth's commitment to understanding its landscapes as the rich material product of both natural and human history. In the first of its three sections, "the country as formed by nature" (*Guide* 170), Wordsworth provides a detailed account of the physical geography of the region, discussing its primary natural features and landforms, such as mountains, valleys, lakes, tarns, rivers, surface vegetation, woods, and climate. This section shows a substantial grasp of contemporary geology, a perspective that would be supplemented in the expanded edition produced by Wordsworth's publisher, Hudson & Nicholson of Kendal, which included both a flora drawn up by Thomas Gough of Kendal and letters on the geology of the region by Adam Sedgwick.[10]

In the most innovative section of the *Guide*, dealing with the "aspect of the country as affected by its inhabitants" (*Guide* 194), Wordsworth provides a historical ecology of the ways in which the people living in the Lake District, through their interaction with their physical surroundings, produced a place whose unique beauty was an expression of their cultural and economic values. Strikingly, Wordsworth emphasizes the material dimensions of landscape formation, basing his history on "a description of the ancient and present inhabitants, their occupations, their condition of life, the distribution of landed property among them, and the tenure by which it is holden" (194). In fact, he argues that "the distribution of lands among . . . tenants" is the primary factor determining "the face of the country materially to the present day." He looks to economics to explain why the region exhibits "such a striking superiority, in beauty and interest, over all other parts of the island" (197). Thus, in discussing the unique intermixing of enclosed and common lands in the valleys and on the hillsides adjacent to them, he observes that the word "dale" derives from the early distribution of land in the valleys among the tenantry; thus, "the several portions, where custom has survived, to this day are called *dale*, from the word *deylen*, to distribute" (198). As I have suggested, Wordsworth understood both nature and the mind as being the historical products of the interaction of human beings with the natural world, each depending on the other. To Wordsworth, the Lake District was as much a historical document of a certain kind of society as it was the embodiment of a certain kind of nature.

The *Guide* is very much a colonial environmental history, for it tells the story of how the land was cleared and settled, first by the Celts, fleeing the invading Romans, then by the Saxons, Danes, and Normans, and finally by an "influx of new settlers," the rich consumption-driven English tourists and other lovers of the picturesque who erected "new mansions out of the ruins of the ancient cottages" (*Guide* 224). The latter receive the brunt of Wordsworth's criticism in the *Guide*, for rather than producing landscapes that express a deep material interaction with the place and its history, these new settlers were introducing new natures to it wholesale. In the second section, as a starting point for this history, Wordsworth takes up the metaphor that he used at the beginning of the book by positioning the reader imaginatively on a cloud somewhere between Great Gavel and Scafell looking down on "the main outlines of the country" (171). This time, however, we are asked to imagine what the Lake District would have looked like "before the country had been penetrated by any inhabitants" (194). In a powerful enactment of historical imagining, Wordsworth asks us to see a nature that no longer exists, the Lake District as a *wilderness*: "He may see or hear in fancy the winds sweeping over the lakes, or piping with a loud voice among the mountain peaks; and, lastly, may think of the primeval woods shedding and renewing their leaves with no human eye to notice, or human heart to regret or welcome the change" (194).

Here Wordsworth was drawing upon Thomas West's *The Antiquities of Furness* (London, 1774), but just as importantly, as his frequent mention of

American travelers in the *Guide* indicates, he also had his eye on the ecological changes that had followed in the wake of the colonization of America. Colonial history, in other words, provided the template for Wordsworth's reconstruction of the past. Writes Wordsworth, "Such was the state and appearance of this region when the aboriginal colonists of the Celtic tribes were first driven or drawn towards it, and became joint tenants with the wolf, the boar, the wild bull, the red deer, and the leigh, a gigantic species of deer which has been long extinct" (194). Here is one of the few places in the book where wild animals are even mentioned. Aside from a passing reference to waterfowl, the *Guide* speaks of a nature in which most of the larger mammals and birds are absent. One does not need to look far, however, for an explanation for this absence: when Wordsworth wrote the *Guide*, all the larger animals that he lists, once the "joint tenants" of the region, had been destroyed. Wordsworth mentions that a desolate part of Ullswater Vale was "formerly haunted by eagles" that nested in "the precipice which forms its western barrier" (167), but this was no longer the case. While the Legh (the Irish Elk or "segh" [*Megaloceros giganteus*]) and the wild bull or Aurochs (*Bos primigenius*) had long ceased to exist, others had been exterminated more recently, for instance, wild boar at the end of the sixteenth century, and gray wolf by the turn of the seventeenth century. The disappearance of woodlands also put an end to deer, so by the end of the eighteenth century, all three types of deer which had traditionally inhabited England—fallow deer (itself an early introduction), roe deer, and red deer or stag—were now extinct in the wild and were to be found only in deer parks (Yalden, *History of British Mammals* 171–73).

In its recognition that human colonization of the Lake District had led to substantial local extinctions, the *Guide* can usefully be compared to the work of the Scottish naturalist John Fleming, whose "Remarks Illustrative of the Influence of Society on the Distribution of British Animals" (1824) represents the first major attempt to write a history of the impact of human beings on the geographical distribution of animals. Although the idea of the extinction of species was generally accepted at the time at which the essay appeared, geologists were primarily interested in those disappearances that had taken place in the distant past, which were then thought to have been caused by a universal deluge. As an early promoter of uniformitarianism, Fleming was unique in using the facts surrounding recent animal extinctions to draw conclusions about the laws governing the extinction of animals in the deep past. He stands apart from other naturalists in believing that the majority of animal extinctions, since human beings first made their appearance on the earth, were attributable either directly or indirectly to human agency. With some element of understatement, given the story that he is telling, Fleming writes that "man must have altered greatly the geographical range of many species, and may even have succeeded in effecting the total destruction of a few" ("Remarks" 291). Focusing on the recent history of British animals, he writes that they can generally be divided into three classes: (1) "those species, the individuals of which are daily becoming scarcer, in consequence of the agency of man"; (2) "those which man has succeeded in extirpating, but which still find an asylum in the more thinly peopled or less cultivated districts of Europe"; and (3) those that were once natives of Britain but "have ceased to exist in a living state on the earth" (302).

Following the lead of the French naturalist Georges-Louis Leclerc, Comte de Buffon, Fleming understands civilization as an ongoing state of warfare on animals. The graminivorous animals are hunted for food, skins, and fat, while animal predators are exterminated in self-protection, to preserve domesticated animals, or because they interfere with our objects of pursuit. In Fleming's view, the ever-increasing intensity and scope of this warfare are inescapably bound up with the ever-increasing wants of civilized human beings: "The war which man thus carries on against the lower animals, is influenced, in a remarkable degree, by the progress of society. The wants of man increase in kind and variety with his advances in civilisation and his means of supplying them become proportionally numerous. The war carried on by rude tribes is limited in its objects, and uncertain in its results. But with the progress of experience and improvement, the objects of the chace cease to be limited, while the methods of capture, and engines of death, become more numerous, complicated, and effectual" ("Remarks" 290). Fleming provides a chilling account of the number of British mammals that in historical times have been exterminated (including wolves, bear, wild boar, oxen, wild horses or tarpan, antelope, and beaver), along with those that were on the verge of disappearing (otters, martins, polecats, wildcats, and badgers). British birds suffered similar pressures. Fleming notes that avian predators such as eagles, ravens, and buzzards "have disappeared from the more cultivated districts" and that the draining of marshlands for agriculture severely affected water birds, such as mallards, snipes, redshanks, and bitterns. Egrets, once common in the fifteenth century, and cranes, abundant until the seventeenth century, could no longer be found in

Britain, while the Scottish capercailzie or wood grouse had disappeared during the previous fifty years. As Fleming looked into the deep geological past, he was also aware that Britain had once been the home of Irish elk, mammoths, tigers, rhinoceroses, hippopotamuses, horses, oxen, cave bears, and hyenas. In an argument that would have an enormous influence on Charles Lyell, he contended that it was unnecessary to claim, as did William Buckland in *Reliquiae Diluvianae* (1823), that the destruction of these animals proved that a universal deluge had occurred. Instead, Fleming found a more plausible explanation for this catastrophe lying closer to hand in mankind's long-standing "destructive warfare against many animals" (288).

Where Fleming saw a dark future in which human civilization would continue to extend the range and depth of its warfare on animals, Wordsworth hoped for "the coming of a milder day" ("Hart-Leap Well," *Poetical Works* 2:249–54, line 175) when human beings, through their recognition of the importance of the natural world to their own well-being, might learn to love the natures that they were destroying. In the "spots of time" of *The Prelude*, for instance, human violence toward nature is the norm. "I was a plunderer then" (1805 *Prelude*, l:336), Wordsworth writes in the egg-stealing episode, while in others, such as the woodcock and boat-stealing episodes and in the mimicking of the owls in book 5, he seeks to show how a self-centered, anthropogenic ethos can give way to a recognition of being part of a larger whole in those moments when our confidence in our capacity to control nature is shaken. In "Nutting," the violent destruction of a hazelnut bower leads to a recognition of the need for "gentleness of heart" in our dealings with nature because "there is a Spirit in the woods" (*Poetical Works* 2:212, lines 55–56). Wordsworthian natural education seems to require this discipline. Elegy emerges from Wordsworth's recognition that our love of nature seems inextricably bound up with a history of violence. The woods that one loves are filled with ghosts. Wordsworth's love of nature and his concept of "natural education" are certainly bound up with nostalgia, but this love and learning also hold on tightly to the ghosts of natures past because they allow us to recognize nature's fragility. Natural history, in other words, should teach us to care for things that are easily lost; it should give us hearts that have learned from the violence of the past, and that is what Wordsworth's narratives seek to do. As Geoffrey Hartman has suggested, they recount "the emergence of the gentle mind out of the haunted mind" (*Unremarkable Wordsworth* 57), a mind, one would argue, that is humanized through its hauntings.

"Hart-Leap Well" is a poem of this kind. It tells the story of an ancient hunt in which a five-year-old stag, known in the past as a "hart," was hounded all day long until it could run no more. On the spot where the animal died, the hunter, a knight named Sir Walter, built a pleasure palace and an arbor to commemorate this glorious chase, placing a stone drinking basin where he found the dead animal and three pillars to mark where its "hoofs the turf... grazed" (*Poetical Works* 2:249–54, line 68) as it leaped down a steep hill trying to escape. When Wordsworth came upon the place, on 17 December 1799, during a walking journey with Dorothy from Sockburn-on-Tees to Grasmere, nothing remained of Sir Walter's mansion, which he boasted would endure until "the foundations of the mountains fail" (line 73), but the stones were still there, and Wordsworth could still discern in the landscape the traces of the arbor: "Here in old time the hand of man hath been" (line 112).[11] Inquiring of a local shepherd the story behind these ruins, he learned of the chase (which is recounted in part 1 of the poem) and that the locals believed that "the spot is cursed" (line 124). Wordsworth drew a somewhat different conclusion, however, arguing that nature leaves these ruins "to a slow decay" so that "what we are, and have been, may be known" (lines 173–74). The headnote to the poem, which indicates that it is based on "monuments [that] do now exist as I have there described them" "near the side of the road that leads from Richmond to Askrigg," aligns "Hart-Leap Well" with the literature of ruins. The poem can be understood as a complex and ambivalent engagement with feudal culture, one that displays a middle-class critique of the pride, exuberance, and waste, as Thomas Pfau observes, of medieval aristocratic culture: "The Knight's aristocratic caprice ('restless as a veering wind,' l. 17) appears self-consuming here. He seems to ravage his own resources, ruining the dogs (the 'noblest of their kind') and soon poisoning the water and land" (*Wordsworth's Profession* 222). The poem provides a simple, though powerful, ecological "lesson... Never to blend our pleasure and our pride / With sorrow of the meanest thing that feels" (*Poetical Works* 2:249–54, lines 177–80).

When Wordsworth came upon the three stone pillars that marked the "Three several hoof-marks which the hunted Beast / Had left imprinted on the grassy ground" (*Poetical Works* 2:249–54, lines 50–51), he was in many ways seeing again Sir Walter's discovery of a "sight... never seen by human eye" (line 54), but whereas the knight saw the hoofprints of an individual deer, Wordsworth was seeing the actual historical traces of one of the last places where a now extinct species had once walked the earth. Indeed, in his book *Romantic Richmondshire* (1897), the local historian Harry Speight, speaking of Hart-Leap Well, comments that it was "on these moors" that "the last wild red deer in Richmondshire were shot" (376) in the 1750s, the last of these being killed by Mr. John Hutton of Marske "a little over a century ago" (372). When, in 1843, Wordsworth commented on the poem to Isabella Fenwick, extinction seems to have been on his mind, for he noted that "both the stones and the well are objects that may easily be missed; the tradition by this time may be extinct in the neighbourhood; the man who related it to us was very old" (*Poetical Works* 2:514n). The poem can thus be said to reflect on the disappearance not only of a feudal aristocratic hunting culture but also of the red deer or stag and the stories associated with it. In preserving in narrative the ecological history of a place that was associated with "the memory of" "a remarkable Chase," Wordsworth is recovering in the landscape the historical presence of both the hunters and the hunted (*Poetical Works* 2:249). In this sense, "Hart-Leap Well" is being associated with other places in England, such as Hunston, Huntercombe, Hunterley, and Huntington, named as hunting locales, and also places such as Hindlip ("deer leap"), Lypiatt, Lypiate, and Lippits Hill, linked to deer enclosures.[12] Fleming did something similar when he used ancient place-names as evidence of the previous geographical distribution of exterminated animals across Britain. He notes, for instance, that there exist "several places, the appellations of which have the prefix Brock or Badger, attesting in the Gothic name the prevalence of these animals, and the Scandinavian power by which it was imposed" ("Remarks" 293–94). In Scotland, he comments that "the parish and family of Swinton owe their name to this animal [the wild boar], the former celebrated for harbouring, the latter for destroying them" (294). For Wordsworth, who wrote a series of poems on "the naming of places," the existence of so many places in Britain named after extinct animals must have provided sobering evidence of "what we are, and have been," places such as Hertford (and the county Hertfordshire), Harthill, Hartwell, Hattersley, Heatherslaw, Hindley, Hindhead, Stagenhoe, and, of course, Hart-Leap Well (named after red deer); Buckhill, Buckfast, Roecombe, Rogate, Reigate, and Rayleigh (for roe deer); Lostford (for lynx); Beverley (for beaver); Barbon and Barford (for bear); as well as over two hundred places named after wolves, with others denoting the previous locales of foxes, otters, and martins.[13]

"Hart-Leap Well" can be said, then, to reflect broadly on the extinction of England's red deer and on the importance of marking their place in its natural history. Although the poem is about the killing of a single deer, rather than of the species, it would have been difficult for Wordsworth to have come upon such a monument without also thinking about the recent disappearance in the wild of what was Britain's largest animal. David Perkins has argued that the poem expresses Wordsworth's contradictory attitude toward hunting and hunters: "In 'Hart-Leap Well' . . . the chase, its practices, and the emotions it evokes in hunters are exhibited and criticized. More broadly, the poem pleads for sympathy with animals. And animals themselves are a figure in the poem for nature in general—as is still the case in discourses today—so that the widest issue the poem raises is that of human relations to the natural world, whether mankind is necessarily egoistically closed and predatory" (*Romanticism and Animal Rights* 80). Perkins's attractive reading of the poem ignores, however, an important fact about the historical relationship between deer and the aristocracy which is central to the poem. Where there are no forests or woods, there are no deer, so their disappearance in England had less to do with hunting than with the extensive deforestation that accompanied the rise of farming. In fact, from the Norman period onward, the continuance of deer populations was integrally bound up with the establishment of Crown forests reserved for the hunting of game and with the harsh laws that were enacted against poaching, the most notable of these being the Black Act of 1723. In a context in which access to forests and game was structured by class, the conservation of deer and the woodlands that supported them was inextricably bound up with aristocratic culture, and for the working class the continued existence of the red deer—a nature reserved for aristocrats—was often seen as an expression of the "Norman yoke." It was only when the aristocracy began to lose interest in hunting and many forests were sold off, enclosed, or replanted with oak for timber that deer populations diminished.[14] Ralph Beilby, who provided the prose for Thomas Bewick's *A General History of Quadrupeds*, lamented neither their loss nor the disappearance of knights:

> The hunting of the Stag has been held in all ages as a diversion of the noblest kind; and former times bear witness of the great exploits performed on these occasions. In our island, large tracts of land were set apart for this purpose; villages and sacred edifices were wantonly thrown down, and converted into one wide waste, that the tyrant of the day might have room to pursue his favourite diversion. In the time of William Rufus and Henry the First, it was less criminal to destroy one of the human species than a beast of chase.

Happily for us, these wide-extended scenes of desolation and oppression have been gradually contracted; useful arts, agriculture, and commerce, have extensively spread themselves over the naked land; and these superior beasts of the chase have given way to other animals more useful to the community. In the present cultivated state of this country, therefore, the Stag is almost unknown in its wild state. (109)

"Hart-Leap Well" points up the barbarity of hunting but also recognizes the degree to which the continuance of red deer and the wilderness areas that preserved them was bound up with aristocratic culture.

By the end of the eighteenth century, as new breeds of sheep, such as the Blackface and Cheviot, were introduced into the Scottish Highlands, the clearances impacted not only traditional croft farmers but also red deer. This situation changed, however, when Queen Victoria's purchase in 1852 of the Balmoral Estate gave royal imprimatur to deer hunting. Red deer were reintroduced to the Scottish Highlands, with the result that rural Scots were forcibly removed from their traditional lands not only to support sheep raising but also for the establishment of game preserves and deer parks. "Are sheep walks, deer forests, hunting parks, and game preserves, so beneficial to the nation that the Highlands must be converted into a hunting desert, and the aborigines banished and murdered?" asked Donald McLeod, in a letter to the editor of the *Northern Ensign* in Caithness. "I know that thousands will answer in the negative; yet they will fold their arms in criminal apathy until the extirpation and destruction of my race shall be completed" (*McLeod's Gloomy Memories* 129). Thus, the fact that it is a shepherd who recounts the tale of the murder of the hart and the cursed natural environment that it has produced complicates our understanding of the poem. In what is otherwise a somewhat cryptic passage in "Hart-Leap Well," Wordsworth seems to point to the consequences of deforestation when he writes that the three "lifeless stumps of aspen wood" that mark the place where the arbor once stood appear not to be recognized *as aspens* by the people living in the neighborhood: "some say that they are beeches, others elms" (*Poetical Works* 2:249–54, lines 125–26). In a society where forests and woodland really mattered, such confusion would not have been possible.

Deforestation is a central theme of the *Guide*. Wordsworth notes that when the "first settlers" (*Guide* 189) or "aboriginal colonists" (194)—the Celts—arrived, during the Roman occupation, the entire region was forested, with a deciduous forest canopy of oak, ash, birch, and the occasional wych elm and an understory of white and black thorn, hollies, and yews. Alders and willows, he notes, were also found in moist areas. Scots pine (*Pinus sylvestris*), which he calls the "native Scotch firs" (188–89), were also abundant, but he adds that "not one of these old inhabitants has existed, perhaps for some hundreds of years" (189). Little of this ancient forest remained when Wordsworth wrote the *Guide*. With the exception of the extensive wildwoods of Lowther, the only traces of "the universal sylvan appearance the country formerly had" (189) were to be found in the woodlands protected by enclosures and in the occasional solitary forest trees that lay scattered across the mountains. Timber trees were rare, the best of these being found near the lakes, and Wordsworth comments that "unless greater care be taken, there will, in a short time, scarcely be left an ancient oak that would repay the cost of felling" (189). Wordsworth attributes the vast deforestation of the Lake District to human agency, but he does not consider it in negative terms. The *Guide* does not make an ecological argument for the preservation of wildwood. Instead, Wordsworth uses aesthetics to argue for the preservation of a unique cultural landscape composed of woodlands, lakes, and pasture, in which plants are valued for their contribution of form, color, and texture to a landscape; they cover the land like paint on a canvas and constantly change with the seasons. "I do not indeed know any tract of country in which, within so narrow a compass, may be found an equal variety in the influences of light and shadow upon the sublime or beautiful features of landscape," he writes (174). The *Guide* is probably the first instance in the history of literature where literary capital and aesthetics have been marshaled in order to mount an argument for the conservation of a unique ecological landscape. Training his readers to appreciate the material, historical, and aesthetic dimensions of this landscape was essential to this task, so Wordsworth devotes much of the *Guide* to bringing his readers "to habits of more exact and considerate observation than, as far as the writer knows, have hitherto been applied to local scenery" (171). As he writes at the end of *The Prelude*, "what we have loved / Others will love; and we may teach them how" (1805 *Prelude* 13:445–46).

In the *Guide*, Wordsworth seeks to show how the aesthetically rich scenery of the region was brought into being by the dynamic material interaction of the inhabitants with their physical surroundings. In explaining, for instance, why hollies continued to thrive in the upland pastures, he observes that the primary responsibility of the upland villains was to preserve the flocks and herds, so they cleared all the upland thickets to protect the domestic animals from the wolves while carefully preserving the holly trees as winter feed. The pastoral economy thus dictated which species of plant and animal were preserved and which ones—in this case, white and black thorns and wolves—were destroyed. The practice of dividing lands into small irregular parcels also contributed to the diversity of the landscape. Even though the ancient forest disappeared, a sylvan appearance was reestablished when the pastoral farmers planted "fences of alders, willows, and other trees" (*Guide* 199). Ash trees were also used as early winter cattle feed. Near the end of the seventeenth century, with the reestablishment of iron forges in the region, it was economically profitable for enclosed land on the steeper and more stony areas to be converted into close woods. As "shepherds and agriculturalists" independently worked on the small parcels of land that "they occupied and cultivated" (201), a flat, homogeneous wilderness landscape was transformed, in Wordsworth's view, into one composed of a rich intermixture of plants and plant forms, the landscape being itself the fullest expression of this agricultural and pastoral economy. As Wordsworth observes, this is "the cause by which tufts of wood, pasturage, meadow, and arable land, with its various produce, are intricately intermingled in the same field; and . . . how enclosures entirely of wood, and those of cultivated ground, are blended all over the country under a law of similar wildness" (201). Elsewhere, he describes the intricate balance of wood and fields created when "the first settlers having followed naturally the veins of richer, dryer, or less stony soil . . . shaped out an intermixture of wood and lawn, with a grace and wildness which it would have been impossible for the hand of studied art to produce" (189).

That Wordsworth would in both cases use the word "wildness" to describe a landscape that he knew had been radically transformed by the agency of man indicates the degree to which his ecology sought to preserve a designed landscape whose beauty lay in its integration of natural elements within a new natural aesthetic order. In writing the history of the role that human beings had played in remaking the landscape of the Lake District, showing in "historic detail . . . the manner in which the hand of man has acted upon the surface of the inner regions of this mountainous country," Wordsworth saw this region as a special place where that agency was "incorporated with and subservient to the powers and processes of nature" (*Guide* 201). Whereas Fleming saw no end in the escalation of species destruction caused by the human colonization of nature, and whereas George Perkins Marsh, in what is often read as the foundation of historical ecology, *Man and Nature* (1864), would write that "man is everywhere a disturbing agent. Wherever he plants his foot, the harmonies of nature are turned to discords" (36), Wordsworth believed that, for approximately four centuries in a certain part of the world, "a community of shepherds and agriculturalists, proprietors, for the most part, of the lands which they occupied and cultivated" (*Guide* 201) had become possessed by a nature that they had half created and that their moral character and commitment to liberty and independence were derived from their interaction with this place.

Of course, when Wordsworth wrote the *Guide*, this society, this "perfect Republic of Shepherds and Agriculturalists" (*Guide* 206), was gone, and after more than five hundred years of isolation from the broader commercial and political interests of England, the region was undergoing a new phase in its history. Wordsworth blamed the ensuing changes on economic factors and on the influx of tourists attracted to the Lake District by the rise of "Ornamental Gardening" and its "relish for select parts of natural scenery" (207). Many who came to see its landscapes decided to stay ("visitors . . . from all parts of England . . . became settlers" [208]), so the rise of the picturesque could be seen as having instigated a new wave of invaders who colonized the area with their own ideas about what nature was or ought to be. At the same time, the traditional pastoral farmers were facing hard times, so it made properties that had been passed from one generation to another for centuries vulnerable to being acquired by wealthy new landowners (called "off-comers" by the locals), who came to the region not to farm but to enjoy its picturesque beauty and who set about "improving" their properties by modeling them on Italian landscapes.[15] In her *Grasmere Journal*, on 19 May 1800, Dorothy notes a conversation that she had with her neighbor John Fisher, who commented that "in a short time there would be only two ranks of people, the very rich and the very poor, for those who have small estates says he are forced to sell, and all the land goes into one hand" (*Journals* 19).

The third section of the *Guide* describes this recent resettlement as both a cultural and ecological invasion, of people and plants, which was reshaping the physical landscapes of the Lake District. Wordsworth devotes much of his attention to the destruction caused by the axe, mentioning "a rich wood of birches and oaks" that once graced a piece of enclosed land on Ullswater Lake and sadly observing that "those beautiful woods are gone, which perfected its seclusion." "Scenes," he writes, "that might formerly have been compared to an inexhaustible volume, are now spread before the eye in a single sheet" (*Guide* 168). Writing to Sir George Beaumont in 10 November 1806, he remarks that "among the Barbarisers of our beautiful Lake-region, of those who bring and those who take away, there are few whom I have execrated more than an extirpator of this beautiful shrub, or rather tree the Holly" (*Letters: Middle Years* 1:93). More deplorable, however, than the disappearances were the introductions. "Much as these scenes have been injured by what has been taken from them, buildings, trees, and woods, either through negligence, necessity, avarice, or caprice," Wordsworth writes, "it is not the removals, but the harsh *additions* that have been made, which are the worst grievance—a standing and unavoidable annoyance" (*Guide* 222). He laments the rise of commercial forestry and the wholesale importation of new trees and shrubs "within these last fifty years, such as beeches, larches, limes, &c. and plantations of firs, seldom with advantage, and often with great injury to the appearance of the country" (189). In a drive for profit, landowners were planting conifers, which disrupted the intermixed patterning of farmland, upland pastures, and enclosed woods. "Small patches and large tracts of larch-plantations [and pine] . . . are over-running the hillsides," Wordsworth writes (219), arguing that an area primarily valued for its beauty was thus being turned into a "vegetable manufactory" (217).

As a gardener, Wordsworth was fully cognizant of the ego-centered pleasure that comes from purchasing and cultivating exotic plants, which "by their very looks, remind us that they owe their existence to our hands, and their prosperity to our care" (*Guide* 218). What concerned him was that these "alien improver[s]" (209), as he called them, in their quest to re-create Arcadian landscapes, were extending these ornamental plantings beyond their doorsteps and changing the face of the entire country. John Christian Curwen, who owned the villa of Belle Isle, built in 1774 on an island in Windermere, with its circular house flanked by "platoons of firs" (209), enclosed and planted five hundred acres of Claife Heights in 1798 with thirty thousand larches at a cost of £2 15s. per acre.[16] Wordsworth was appalled by these activities: "What shall we say, to whole acres of artificial shrubbery and exotic trees among rocks and dashing torrents, with their own wild wood in sight—where we have the whole contents of the nurseryman's catalogue jumbled together—colour at war with colour, and form with form?—among the most peaceful subjects of Nature's kingdom, everywhere discord, distraction, and bewilderment!" (219). Wordsworth's recourse to the same language that he had previously used in *The Prelude* to describe the heterogeneity and disorder of London streets should not surprise us, for he recognized that the globalized transport and mixing of natures—"birds and beasts/Of every nature, and strange plants convened from every clime" (7:230–32)—were becoming just as much an aspect of the Lake District as they were of metropolitan life. In *The English Landscape in the Twentieth Century*, Trevor Rowley shares this perspective, noting that, because local building materials and craft traditions no longer play a major role in reshaping our environments, a contemporary landscape "owes little or nothing to what went before. In many respects, the landscape is no longer directly linked to the land on which it lies" (1); "Much of our contemporary landscape is rootless—as a society we have not understood this, let alone begun to come to terms with it" (9). In the latter part of the *Guide*, Wordsworth seeks to articulate model landscape practices that would retain the historical relationship between the landscapes of the Lake District and the land. In arguing that houses should appear "to have grown than to have been erected," as if "risen, by an instinct of their own, out of the native rock" (202), Wordsworth was not just naturalizing culture, or insisting on an organicist aesthetic, but also arguing for landscapes that displayed the shaping powers of human agency yet were nevertheless grounded in material environments that preserved their historical and local roots.

Wordsworth's extensive employment of the language of the picturesque in the *Guide* is thus structured by contradiction: even as he recognized the degree to which the rise of the picturesque and the tourist economy that it supported had supplanted the pastoral farming economy that had produced the unique landscapes of the Lake District, he also recognized that it was a powerful discourse that might be channeled to support his preservationist concerns.[17] As Scott Hess remarks,

> The *Guide* presents the area in overwhelmingly aestheticized terms. It gives detailed and authoritative prescriptions on how best to appreciate everything from the effects of light and shade in the valleys to the forms, geological composition, and coloring of mountains; the changing effects of vegetation in various seasons; the shapes of lakes and their reflective qualities; the visual qualities of the native birds; the forms of mountain tarns and streams; the composition of woods and the forms and coloring of the various trees within them; the aesthetic effects of climate, such as rain, mist, clouds, and varying sky colors; the impact of human habitation and agriculture on the patterns of fields and woods; and the aesthetics of roads, cottages, churches, and mountain bridges. (*William Wordsworth* 87–88)

By valuing the Lake District exclusively in aesthetic terms, however, the poet was taking up a very different relationship to the land than that of the "statesmen" who originally produced its landscapes. In the *Guide* the material agency of labor and livestock are replaced by the pleasure of seeing and describing landscapes. With little in the way of manufacturing, mines, or quarries and with poor soil, Wordsworth writes, "the staple of the country is its beauty and its character of retirement" (*Prose Works* 3:347). By promoting the aesthetic dimensions of the region, Wordsworth no doubt hoped that the British nation might come to its rescue. Famously, he suggested that the Lake District be seen as "a sort of national property, in which every man has a right and interest who has an eye to perceive and a heart to enjoy" (225). As Hess further observes, Wordsworth symbolically transformed the Lake District into "a kind of open-air museum . . . a museum of nature, an autonomous sphere set apart from the ordinary commercial and social world for rituals of imagination" (*William Wordsworth* 173).

Silence reigns throughout the *Guide*, because Wordsworth sought to freeze his landscapes in time and because they are unpeopled. In *The Dark Side of the Landscape*, John Barrell notes the manner in which the laborers who created the landscapes painted by Constable and Turner progressively disappear into the background of their paintings. In the *Guide*, Wordsworth read the land as telling the powerful story of how the pastoral farmers who had created its unique landscapes and brought a certain kind of beauty and appreciation of beauty into the world were forced from their land. Like Bartram, he was providing images of what a unique historical nature looked and felt like at the moment in which it was disappearing from the earth. Both men hoped that their writings might slow the destruction of these places, while providing their readers with an understanding of the nature that was being lost in the displacement of the people whose loves and lives created it.

Wordsworth's writing also marks a fundamental cultural shift in the temporal understanding of the natural world, for in his work nature is fundamentally associated with the past and exists outside and in opposition to modern time. Wordsworthian nature is deeply linked with memory, nostalgia, childhood, past modes of being, and loss. *Nature present* is understood as being already a part of *nature past*. In *Time and the Other*, Johannes Fabian discusses the scientific "denial of coevalness," by which he means "a persistent and systematic tendency" in anthropology "to place the referent(s) of anthropology in a Time other than the present of the producer of anthropological discourse" (31). In Wordsworth's writing, nature also undergoes a similar process of anachronization, as the time that nature inhabits is understood as being different from the time that Wordsworth occupies as a narrator. It is *in* the same time, but not *of* it. This tendency to separate nature from modern time would shape much of the nineteenth- and twentieth-century understanding of the natural world. In "To the Cuckoo," nature does cling to the present, but it primarily speaks of the past. That is why the bird's voice expresses a doubled temporality: "*I have heard / I hear thee* and rejoice" (*Poetical Works* 2:207–8, lines 1–2; my emphasis). Throughout the poem, the cuckoo's "wandering Voice" (line 4) powerfully links past and present. Whatever changes the speaker has undergone, he rejoices in a song that returns him to a listening that took place in the past: "the same whom in my School-boy days / I can listen to thee yet . . . till I do beget / That golden time again" (lines 17–18, 25, 27–28). Unfortunately, the tense that is lacking in Wordsworth's poetry of nature, the tense that largely disappears during the nineteenth century, is that of "nature future."

CHAPTER SEVEN

John Clare and the Ghosts of Natures Past

In 1841 John Clare came to the stark realization that, although he had been a resident of England all his life, he was now living in a state of exile. Being a patient for almost four years at Dr. Matthew Allen's High Beach Asylum in Epping Forest precipitated this recognition, but the feeling was not new. Deep down, Clare had always known that enclosure had turned his claim that he belonged to a place into a poetic fiction. He no more owned the fields in which he worked than mechanical operatives owned the manufactories in which they labored or the crowded streets on which they walked. Clare escaped from the asylum in July and returned to his cottage in Northborough, only to feel "homeless at home." "My home is no home to me," he would declare in a letter written to his childhood love, Mary Joyce, who was dead at the time (27 July 1841, *Letters* 649). In December of that year he was committed to Northampton Asylum, and thus his feelings of exile would be fully and permanently realized as he became the equivalent of "a memory lost" ("I Am," *Later Poems* 1:396–99, line 2).[1] In *Child Harold*, the ambitious unfinished poem written during this tumultuous year, Clare directly confronted what it means to lose one's place in the world, to be exiled from a place not because you have left it, but because its nature has left you. Clare's sense of exile thus inherently speaks of the discovery that even nature, that most rooted of things, has no more claim to place in the modern world than a poor agricultural laborer. Both can suffer displacement and exile.

If Clare had lived in the twentieth century, he would have been able to look to other poets to help understand this condition, writers such as the Palestinian poet Mahmoud Darwish, whose poetry engages with the Israeli occupation of Palestine, which forced him to flee to Lebanon. When Darwish returned to Palestine, his place was gone: his village, al-Barweh, had been leveled, and he could no longer claim citizenship. "I had been a refugee in Lebanon, and now I was a refugee in my own country," he says in a 1969 interview (Darwish, al-Qasim, and Adonis, *Victims of a Map* 11). Darwish's experience of exile is different from Clare's, and one does not want to collapse historically different circumstances, yet both poets, in unique ways, speak about what it means to lose one's sense of geographical belonging. Theirs is the exile that comes from discovering that the place that matters most to you has been taken away and permanently erased from the earth and that you now stand on alien ground. "Is it true, good ladies and gentlemen, that the earth of Man is for all human beings / as you say?" Darwish asks; "In that case, where is my little cottage, and where am I?" ("I Talk Too Much," in *Unfortunately, It Was Paradise* 13). These words might easily have been penned by Clare, who was reduced to being just an abstract "I am," not a *dasein* (a "being there"), and to seeing his life as "a dream that never wakes" (*Child Harold, Later Poems* 1:40–88, line 255). Life had become a road that got longer with every step: "Night finds me on this lengthening road alone" (line 256); "In this cold world without a home / Disconsolate I go" (lines 934–35). In 1841 Clare was suffering from delusions, so it has been easy for critics to treat his sense of alienation and displacement as symptoms of psychological distress. Married in his mind to two women—to Patty Turner, with whom he had seven children, and to Mary Joyce—Clare was also seeing himself, at various times, as Shakespeare, Lord Nelson, and the boxer Jack Randall. In adopting the persona of Byron, by writing a supplementary canto to *Childe Harold's Pilgrimage* and his own version of Byron's *Don Juan*, however, Clare was not just engaging in mad mimicry. Instead, he was attempting to use the writings of the great English poet of exile in order to understand the unique form of exile that he was living, one in which neither people of the working class nor creatures in nature could claim their own place.

No Romantic poet wrote more passionately about the joy of experiencing nature in all its immediacy than John Clare, and no poet argued more strongly for its permanence and continuity across generations as part of the original design of Creation. Yet few poets have conveyed in more poignant terms what it means to lose one's nature for good. Clare grew up knowing only one nature: the mixture of arable land, woodland, limestone heath, meadows, and fen that made up his native rural parish of Helpston. By the age of fifteen, he writes, "I had never been above eight miles from home in my life" (*Prose* 20). In an age in which Britons increasingly associated themselves with mobility, with a world of moving people and things, Clare's commitment to the stationary and to thinking about nature in traditional terms, as the place in which he lived and belonged by birth, sets him apart from most of his contemporaries. Gilbert White also felt that he was living in a stable unchanging nature, but he died in the same year that Clare was born, so he was not forced, as Clare was, to deal with the massive changes in rural life introduced by agricultural improvement and the enclosure and privatization of land which made it possible. Margaret Grainger is certainly right in claiming that Clare's "rootedness . . . is one of his greatest strengths" (Clare, *Natural History Prose Writings* l). As Simone Weil argued, "To be rooted is perhaps the most important and least recognized need of the human soul" (*Need for Roots* 41). At the same time, it left him vulnerable to being permanently displaced when he lost his roots and was unable to put down others elsewhere.

Clare and the Mobilization of Natures

For modern readers, then and now, it is difficult to understand Clare's inability to adapt to a world of circulation, migrancy, and exchange. In a famous passage from his autobiography, Clare recounts how, as a young child, he journeyed across Emmonsales Heath beyond the very limits of his knowing: "I got out of my knowledge, when the very wild flowers seemd to forget me & I imagind they were the inhabitants of new countrys the very sun seemd to be a new one & shining in a different quarter of the sky. . . . I was finding new wonders every minute & was walking in a new world & expecting the world's end bye & bye but it never came." This anecdote insists on the intensely local aspects of Clare's nature and his simultaneous recognition of its limits. "When I got back into my own fieldds," he writes, "I did not know them everything lookd so different" (*Prose* 13). In this passage, Clare sees journeying as the cause of his loss of a sense of familiarity with his world, yet he did not replace this defamiliarized world with another. Travel did not lead Clare, as it did so many of his contemporaries, to desire to know more about other places or to compare his natural locality with others. Instead, it produced disorientation, estrangement, dislocation, and, ultimately, a homelessness that was of disabling intensity even as it enabled him to grasp critically the dark side of modernity.

The landscape of Helpston may have "'made up' his 'being,'" as Clare declared, but his poetry responds to a very different situation: his discovery that his identity was rooted on moving ground (Tibble and Tibble, *John Clare* 1). Growing up, he certainly believed that the nature that he knew as a child would never change, and throughout his poetry Clare identifies his childhood relationship with nature with permanence and joy. Yet the 1809 act enclosing Helpston proved him wrong. For Clare nature was composed of a mixture of forest, "waste" lands, and commons that were sustained by traditional open-field agriculture. As an adult, he watched this nature slowly disappear into the past, as it was enclosed, carved up, fenced in, and reduced to "little parcels little minds to please" ("The Mores," *Middle Period* 2:347–50, line 49). Clare's life as a writer was inextricably bound up with this loss, for his first poem, "Helpstone," was begun in the same year that the parish was enclosed. Clare is thus best understood as a rural laboring-class poet who consciously struggled to maintain the idea of the local nature he knew as a child in the face of the changes that were taking place around him. His poetry documents the social and ecological cost of modernity by addressing the dislocation caused by the destruction of an English rural nature that had been deeply bound up with the long-standing traditions of English rural life.

Clare found his voice in the poetry of social and ecological protest, in a struggle for lost ground, and his poetry displays his changing understanding of what that ground was. In the early poetry there is the glimmer of hope that his writing might win over others to value and love what he did. By the 1830s, however, Clare knew that he had lost this nature for good. Consequently, the voice that emerges in his best poetry does not so much struggle for lost ground as stand on it. In this extraordinary period, Clare looked to his poetry to provide a nature that now appeared to him in ruins. He writes about what it means to live and to write poetry in the face of the loss of the traditional nature that had sustained English rural life. Clare's poetry provides a glimpse of the ecological and social impact that this change had on the lives of English rural laborers, but we need not limit the context of our understanding of Clare to England. English rural laborers were not the only people who lost an ancestral or traditional nature during this period. Read in light of what was happening elsewhere, Clare's poetry can also give us a better understanding of what it might have meant to native and indigenous peoples in other parts of the world who were also grappling with the catastrophic loss of their own local natures and the ways of life which they sustained.

John Barrell's seminal 1972 study *The Idea of Landscape and the Sense of Place, 1730–1840* provided the first major theorization of the importance of "place" and "locality" in Clare's poetry. For Barrell, Clare's intensely local commitment to Helpston was not simply an expression of a personal attachment to his birthplace, but instead reflected a certain way of thinking about place which Barrell sees as characteristic of the English rural laboring class. At a time when knowledge gained from being inside a place was being replaced by newer forms of knowledge that operated across distances, making comparisons across space, Clare's ways of knowing, based on the appreciation of the particularity of localities, set him apart from the landowning and professional classes, who were less interested in preserving traditional local knowledges, cultures, and vernaculars than in integrating places into larger networks of trade and communication. Barrell writes, "Mobility was an essential condition of the attitude we have been examining: it meant that the aristocracy and gentry were not, unlike the majority of the rural population, irrevocably involved, so to speak, bound up in, any particular locality which they had no time, no money, and no reason ever to leave. It meant also that they had experience of more landscapes than one, in more geographical regions than one; and even if they did not travel much, they were accustomed, by their culture, to the *notion* of mobility, and could easily imagine other landscapes" (*Idea of Landscape* 63). Where Clare wrote about and sought to preserve the one nature that he knew and valued, the gentry and professional classes knew many, and they were intent on translating them and remaking them to suit their specific needs. Informed by ideas of improvement, wealthy landowners displayed their power and control over nature, as I have already indicated, by refashioning their properties to conform to new ideas of landscape drawn from the paintings of Nicolas Poussin and Claude Lorrain, or from travelers' accounts of the American wilderness or Chinese gardens, or from the categories of taste promulgated by writers on the picturesque. Meanwhile, a whole new class of farmers and professional agriculturalists, surveyors, and land agents—people whose geographical experience was shaped by newspapers, books, and roads—began to think of rural villages and towns no longer as isolated, self-contained entities, but instead in terms of their "relations with other places" (92).

Barrell's argument for the difference between Clare's stationary local knowledge and the more mobile and integrative ways of knowing places which developed during the eighteenth and nineteenth centuries has been justifiably influential in Clare studies. Clare's commitment to a local understanding of nature should not be seen, however, as a straightforward expression of class, for Clare strongly differentiated himself from other laborers who showed little appreciation of the natural world around them. Also, there is little question that the British rural poor embraced, either willingly or by necessity, that world of movement and change that led them to seek a new life in cities and often in the far reaches of the earth. Clare's commitment to place and to an idea of English nature inseparably bound up with a traditional rural past was not inherently a position thrust upon him by class, but instead one that he self-consciously took up as a poet from that class.[2] To cite Alan D. Vardy, Clare's defense of the "old occupations" and "of dialect, vernacular speech and orality in his poetry constituted a cultural intervention fully aware of its social and political implications" (*Politics and Poetry* 1–2). Clare rejected the mobilization of nature and place not because he could not move or did not understand mobility, but because he understood all too well their social and ecological consequences.

As a writer engaged in the broader cultural project of putting nature into print, Clare was not denied access to other natures, and he was fully capable of providing his readers with exactly the kind of "knowledge at a distance" which was being developed in literary and scientific circles. In fact, between September 1824 and January 1826, as his publishing firm, Taylor & Hessey, was on the verge of collapse, Clare was planning to write a "'Natural History of Helpstone,' in a series of letters to Hessey," modeled on White's *Natural History of Selborne* (*Prose* 104).[3] By March 1825, Clare had a title: "'Biographys of Birds and Flowers,' with an appendix on Animals & Insects" (*Journal*, 11 Mar. 1825, *Prose* 139). An avid gardener, Clare went on botanizing expeditions with Edmund Tyrell Artis and Joseph Henderson, consulted natural history books at Milton House, and collected ferns and orchids. By January 1826, however, with the delay in the publication of *The Shepherd's Calendar* and the collapse of Taylor & Hessey, Clare indicated to John Taylor that he had relinquished the idea of writing a prose natural history of Helpston (24 Jan. 1826, *Letters* 356). Other factors had also come into play. With limited access to other natural histories, Clare's knowledge was largely limited to the plants and animals that he could observe. In addition, he refused to adopt Linnaean nomenclature, which he saw as a "hard nicknaming system of unuterable words" that overloaded botany "in mystery till it makes it darkness visible" (*Prose* 117). But the decision not to write in the vein of contemporary natural history did not mean that Clare was any less committed to studying and writing about nature. Instead, it reflected his belief that poetry was the proper medium for its representation. The richness of the natural history poetry that followed is one of Clare's great legacies. As M. M. Mahood comments in her superb study of Clare and botany, the

"370 plants that Clare actually names in his poetry and prose" are "an astonishing tally" (*Poet as Botanist* 112).

The difference between how Clare and contemporary naturalists represented nature had little to do with questions of attention to detail, but instead with how this knowledge was to be used. Naturalists were not interested in observing the individuals of a species as individuals, nor were they interested in the literary, cultural, or social associations connecting a plant or animal to a place: for them, a marsh marigold (*Caltha palustris*) in Kent was the same plant when found in Northborough, and calling the latter a "horse blob" only confused matters. Clare, in contrast, wanted to see nature as an integral part of localized communities. Helpston, for Clare, was a neighborhood in the fullest sense of the word. As Douglas Chambers argues, "[Clare's] poetry demonstrates that 'just representations of general nature' make no sense apart from the particular names of individual trees and flowers: names authenticated by the oral tradition that has transmitted them" ("A love for every simple weed" 238). Clare excels in the representation of nature as a habitation and in his close observance of the relations among creatures in their environments and with the human beings who share the world with them.[4] For Clare, natural history is a means of acknowledging and sharing one's place and life with nonhuman beings. That is why he often personifies plants, referring to botany as the means by which one can "know their names as of so many friends & acquaintance," and why he hoped that botany would become popular among "the future shepherds & ploughmen of [his] country," so that they might become "acquainted with the flowers of their own country that make gardens in summer of the spots where they live & labour" (Clare to Elizabeth Kent, Aug. 1823, *Letters* 284). Clare uses his natural knowledge to insert or to embed himself and his readers in the particularity of the world he observed, and he was profoundly critical of the manner in which contemporary natural history forcibly ripped living beings from their socioecological communities. In "Shadows of Taste," he writes,

> Take these several beings from their homes,
> Each beauteous thing a withered thought becomes
> Association fades & like a dream
> They are but shadows of the things they seem
> Torn from their homes & happiness they stand
> The poor dull captives of a foreign land. (*Middle Period* 3:303–10, lines 147–52)

Here Clare clearly recognizes the manner in which contemporary natural history mobilized natures by translating them into "shadows" of what they once were, both a ghost-making activity and a forced removal that he equates with slavery, tearing living beings from their "homes & happiness" and relocating them as the "captives of a foreign land," where, to add insult to injury, their original names were taken away and they were given new Latin ones by metropolitan scientists.

The Eternity of Nature in a Time of Change

For Clare the rights and dignity of nature and rural folk were inseparably bound up with each other and with the local traditions of rural communities. In the poetry of the Helpston period, he depicts the concrete and enduring bond between rural laborers and the natural world. His goal is not to present an individual subjective experience of nature (as one finds in the poetry of Wordsworth), but instead to recover, and thus to preserve, common traditions of experience which had developed over centuries. Tradition is not something that we inherit from the past; instead, it draws its vitality and strength from being constantly renewed or found again in the present. In these poems Clare presents a nature that is continuous with the present, a nature that does not change, even though, in the modern world, it struggles to endure. It is worth stressing that Clare cannot imagine one nature giving way to another, as often happens in colonial nature writing; he does not see nature in general as a historical phenomenon, though every plant and bird in an ecological community has its own history as an individual. Consequently, in his poetry either nature is present in all its immediacy, or it is absent, a loss that memory seeks to recover. Clare's nature present does not change, nor do its joys.

In "The Eternity of Nature," Clare finds a powerful figure for this continuity of tradition in the image of a daisy that "lives & strikes its little root / Into the lap of time" (*Middle Period* 3:527–31, lines 4–5). For Clare, culture and tradition are the very soil in which nature roots itself. The same joy that urges a present-day child, and that will urge a child "many thousands" (line 12) years hence, to pluck a daisy was also felt by Eve, "when all was new" (line 17) and she did "stoop adown & show / Her partner Adam in the silky grass / This little gem that smiled where pleasure was" (lines 20–22). Every living field is a version of the first garden, and the pleasures that spring from nature are free and available to anyone wherever a flower is found. The attendant joy is what links all human beings to the earth and gives them a sense of belonging. Whereas William Blake in "The Marriage of Heaven and Hell" emphasized the historically produced character of the flower ("To create a little flower is the labour of ages" [plate 9]), Clare sees the plant and its joys as being continuous across time. Still, there is trouble in paradise: whereas the daisy once bloomed in Eden "under heavens breath" (line 25), now, in its contemporary setting, it is "trampled under foot" (line 3) and blooms "with sorrow" (line 24). And yet, despite its changed circumstances, growing "on blighted earth & on the lap of death / It smiles for ever" (lines 26–27).

"The Eternity of Nature" presents a radical faith in the permanence of nature despite the changes that Clare saw happening around him. Both the daisy and the pleasure it provides are indestructible because they are guaranteed by Creation. Time may destroy the works of man, but nature remains the same forever. "When kings & empires fade & die," cowslips, "as times partners," will still be "As fresh two thousand years to come as now" (lines 29–31). "Endless youth / Lives in them all unchangeable as truth" (lines 75–76), and this quality of immutability extends to their individual forms and behavior, which were also permanently fixed in that first moment of creation: "the cows lap peeps this very day / Five spots appear which time neer wears away" (lines 79–80). The hum of the bee, the tune of the nightingale, the song of the robin—all nature's songs are "a music that lives on & ever lives" (line 46), "for time protects the song" (line 40). And like the smiling flowers of "Songs Eternity," a poem that Bridget Keegan calls Clare's *ars poetica* (Keegan, "'Camelion' Clare"), these songs, first "sung to adam & to eve," will remain "evergreen" until the end of time:

> Songs is heard & felt & seen
> Every where
> Songs like the grass are evergreen
> The giver
> Said live & be & they have been
> For ever (*Middle Period* 5:3–5, 5n).

Clare famously remarked that he did not compose his poems, but instead "found [them] in the fields, / And only wrote them down" ("Sighing for Retirement," *Later Poems* 1:19–20, lines 15–16). No doubt he hoped that the time that protected nature's song would also protect his poems too.

And yet time and tradition failed both Clare and the nature that he sought to preserve. Though committed to the idea of the permanence of nature and its songs, Clare lived in a world that was changing rapidly. Where many of his contemporaries saw mobility as an opportunity, Clare approached it with dread, equating it with the loss of place, rights, identity, and community. This close connection between movement and loss can be seen in "Helpstone," the first of many poems to address the destructive impact of enclosure on his "native place" (*Early Poems* 1:157–63, line 71). Adopting a conventional eighteenth-century topographical model, in which the *genius loci*, the spirit of a place, inspires the genius of the poet who depicts it, "Helpstone" announces the appearance in literature of a previously "unletterd spot" and of a new poet who will sing its glories, promising to "advance [the] name" of a "mean Village ... / Unknown to grandeur & unknown to fame" even as it brings to public attention the "dawning genius" of its first "minstrel" (lines 2–5). Yet, ironically, the poem is less about re-marking a place than about recognizing its erasure. Instead of describing Helpston as it *is*, Clare represents the parish as it *was*, for the place is associated with "past delights" that now exist only in memory: a "vanish'd green," an absent brook, "many a bush & many a tree" that were leveled, and the abundance of flowers—the "golden kingcups," "silver dazies," "silver grasses," "lilac," and "Cows laps"—now all part and parcel of a "long evanish'd scene" (lines 73–103). The "golden days" of pastoral are evoked, but they are seen as being "long vanish'd from the plain" (line 55). Clare does not mince his words about the economic greed and violence that have produced this destruction; under the banner of "improvement," "A tree [is] beheaded," "a bush [is] destroy'd" (line 88). What is striking is that he sees this destruction as a kind of forced removal—an eviction, but one that applies to plants rather than people. It "Griev'd [him] at heart," Clare writes, "to witness their removes" (line 94).

John Barrell has noted the degree to which Clare's criticism of the destruction of the local nature of Helpston is confusing because it is overlaid with his simultaneous adoption of the eighteenth-century convention, basic to Oliver Goldsmith's "The Deserted Village," of an older person returning to the scenes of his childhood only to find them forever changed (*Idea of Landscape* 110–20). The reader is thus left wondering whether the losses mourned by the poet are those that inevitably come with age, or are indications of real, material changes in the ecology of the place. Both Goldsmith and Clare are arguing the latter, but they clothe this new political and ecological message in the sentimentalist conventions of a poetry that nostalgically mourns the lost Eden of childhood.[5] Goldsmith's criticisms of enclosure and of the manner in which it was forcing rural laborers to emigrate to America seem to weigh less with the poet than his disappointment that the desertion of the village has spoiled his retirement plans, the long-held hope "to return— and die at home at last" (*Collected Works* 90). Clare echoes Goldsmith in the conclusion of "Helpstone" ("Find one hope true to die at home at last" [*Early Poems* 1:157–63, line 178]), but the pain of loss is much greater because Clare is considering what enclosure means not for those who have left the village, but for those who have stayed behind.[6]

Gary Harrison's argument that, instead of reading Clare as "the poet of place," we should see him as "the poet of between places" helps clarify Clare's social and literary self-positioning, for whether he was attempting to occupy the contradictory social category of a "peasant poet" or seeking to recover a relationship to place which he felt was already in the past, Clare stood between worlds ("Hybridity, Mimicry" 149).[7] In describing his "Dear native spot" (*Early Poems* 1:157–63, line 51), Clare moves through a divided landscape, one in which he sees in the enclosed Helpston of the present the absent landscape of the Helpstone that once occupied its place. Clare's poetry is filled with these doubled landscapes, in which memory dwells on— even dwells in—the spectral traces of what has been lost in what remains. "Now alas those scenes exist no more," he writes (line 115), so instead of being introduced to the Helpston of the present, the poem is about "Perishd spots" and "ruind scenes," the "well known pastures oft frequented greens / Tho now no more" (lines 145–47). Helpston was once an Eden, he suggests, but now the "fates" have chosen to lay its "beauties bye / In a dark corner of obscurity" (lines 119–20). Clare searches these "dark corners," so that the poem is less about mapping the features of an overlooked place than about recovering the nature that it has lost. Whereas tourist poems celebrate those authentic places that must be visited before the tourists destroy them, "Helpstone" is already a ruin whose glory is only visible to the eye of memory provided by Clare. The poet essentially wanders through an absent parish. This is why the poem is actually less about being at home than about homelessness, about being exiled from his native place. For Clare, enclosure was an act of "removal," one that permanently uprooted him and the natural world that was his neighbor from their traditional place. Speaking of Clare's later poetry, Tim Chilcott sees "a poetry of presence . . . evolving into a poetry of absence" ("*Real World & Doubting Mind*" 118). From the beginning Clare—the great poet of local nature—was already writing road poems dealing with exile, loss, and absence. The conclusion of the poem speaks not just of loss but also of being lost on roads going nowhere: "So when the Traveller uncertain roams / On lost roads leading every where but home," he "makes for the home which night denies to find" (*Early Poems* 1:157–63, lines 179–80, 184). For Clare, mobility was not an expression of freedom, but rather the equivalent to exile.

Tradition and the Rights of Nature and Rural Laborers

A comment that Clare made in his journal on 29 September 1824 helps clarify why he understood mobility as a form of dispossession:

> Took a walk in the fields saw an old wood stile taken away from a favourite spot which it had occupied all my life the posts were overgrown with Ivy & it seemd so akin to nature & the spot where it stood as tho it had taken it on lease for an undisturbd existance it hurt me to see it was gone for my affections claims a friendship with such things but nothing is lasting in this world last year Langley Bush was destroyd an old whitethorn that had stood for more than a century full of fame the gipsies shepherds & Herdmen all had their tales of its history & it will be long ere its memory is forgotten. (*Prose* 109–10)

Clare's complaint about the removal of the stile is part of his ongoing criticism of enclosure, because he could not but recognize that it had once provided pedestrian access to land that was now being claimed as private property. Here Clare fuses nature and tradition, claiming that the wooden stile, now "overgrown with Ivy," has become "akin to nature" through its long endurance in place. "Akin" suggests likeness, but also kinship, a belonging that includes the poet who "claims" a "friendship" with it because it has been in "a favourite spot" for "all [his] life." In speaking of the stile's removal, Clare explicitly employs legal language, remarking that it had "occupied"—that is, it had taken possession of—this spot "as tho it had taken it on lease for an undisturbd existance." Where traditionally tenants were accorded specific rights through occupancy, under the new forms of ownership epitomized by enclosure, all natures, human and nonhuman (even those that had occupied the land since the beginning of time), were subject to the goodwill or caprice of those who now claimed absolute ownership over their property.[8] That is why the circumstances of the "stile" were not substantially different from those of the "old whitethorn" called Langley Bush.

In eighteenth-century and nineteenth-century literature, it is not uncommon to see agricultural laborers and indigenous peoples being identified with nature—as being wild, uncivilized, rooted in place, and incapable of change. Clare makes a similar association, but for very different purposes. Instead of emphasizing the ahistorical dimensions of the rural laboring class and of nature, Clare sees both as being long-standing tenants of the earth who are now suffering the indignity of being uprooted and removed from their places. The creatures of nature, in Clare's view, claim their belonging to the earth they inhabit by virtue of their long-standing occupation of it. Ecology and social protest are brought together in his belief that laborers and rural English nature were suffering under a new economic regime that made them landless by abrogating their traditional rights of tenancy: in claiming that "nothing is lasting in this world," Clare may sound like he is adopting the ancient language of mutability, but he is really making a political and ecological critique of his times. In "To Wordsworth," after singling out Simon Lee "grubbing up the root" and remarking that Wordsworth's poems are like fields in which one comes upon different flowers, Clare declares, "I love them all as tenants of the earth" (*Later Poems* 1:25, lines 7–8).

Clare's use of the idea of tenancy to understand place and belonging in nature is very different from how Wordsworth uses the term in the *Guide* to denominate the extinct animals that were once "joint tenants" with the Celts when the Lake District was a primordial wilderness. Wordsworth uses the word "tenant" because he understands these animals as having had no real claim to the places that they occupied. The deer, bear, wolves, and boar are temporary lodgers in a place that will ultimately displace them. Clare rejected the idea that nature has no property in itself and that nonhuman beings have no right to the land they occupy. Against the modern legal definition of land as a private freehold possession, he reasserted the moral economic claim that nature and country folk, by virtue of their ancestors having lived in their places for centuries, could claim traditional tenants' rights, at the very least "for an undisturbd existance." Whereas traditionally land had been shared among different social groups, enclosure represented the triumph of a capitalist system in which property rights trumped customary entitlements and obligations. As Nancy Fraser and Linda Gordon remark, "When land became a commodity, rural populations lost their age-old rights of tenancy and use" ("Contract versus Charity" 122). Discussing the eighteenth-century struggle between traditional and capitalist conceptions of property, E. P. Thompson observes, "What was often at issue was not property, supported by law, against no-property; it was alternative definitions of property-rights: for the landowner, enclosure—for the cottager, common rights; for the forest officialdom, 'preserved grounds' for the deer; for the foresters, the right to take turfs" (*Whigs and Hunters* 261). Clare recognized that the new conception of property rights embodied in the act of enclosing common lands dispossessed not only the rural laboring class but also the nonhuman beings that had long inhabited the rural countryside. Having seen the landscape of Helpston stripped of the great elms and oaks that had once occupied a place of honor in its landscape, he recognized that the English nature around him was now no more rooted in place than he was. Its hold on place had become tenuous. In "Remembrances," the destruction of Langley Bush is once again figured as a forced move: "by Langley bush I roam but the bush hath left its hill" (*Middle Period* 4:133, line 61). "Langley bush" has become a name only, serving as a ghostly stand-in for a nature that has "left *its* hill" (my emphasis). In the new economy signalized by enclosure, a part of nature that had once claimed a preeminent right of place on a hilltop was now just as liable to being made homeless as a rural laborer. In a journal entry for 26 November 1824, Clare recounts a visit to Lea Close to see whether the old hazelnut tree, from which he had collected "a half peck of nuts" when he was a boy, was still standing. It was, but all of the other trees were gone: "the Inclosure has left it desolate its companion of oak & ash being gone" (*Prose* 126). For the professional classes of imperial Britain, the capacity and legal right to disembed plants and animals from their traditional localities made it possible to use nature in new ways, integrating natural beings into new forms of knowledge and transferring them to new locales. For Clare this mobilization of nature was a greed-driven act of dispossession.

In his poems on bird nests, Clare focuses on nature's claim to the rights of tenancy and seeks to provide an alternative to the destructive, appropriative relationships that normally characterize human beings' relations with the animal world. As Elizabeth Helsinger remarks, these are poems in which Clare searches for the "remnants of unowned land in an enclosed landscape" (*Rural Scenes and National Representation* 151), seeking the freedom that he associated with community.[9] In "The Robins Nest," Clare describes a part of the woods which has been untouched by human improvement, its wildness a sign of its status as an "old spot" (*Middle Period* 3:532–36, line 23), an "ancient place" (line 26), where human beings still share their place with other organic beings,

> Where old neglect lives patron & befriends
> Their homes with safetys wildness—where nought lends
> A hand to injure—root up or disturb
> The things of this old place. (lines 50–53)

Here "the very weeds as patriarchs appear" (line 63), and their continuing presence in this place indicates that it has not yet been touched by the "war with nature" (line 55). The interaction with the natural world is understood as being based on the rights of cohabitation. Come "ten years hence," Clare declares, and the same plants will greet your eyes, "like old tennants peaceful still" (lines 66–67). Here too the wood robin displays little fear of human beings, and "With fluttering step each visitor recieves" (line 69). Like his progenitors, the robin rarely leaves this spot, for all of his needs are satisfied here:

> In heart content on these dead teazle burs
> He sits [&] trembles oer his under notes
> So rich—joy almost chokes his little throat
> With extacy & from his own heart flows
> That joy himself & partner only knows
> . . .
> And there these feathered heirs of solitude
> Remain the tennants of this quiet wood
> & live in melody & make their home
> & never seem to have a wish to roam. (lines 71–90)

For Clare mobility was a symptom of homelessness and dissatisfaction. Ancient natures, unlike their modern counterparts, seek to stay put. Appropriately, he ends the poem with a description of the robin's nest, which snugly shelters five brown-colored eggs from which will come "bye & bye a happy brood," "The tennants of this woodland privacy" (lines 100–101).

The most powerful fusing of natural and human tenancy is to be found in "The Lament of Swordy Well," where an ancient stone quarry, "a piece of land" personified as a member of the parish poor, comes before the reader, not to beg, but to appeal for his customary rights, to "pray to keep [his] own"

(*Middle Period* 5:250–56, lines 21, 12). For the first time in literature, nature appears as a homeless person. Granted to the overseers of Helpston's roads by the 1809 Enclosure Act, Swordy Well complains that he has been stripped of all his belongings by vultures, a "grubbling geer" (line 45) that "claim[s] [his] own as theirs" (line 65). "Ive scarce a nook to call my own / For things that creep or flye," he complains (lines 113–14).

Denied access to any legal court of appeal and seeking, in Helsinger's words, to "own" and to "keep" what is his, in the sense of claiming and supporting what belongs to his identity (*Rustic Scenes and National Representation* 149–50), Swordy Well brings his case before the reader, hoping that, in the absence of legal or social justice, poetry can provide him with a hearing:

> Though Im no man yet any wrong
> Some sort of right may seek
> & I am glad if een a song
> Gives me the room to speak (*Middle Period* 5:250–56, lines 41–44).

The Swordy Well who appears in the poem is hardly recognizable, even to himself, and yet he has been lucky, for he says, "Of all the fields I am the last / That my own face can tell" (lines 251–52). Once a rich ecological community, home to rare orchids ("flowers that blo[o]med no where beside" [line 135]), to great crested newts, and to the rare (but now extinct) large copper butterfly (*Lycaena dispar dispar*), Swordy Well has been robbed of his streams and soil cover: "The butterflyes may wir & come / I cannot keep em now" (lines 93–94). Just as his diked springs can "Scarce own a bunch of rushes" (line 58), so too

> The muck that clouts the ploughmans shoe
> The moss that hides the stone
> Now Im become the parish due
> Is more then I can own (lines 37–40)

The "Lament of Swordy Well" is, indeed, a poem about *extinction*, for it speaks of changes in the land which are permanent and enduring. His is the story of how a "waste" land, rich in ecological diversity, has been transformed into a wasteland.[10] Swordy Well has learned, to his despair, that nature is not permanent and that it can be destroyed. Looking into a dark future, he sees a time when his "name will quickly be the whole / Thats left of swordy well" (*Middle Period* 5:250–56, lines 255–56). In the poem, Clare gives poetry the important role of documenting the true dimensions of an ecological catastrophe. Plants and animals do not easily communicate with us, so it is all too easy for human beings to ignore their well-being and to forget what they are doing and have done to nature. In "Obscurity" Clare writes,

> Old tree, oblivion doth thy life condemn
> Blank & recordless
> . . .
> So seems thy history to a thinking mind
> . . .
> Thou grew unnoticed up to flourish now
> & leave thy past as nothing all behind" (*Middle Period* 4:256, lines 1–9).

Clare's recognition that the nature that he valued would never return, that "The past is past—the present is distress" ("Here after," *Middle Period* 2:200, line 8), changed the ways in which he invokes tradition and communes with nature.

The intense alienation that Clare felt when he moved the three miles from his family cottage in Helpston to Northborough has been famously misinterpreted as a sign of his incapacity to advance beyond localized conceptions of identity. For Clare, however, the issue had nothing to do with distance, but with the move itself, which meant that he could no longer claim that he belonged to a specific place. Legally alienated from his native parish, he was now a stranger moving among strangers in an alien space. "Here every tree is strange to me / All foreign things where eer I go," he writes in "The Flitting," a poem in which he grappled with his uprootedness (*Middle Period* 3:479–89, lines 97–98). In northern English and Scottish dialect, the word "flitting" refers not just to a kind of flickering movement, but to moving from one place of habitation to another. It also can refer to the removable goods and furniture, the "flitting," that one carries on such a move. "The Flitting" is a poem about nostalgia, the pain that comes with losing one's "own old home of homes" (line 1), but Clare self-consciously uses his feelings of psychological dislocation in order to comprehend what it means to live in a world where nativity no longer constitutes a claim to a place and where all human and nonhuman beings are subject to removal. In the poem Clare asks, how is one to live in a world where everything is flitting, and what is one's relationship to an environing nature when one can "own the spot no more," in Clare's inclusive sense of "owning" as a form of "belonging"?

John Lucas has argued persuasively that "'The Flitting' is about dispossession. It is a grieving, eloquent utterance of a sense of being denied ownership of, or relationship with, all that you feel most intensely to be yours, all that feels so intimately connected with you that it is integral to your sense of selfhood" ("Places and Dwellings" 91). It is also a poem about exile, about being "Alone & in a stranger scene / Far far from spots my heart esteems" (*Middle Period* 3:479–89, lines 49–50). Struggling to explain his dependence on a nature now alienated from him in both legal and psychological terms, trying to imagine how to write poetry about a nature from which he was legally alienated, Clare compares his thoughts to weeds: "& still my thoughts like weedlings wild / Grow up to blossom where they can" (lines 59–60). Weeds are important symbols in the later poetry, for he believed that they constituted a living link to Adam's original open garden ("weeds remain / & wear an ancient passion that arrays / Ones feelings with the shadows of old days" ["The Robins Nest," lines 55–57]). For this reason Clare rejected the accepted view that "disorder is an ugly weed" ("Shadows of Taste," line 154), that a weed is, in Samuel Johnson's dictionary definition, "an herb noxious or useless" (*Dictionary*). In the same poem in which Clare recounts how he came to see poetry as a vocation—"I'd a right to song / & sung"—he also insists that there "could not be / A weed in natures poesy" ("The Progress of Ryhme," *Middle Period* 3:492–504, lines 80–81, 91–92). Still, under the driving force of improvement, these ancient neighbors of humankind continued to be seen as enemies of progress, as "plants growing out of place." Behind Clare's use of the word "weeds" is its association with clothing and thus with ideas of protection and vulnerability. Alfred Tennyson, in *In Memoriam*, declares, "In words, like weeds, I'll wrap me o'er, / Like coarsest clothes against the cold" (5.9).

In "The Flitting," Clare sees his poetry and thoughts as "weedlings wild," as displaced plants rooting themselves wherever they can. Being out of place thus forms the basis of a new conception of identity grounded in mobility. Seeing a "shepherd's purse" at his feet leads him to imagine another place and time, but it does not take him away from the here and now:

& why—this 'shepherds purse' that grows
In this strange spot—In days gone bye
Grew in the little garden rows
Of my old home now left—And I
Feel what I never felt before

> This weed an ancient neighbour here
> & though I own the spot no more
> Its every trifle makes it dear. (*Middle Period* 3:479–89, lines 193–200)

In this "ancient neighbour," whose name harkens back to the pastoral world, Clare "historicises and enculturates this weed as a text not only of himself but of a lost society" (Chambers, "A love for every simple weed" 243). He does not use this association, however, to translate himself to another place, but instead to seek to ground himself in "this strange spot," a spot that he does not own.

Conscious of the fragility of such weedy rootings (they are "trifle[s]"), Clare nevertheless makes in this poem his most radical claim for the continuity of nature and the enduring importance of a "verse that mild & bland / Breaths of green fields & open sky" (*Middle Period* 3:479–89, lines 161–62). Every weed and blossom claims an Edenic pedigree:

> All tennants of an ancient place
> & heirs of noble heritage
> Coeval they with adams race
> . . .
> & still they bloom as in the day
> They first crow[n]ed wilderness & rock" (lines 129–46).

This Edenic pedigree also allows Clare to posit that in the dying day of empires "Poor persecuted weeds [will] remain" (line 212), as Adam's open field garden continues in the "waste" lands. Yet this hope was itself a tenacious weedling thought, seeking "to blossom where [it] can" (line 60), for Clare very well knew that both poetry and the nature that mattered to him were disappearing from the earth. In "Decay A Ballad," he writes that "Nature herself seems on the flitting," and with it, his poetry:

> spots where still a beauty clings
> Are sighing 'going all a going'
> O poesy is on the wane
> I hardly know her face again. (*Middle Period* 4:114–18, lines 4–10)

There were many good reasons for Clare to move to Northborough, so his feelings of extreme dislocation can be seen as symptoms of the psychological difficulties that would increasingly plague him. Nevertheless, we should be careful not to narrow the implications of his work to biographical considerations, no matter how poignant they might be, for Clare's experiences of migration and exile were not unique, and he was not the only person during the nineteenth century who had to build his hopes on hopeless circumstances. The forced migration of slaves, convicts, and indentured laborers and the displacement of many indigenous peoples from their native lands parallel the experience to which Clare gave intensified expression in his poetry. Furthermore, British rural villagers were not the only people to find themselves displaced and their arable and non-arable lands transformed by self-serving legalistic definitions of property rights; fences, plows, and English property law left their mark in other places too. What makes Clare's later poetry so important is that he writes about what it means to irrevocably lose one's nature and his poetry questions the ethics of European conceptions of land ownership.

Radical Hope

In *Radical Hope* (2006), Jonathan Lear explores the ethical question of how one should live in the face of the total collapse of one's culture. Lear draws upon the case of the last great Crow chief, Plenty Coups, whose people suffered the catastrophic loss of the northern plains buffalo. Plenty Coups "refused to speak of his life after the passing of the buffalo.... After this nothing happened. There was little singing anywhere. 'Besides,' he added sorrowfully, 'you know that part of my life as well as I do. You saw what happened to us when the buffalo went away'" (*Radical Hope* 2). With no buffalo, time and history came to an end for Plenty Coups, and yet he found meaning for himself and his tribe in the radical hope conveyed by a dream he had as a young man. This dream told him how he might help his people to survive and to keep their land, as another Crow, Yellow Bear, interpreted it: "The tribes who have fought the white man have all been beaten, wiped out. By listening as the Chickadee listens we may escape this and keep our lands" (72). In his later years, sitting under the same tree that appeared to him in his dream sixty years earlier, Plenty Coups found some solace in the fact that he had guided his people through the terrible times that had destroyed other tribes: "And here I am, an old man, sitting under this tree just where that old man sat seventy years ago when this was a different world" (143). Plenty Coups's world came to an end with the loss of the buffalo, but he could still turn to other aspects of the land to reaffirm the continuity between his people and their past. Clare's turn to the "shepherd's purse" does something similar, but rather than speaking of the continuity of a place through time, Clare's poetry focuses on natures lost. For Clare there was no Langley Bush left to ground his visions. In this regard, his experience comes closer, in many ways, to that of the Sioux chief Sitting Bull and the many other tribes who lost their traditional lands to the white man and were forced to relocate elsewhere. Left without a local nature on which to build his hopes, Clare was left only with his dreams. Edward Said writes that "in a very acute sense exile is a solitude experienced outside the group: the deprivations felt at not being with others in the communal habitation" ("Reflections on Exile" 177). Clare wrote to re-create this community, one that was composed not only of people but also of the other organic beings that had once inhabited this place.

Although Clare had once hoped that nature would safeguard his poetry from oblivion, he came to the realization, during the 1830s and 1840s, that the only place where the nature that he knew and loved would continue to exist was in his poetry. It gave a new urgency to his writing and made language the basis for his radical hopes. That is why his most powerful evocations of the enduring importance of a "harmless . . . song" or of "A language that is ever green" often emerge within contexts of loss and despair ("Pastoral Poesy," *Middle Period* 3:581–84, lines 112, 13). While in the Northampton Lunatic Asylum, Clare recalled in vivid detail his childhood visits to Round Oak, "the Apple top't oak" above Round Oak Springs, which was cut down with the enclosure. The poem concludes with the poet's recognition:

> All that's left to me now I find in my dreams
> . . .
> Sweet Apple top't oak that grew by the stream
> I loved thy shade once—now I love but thy name. (*Later Poems* 1:450–51,
>   lines 1, 29–32)

While Plenty Coups found a confirmation of the truth of his dream in the shade of a tree that was there for him all his life, Clare had to ground the truth of his visions in a different kind of shade.

James McKusick suggests that Clare's "decision to retain certain features of his own regional dialect was motivated in large part by his need to preserve a language that evoked with concrete immediacy the natural phenomena of his native place" (*Green Writing* 89). Clare's poetry represents a significant advance in ecowriting, McKusick argues, because he created an "ecolect" that would capture the unique qualities of the nature of Helpston. Yet Clare rarely spoke of his poetry in terms of individual poetic creation, instead arguing that the songs and words were already there; he was not attempting to represent a nature that existed outside of language, but instead one that was already embedded in it, in the everyday words spoken by the villagers of Helpston. That is why preserving the words by which this nature had traditionally been known, experienced, and understood was just as important to him as preserving this nature itself. Here it is worth stressing Douglas Chambers's view that Clare sought to ground his poetry "in natural speech and native vocabulary" ("A love for every simple weed" 238). Linnaeus famously argued that the task of natural history was to name the Creation. "If you do not know the names of things," he wrote in the *Philosophica Botanica* (1751), "the knowledge of them is lost too" (169). Clare would have agreed, but rather than seeking to establish a universal language that would allow species to be compared across time and space, Clare wanted to preserve the knowledge and historical associations of a very local nature that was embedded in the language of the people who lived in Helpston. "The vulgar," Clare remarks in his First Natural History Letter, "are always the best glossary" for plant names (*Prose* 117). Ecolect is thus inseparably fused with idiolect in his poetry, and, in resisting his editor John Taylor's efforts to rid his poetry of dialect and provincialisms, Clare was struggling for the continuance not just of a nature but also of the unique language in which that nature had long been experienced and understood. Clare recognized that language was not simply a means of communication, but that it also contained the history and the culture of the place and the people to whom it belonged, not in the sense of *owning*, but *being*. The same forces that were obliterating local natures were also destroying the languages by which they were known and particularized. In an age in which the goal of translation was to mobilize natures, Clare's translation of Helpstone is much closer to Walter Benjamin's idea that translations constitute the "afterlife" of a text. Clare's poetry represents the "afterlife" of a nature whose disappearance he witnessed during his life. In insisting on the literal connection between this nature and the songs by which it continued, Clare the poet was Clare the translator: a writer who explored in his poetry the strange afterlife of this nature. Inhabiting its afterlife, inhabiting his translations, Clare recognized how much this nature owed its continued existence to a language that was itself on the verge of disappearing.

    Much has been said about Clare as a memory poet, nostalgically dwelling on the joys of childhood and the past. Of greater importance is the manner in which he uses local history to encourage his readers to consider the claims that the natural world has on us. While his poems can be read, in Tim Chilcott's terms, as Proustian "litanies for the restoration of lost time" ("*Real World & Doubting Mind*" 108), they are more productively seen as poems that are less about nostalgia than about a critical remembrance of what took place in a small, out-of-the-way part of rural England early in the nineteenth century. They ask us to see, and thus to recognize, what has been lost and the injustices that changed the nature of this place.

Clare's poem "Remembrances" exemplifies this critical strategy. As a poem about the disappearance of "summer pleasures," "Remembrances" can easily be read as a conventional elegy for the passing of youth: "I thought them all eternal when by Langley bush I lay / I thought them joys eternal when I used to shout & play" (*Middle Period* 4:130–34, lines 1, 6–7). What gives the poem its critical dimension is that in recalling a host of lost childhood pleasures, it also speaks of the places in the landscape, now gone, where those pleasures took place. Eric Robinson, David Powell, and P. M. S. Dawson remark in their commentary that the poem provides a "map of Clare's boyhood, bring[ing] together many of the favourite places that are mentioned elsewhere in his work, many of which can be located on enclosure and ordnance maps" (*Middle Period* 4:595). The metaphor of a map is appropriate, because "Remembrances" constitutes a counter-history of the Helpston countryside which seeks to map and protest the changes that took place in the land and to recognize the plants, traditional names, local associations, and experiences that were being erased from the landscape. Performing his own form of ordinance mapping, Clare seeks to preserve the names of the places in Helpston which are gone or will soon be gone: "Langley bush," "old east wells," "old lea close oak," "old cross berry way," "swordy well," "round oaks," "little field," and "cowper green." Yet here he recognizes that language is insufficient to compensate for what has been lost: "O words are poor receipts for what time hath stole away / The ancient pulpit trees & the play" (lines 29–30). He is appalled by the greed that even denies poor moles their small share of the earth:

On cowper green I stray tis a desert strange & chill
& spreading lea close oak ere decay had penned its will
To the axe of the spoiler & self interest fell a prey
& cross berry way & old round oaks narrow lane
With its hollow trees like pulpits I shall never see again
Inclosure like a Buonaparte let not a thing remain

> It levelled every bush and tree & levelled every hill
> & hung the moles for traitors—though the brook is running still
> It runs a naker brook cold & chill. (lines 62–70)

The poem ends with an image drawn from the traditional rituals of the May, as Clare portrays himself as a jilted lover, who would have used his "poesys all cropt in a sunny hour" (line 77) to win her love so that she would stay. Yet the offers of poems and flowers in the pun embedded in "poesys" and in the structuring device of *The Midsummer Cushion* (1832), to which "Remembrances" belongs, are unable to convince her to stay: "love never heeded to treasure up the may / So it went the common road with decay" (lines 79–80).

"I love with my old hants to be," Clare wrote in the Pforzheimer manuscript version of "The Flitting" (line 91), suggesting another dimension of dwelling that especially appears in the later poems. "O Poesy is on the wane / I cannot find her haunts again" (*Middle Period* 4:114–18, lines 49–50), he writes in "Decay A Ballad." Here Clare's nature poetry verges on "ghost-writing," for the present is seen as being haunted by the natures it has displaced, natures that have been violently uprooted yet refuse to leave. Dispossessed of their traditional locales, they still occupy them in a new way. Whether these ghosts of nature past were really there or not, Clare saw them. Walking through what was once Helpston Green, now "all desolate" (line 6) because of the destruction of every tree, Clare saw another nature:

> When ere I muse along the plain
> And mark where once they grew
> Rememb'rance wakes her busy train
> And brings past scenes to view
> The well known brook the favorite tree
> In fancys eye appear
> And next that pleasant green I see
> That green for ever dear. ("Helpston Green," *Early Poems* 2:11–14, lines 17–24)

In "mark[ing]" this nature in at least two senses, in seeing it and producing inscriptions of it, Clare describes a natural ecology linked to rural labor, and both are now gone:

> Both milkmaids shouts and herdsmans call
> Have vanish'd with the green

> The kingcups yellow shades and all
> Shall never more be seen
> But the thick culterd tribes that grow
> Will so efface the scene
> That after times will hardly know
> It ever was a green. (lines 41–47)

"Milkmaids," "herdsman," and "kingcups" will disappear, having been replaced by fields of grain, "the thick culterd tribes" brought by the plow. Contemplating this wholesale erasure of a way of life from the landscape, Clare prophesies that few people in "after times" will even know that this place "ever was a green." Ghosts make themselves visible for many reasons, but often it is because they are seeking justice for a crime committed against them. As Jeffrey Andrew Weinstock suggests, "the ghost is that which interrupts the presentness of the present, and its haunting indicates that, beneath the surface of received history, there lurks another narrative, an untold story that calls into question the veracity of the authorized version of events" (*Spectral America* 5). Clare's ghostly natures play this critical role. They seek to recover the ground from which they have been removed by unsettling the reader, but more importantly, they want to be accorded a kind of justice, if only in our recognition that the present is deeply implicated in their unrest. For Clare the nature of the present was often silent and empty: "Silence sitteth now on the wild heath as her own / Like a ruin of the past all alone" (*Middle Period* 4:130, lines 9–10), he writes in "Remembrances"; "I never dreamed . . . / . . . that pleasures like a flock of birds would ever take to wing / Leaving nothing but a little naked spring" (lines 18–20).

In a letter written to Mary Joyce announcing the composition of *Child Harold*, Clare remarks that "nature to me seems dead & her very pulse seems frozen to an icicle in the summer sun" (*Letters* 646). In the silence and emptiness of a nature come to ruin, in a landscape where the language of rural labor and its close relationship to the land were no longer heard, Clare saw and heard the spectral presence of the past, and he used his language and his poetry to preserve these shadowy presences. Why these legions of natural ghosts had chosen to make themselves visible to Clare was probably not evident to him when he first began documenting in poem after poem the ruthless ecological destruction that was taking place in his neighborhood. By the 1830s, however, Clare must have sensed that his ability to communicate with these ghostly natures had something to do with the fact that he himself was one of the dispossessed. Clare communicated with ghosts and walked among shadows of the past because he shared in their flitting, living on long after the time that mattered most to him had come to an end. He, like his poetry, had become an afterlife.

Mahmoud Darwish would come to realize that "language and metaphor are not enough to restore place to place" (*La Palestine comme métaphore* 25). For Clare these were all he had. An exile in his own country, exiled not by having left his country but by having witnessed it leave him, Clare haunted Helpston as he continues to haunt his readers today. At High Beach Asylum and later at Northampton Asylum, Clare feared that he had become one of the forgotten, inhabiting "glooms & living death / A shade like night forgetting & forgot" (*Child Harold, Later Poems* 1:40–88, lines 621–22). He worried that he would end up as "nothing but a living-dead man dwelling among shadows"; in relation to these identities, the poet would be a kind of bell crier "to own the dead alive or the lost found" (*Prose* 239). Perhaps seeing Clare as a translated being, seeing him as a ghost speaking for a lost nature, or viewing him as a "Wordsworthian shadow" (*Visionary Company* 445) in a very different sense than Harold Bloom originally intended, is the first step in coming to terms with what his poetry can teach us.

CHAPTER EIGHT

Of Weeds and Men
Evolution and the Science of Modern Natures

In recent decades, historians of science have tended to downplay the importance of Charles Darwin's voyage on the *Beagle* for the development of evolutionary theory. We have learned that the idea of evolution was not a "eureka" moment, with Darwin suddenly seeing "into the life of things" (W. Wordsworth, "Tintern Abbey," *Poetical Works* 2:259–63, line 50) on the Galapagos Islands. As Frank J. Sulloway has argued, Darwin's "conversion to the theory of evolution did not spring full-blown as the result of his voyage, but emerged gradually in intimate cooperation with the numerous systematists who helped to correct many of his voyage misclassifications" ("Darwin's Conversion" 388).[1] The main strands of Darwin's theory of transmutation were brought together after he had returned home, in a series of notebooks begun in July 1837, culminating in his reading of Thomas Malthus's *Essay on the Principle of Population* (1798) in late September 1838. "I had at last got a theory by which to work" (120), Darwin would later declare in his *Autobiography*.

Still, the idea that evolution primarily emerged in Darwin's study creates its own kind of blindness, for it minimizes the profound impact that the experience of the colonial world had on his understanding of nature and the questions that preoccupied him during and after the voyage. When Darwin left England in 1832, he was orthodox in both his religious and scientific beliefs. Like his contemporaries, he believed in the stability of nature, with each creature unchanged in its organic form, occupying its ordained place within a designed order. Nature was stationary, and travel and the colonization of new territories were essentially human privileges. When Darwin returned in 1836, having witnessed the wholesale ecological transformation of colonial natures, he had a very different idea of what nature was: instead of being an order of things fixed and rooted in place, he now saw it as being profoundly linked with mobility and change, and this shift in perspective laid the foundations of evolutionary theory. Darwin was thinking about nature in a new way, and this perspective gave rise to new questions and new ways of framing answers. Whereas Darwin's predecessors had interpreted the enormous changes taking place in colonial environments as being anomalous, as temporary disruptions of the "balance of nature" which would eventually be rectified in time, Darwin's revolutionary gesture, drawing on the work of Charles Lyell, was to reverse this perspective, by seeing colonial natures as a model for how nature had always functioned. Rather than treating their destabilized and contestatory aspects as being *exceptional*—because they were either directly or indirectly caused by human agency—Darwin used them as a contemporary lens for seeing nature in a new way. By seeing colonial natures as *exemplary*, he was able to discern the dynamics of biological change occurring in fast motion. Instead of seeing natural environments as expressions of a single original Creation (or of multiple creations), Darwin increasingly understood the geographical distribution of species as being the expression of a long and complex history of conflict and settlement. He thus returned from the voyage on the *Beagle* with an idea of nature as being thoroughly *modern*, as constantly emerging from the crucible of variation, difference, mobility, migration, conflict, territorial expansion, and settlement. That is why, in reflecting on this voyage, he would claim that it "has been by far the most important event in my life, and has determined my whole career" (*Autobiography* 76–77). Evolutionary theory could not have been conceptualized apart from the world of dramatic mobility and change that Darwin had witnessed on his voyage.

Lyell and the Uniformitarian Rethinking of Colonial Ecologies

When Darwin boarded the *Beagle*, he brought with him the dominant ideas of British natural theology, which saw all species as having been created for a purpose and for the places that they were intended to occupy in Creation, and the first volume of a book that would revolutionize that understanding, Charles Lyell's *Principles of Geology*. He first encountered natural theology as a child by reading Gilbert White's *Selborne*, a book that made him want to become a natural history collector and led him to ask "why every gentleman did not become an ornithologist" (*Autobiography* 45). At Cambridge, Darwin read Paley's *Evidences of Christianity* (1794) (for the BA examination) and *Natural Theology* (1802), interestingly, while occupying Paley's former room in Christ's College. Long before ecology emerged as a scientific discipline, it was an important aspect of natural theology, which celebrated the intricate ecological relationships existing among plants and animals as an expression of the wisdom and design of Creation. According to this view, every organism was created to occupy its own place or "station" in the design of Creation, each serving a unique purpose in a stable natural order that exhibited divine wisdom in the intricacy, economy, harmony, and balance of its parts. In this view, every species was "stationary" by virtue of its form or design: because of the perfect alignment that existed between an organism's form and its station, no organism had a reason either to move or to change, or even to want to do so. The largely stable distribution of species across the globe was seen as a reflection of the fact that every species was designed to stay put, each having been adapted, or "fitted," to the different climates and physical conditions in which it was found.

Early in the voyage, Darwin draws upon this idea to explain why newly introduced plants often lacked natural enemies: "I could not help noticing how exactly the animals & plants in each region are adapted to each other.— Every one must have noticed how Lettuces & Cabbages suffer from the attacks of Caterpillars & Snails.—But when transplanted here in a foreign clime, the leaves remain as entire as if they contained poison.—Nature, when she formed these animals & these plants, knew they must reside together.—" (*Zoology Notes* 65). This passage valuably illustrates how initially on the voyage Darwin could observe relationships that actually contradicted the notion that natures were perfectly adapted to their place, by interpreting them as exceptions proving the rule, cabbages and lettuces having been created to serve the dietary requirements of caterpillars and slugs. Darwin wants to believe that "Nature" created lettuces, cabbages, caterpillars, and snails, knowing that "they must reside together," and yet it was just as obvious that these vegetables were thriving in the New World. In this view, colonial environments could be seen as exhibiting a breakdown in the natural order, as the proper ecological balance among creatures was thrown out of kilter by human beings, through their having transported plants and animals beyond their proper stations. Thus, in the *Journal of Researches*, Darwin comments that "when man is the agent in introducing into a country a new species, this relation [of one organism to another] is often broken" (30). Colonial natures were thus natures that suffered by virtue of the presence of species that were out of place.

Dung beetles were high on Darwin's list of species to study, because these insects were clearly dependent on other creatures for their existence. Furthermore, they were frequently seen as dramatic proof that all creatures served a useful purpose within an economy of nature where nothing, not even waste, was wasted. Gilbert White had famously used the example of cattle dropping their dung in a stream "in which insects nestle; and so supply food for the fish" to argue that "Nature, who is a great economist, converts the recreation of one animal to the support of another" (*Selborne* 26). Kirby and Spence, in their *Introduction to Entomology*, speak in similarly honorific terms about the high ecological importance of stercoraceous insects: "How disgusting to the eye, how offensive to the smell, would be the whole face of nature, were the vast quantity of excrement daily falling to the earth from the various animals which inhabit it, suffered to remain until gradually dissolved by the rain or decomposed by the elements! That it does not thus offend us, we are indebted to an inconceivable host of insects which attack it the moment it falls" (1:253). Kirby and Spence envision a world steadily raining crap, which would quickly become noxious if not for the unflagging custodial services of a legion of bugs.

Darwin was pleased to see that the relationship between carrion beetles and stinkhorn mushrooms was the same in Rio de Janeiro as in England ("a similar relation between plants and insects of the same families, though the species of both are different" [*Journal of Researches* 30]). Imagine, then, his astonishment when, on visiting the vast livestock raising region of Maldonado, he discovered that there was only one species of dung beetle. Consequently, "the ample repast afforded by the immense herds of horses & cattle" raised there was "almost untouched." How could the idea of design be used to explain waste on such a scale? Once again, however, rather than calling into question his belief in the perfect adaptation of species to place, Darwin marshaled the evidence to support it:

> This absence of Coprophagous beetles appears to me to be a very beautiful fact; as showing a connection in the creating between animals as widely apart as Mammalia & ~~Cole~~ Insects. Coleoptera, which when one of them is removed out of its original Zone, can scarcely be produced by a length of time & the most favourable circumstances.—The same subject of investigation will recur in Australia: If proofs were wanting to show the Horse & Ox to be aboriginals of great Britain I think the very presence of so *many* species of insects feeding on their dung, would be a very strong one.—(*Zoology Notes* 175)

As he later remarked on Saint Helena, "observing what a quantity of food of this kind is lost on the plains of La Plata, I imagined I saw an instance where man had disturbed that chain, by which so many animals are linked together in their native country" (*Journal of Researches* 438). His choice of the word "imagined," however, suggests that by 1836 his views had already changed. In the early notes of the voyage, Darwin understands the travel of colonial natures as disrupting the complex ecological economies exhibited by native environments. These breaks in the ecological links among species, understood as exceptions, confirmed his belief in the rich fabric of design found in those natures unaffected by colonial introductions. Nature in the New World thus provided witness to the excesses, waste, disorder, and loss produced when a creature is "removed out of its original Zone" and introduced into another. No "length of time" or "favourable circumstances," he thought, could bring an absent ecological partner into being (even when he was writing these words, naturalists were introducing, often with disastrous results, new species to counter the spread of invasive plants and animals). At the same time, Darwin must have taken some pleasure in grounding England's national history in scatology, claiming that the ancient right of place of its horses and oxen was evidenced by the multitude of insects that over the ages had been feeding on their dung.

Darwin's understanding of ecology in terms of an original design was changed by his reading of the second volume of Lyell's *Principles*, which he obtained in Montevideo late in 1832. Through Lyell, he was forced to reconsider how ecological partnerships might function in a world characterized by the mobility of plants and animals. Lyell's impact on Darwin, who dedicated the *Journal of Researches* to him, has long been recognized.[2] In an 1835 letter to Will Fox, he confessed himself "a zealous disciple of Mr Lyells views" (9–12 Aug. 1835, *Correspondence* 1:460), and a decade later he would write that "I always feel as if my books came half out of Lyell's brains & that I never acknowledge this sufficiently.... I have always thought that the great merit of the Principles, was that it altered the whole tone of one's mind & therefore that when seeing a thing never seen by Lyell, one yet saw it partially through his eyes" (Darwin to Leonard Horner, 29 Aug. 1844, *Correspondence* 3:55). Whereas the first volume of *Principles* dealt with geology, the second demonstrated that "incessant change" (2:2) governed the organic world, with species continually coming into being and disappearing, a point emphatically made by the volume's epigraph, drawn from John Playfair's *Illustrations of the Huttonian Theory* (1802): "The inhabitants of the globe, like all the other parts of it, are subject to change. It is not only the individual that perishes, but whole species" (i). In Lyell's view, the organic world, like the physical globe, was subject to constant change, with new species appearing and others disappearing when the physical environments that they depended on changed. Lyell believed in the reality of species, seeing them as stable biological units, created to suit the environments in which they were found, that could vary only within certain limits, so without a theory of speciation, he could not explain how new species came into being at a given time or place. However, he clearly believed that he knew why species disappeared. Much of volume 2 is devoted to showing that "the successive destruction of species" was "part of the regular and constant order of Nature" (141).

Lyell's explanation for why species disappeared was relatively straightforward, though revolutionary. Created to be perfectly adapted to their environments, species of plants and animals disappeared when their physical habitats changed through alterations in climate, soil, humidity, surface elevation, and so on. Lyell also believed that extinct species were replaced by new ones, brought into being in different places at different times. Strikingly, though Lyell emphasized the geobiological dimensions of extinctions, he also recognized that the ecological changes wrought by the introduction or disappearance of a single species from an environment also needed to be considered. Even the slightest alteration in the complex and myriad organic relationships constituting a habitat, whether from physical causes or from the introduction of a new species, could have enormous consequences over time: "every new condition in the state of the organic or inorganic creation, a new animal or plant, an additional snow-clad mountain, any permanent change, however slight in the comparison to the whole, gives rise to a new order of things, and may make a more material change in regard to some one or more species" (*Principles* 2:146). Biological Malthusianism, the idea that habitats were structured by the competition of species for limited resources and that "the increase of one animal necessarily offers" a "check" to "that of another" (154), provided the mechanism by which slight changes in an environment would become magnified over time.[3]

In laying the groundwork for a new conception of ecology, focused on the dynamics of species populations, Lyell placed less emphasis on individual species than on interspecies partnerships and competition. "In this continual strife," he writes, "it is not always the resources of the plant itself which enable it to maintain or extend its ground. Its success depends, in a great measure, on the number of its foes or allies among the animals and plants inhabiting the same region" (*Principles* 2:131). In arguing that extinctions could occur through the introduction of a new species, Lyell was the first to theorize the ecological consequences of biological invasions. Where

Fleming had equated extinctions with the ongoing warfare between human beings and animals, Lyell provided a more complex account in which extinction was caused by changes in the relationships among competing and allied species within an environment. In a *tour de force* example, Lyell sketched out in broad strokes the impact that the appearance of Greenland bears would have on the ecology of Iceland: first, it would lead to the destruction of deer, foxes, seals, and birds, but this, in turn, would give rise to a corresponding increase in the plants, birds, and fish that these animals fed on; more plants would support more insects, while the increase in waterfowl would cause a decrease in the fish that they preyed on. Lyell thus concluded that "the numerical proportions of a great number of the inhabitants, both of the land and sea might be permanently altered by the settling of one new species in the region, and the changes caused indirectly might ramify through all classes of the living creation, and be almost endless" (144). Since most parts of the globe were already sufficiently stocked with as many species as these environments could support, the spread of a species into a new territory must inevitably destroy the entire ecological balance of a habitat and "must always be attended either by the local extermination or the numerical decrease of some other species" (142).

Lyell was a uniformitarian, believing, as he wrote in a letter to Robert Murchison, that "*no causes whatever* have . . . ever acted, but those *now acting*, and that they never acted with different degrees of energy from that which they now exert" (*Life, Letters, and Journals* 1:234). Nature past and nature present were governed by the same laws and processes, so Lyell was able to explain the past by referring to laws and processes that he could observe operating in the present. In arguing that extinctions were caused not only by physical changes in the earth but also by the emergence of new ecological relationships among species, through the "appearance of an animal or plant, in a region to which it was previously a stranger" (*Principles* 2:145), Lyell was well aware that he had no historical evidence for this claim. "Our knowledge of the history of the animate creation dates from so recent a period," he writes, "that we can scarcely trace the advance or decline of any animal or plant" (143). Having apparently come to a methodological dead end, he solved it by adapting Fleming's position that recent history provided numerous examples of extinctions "in those cases where the influence of man has intervened" (143). In turning to human agency to explain the role that the introduction of new plants and animals might play in the ecology of extinctions, Lyell certainly anticipated Darwin's similar use of artificial selection in order to explain natural selection. More importantly, however, by so doing he gave his biogeography a strongly colonialist slant, as he argued both that colonialism should be understood in biological terms and that the power of human beings to change environments might illustrate the fundamental laws of nature. Lyell made the colonial world (the present for a nineteenth-century uniformitarian) the template for explaining the past and establishing the laws of nature. The struggle for existence among species was recast as the often highly mediated colonial struggle between a foreign species seeking to expand its range (along with that of its biotic allies) and a native or indigenous assemblage of biota. Rather than considering colonial ecologies as being exceptional, therefore, Lyell saw them as illustrating the fundamental laws and processes governing nature at all times and in all places. The parallel between present-day colonial realities and Lyell's theory of nature is explicit in Lyell's use of the "periodical invasion" of Greenland bears in Iceland to illustrate what happens when "some new colony of wild animals or plants enters a region for the first time, and succeeds in establishing itself" (143).[4] It is not by accident that Lyell compares these invasions to the "marauding expeditions" of "the Danes of old" (143).

Lyell nevertheless still faced another challenge, for the speed, scope, and degree of change being wrought by human agency would seem to call into question any idea of the uniformity between colonial natures and nature itself. "Is not the interference of the human species . . . such a deviation from the antecedent course of physical events, that the knowledge of such a fact tends to destroy all our confidence in the uniformity of the order of nature, both in regard to time past and future?" he asked (*Principles* 1:156). His answer was that in seeking to dominate nature, human beings, though of relatively recent origin, were actually acting like any other animal species. Reason had made it possible for them to travel and to colonize new territories to a far greater degree than other creatures, but in introducing ecological revolutions they were acting as any other animal in a similar situation would if it could. Migration, colonization, and settlement were not unique to the human species, but were instead expressive of a drive within all species to colonize and populate the earth with their kind. Human beings might be more successful at expansion and settlement than other species, but all animals were would-be settlers. Furthermore, Lyell's uniformitarianism allowed him to understand modern colonialism as simply a more dramatic expression of what human beings have been doing since they first appeared on earth. "When a powerful European colony lands on the shores of Australia, and introduces at once those arts which it has required many centuries to mature; when it imports a multitude of plants and large animals from the opposite extremity of the earth, and begins rapidly to extirpate many of the indigenous species, a mightier revolution is effected in a brief period, than the first entrance of a savage horde [but] ... assuming that the system is uniform ... we can with much greater confidence apply the same language to those primeval ages" (157). Colonialism, in other words, was not a new kind of activity, but was as old as the human species, or any other species for that matter.

Principles of Geology extended and deepened Fleming's recognition of the role that human agency had played in determining the global distribution of plant and animal species, yet Lyell was under no illusion that human beings were "improving" global ecologies. In his view, we "impoverish the lands which we occupy": "Man is, in truth, continually striving to diminish the natural diversity of the *stations* of animals and plants in every country, and to reduce them all to a small number fitted for species of economical use" (*Principles* 2:147–48). Where Fleming questioned the ethics of local exterminations and species extinctions, Lyell seems to have viewed them with pride: "these changes ... cannot fail to exalt our conception of the enormous revolutions which, in the course of several thousand years, the whole human species must have effected" (150).

Lyell thus presents a dark imperial vision of a future earth colonized by man and the "limited number of plants and animals which he has caused to increase":

> If we wield the sword of extermination as we advance, we have no reason to repine at the havoc committed.... Every species which has spread itself from a small point over a wide area, must, in like manner, have marked its progress by the diminution, or the entire extirpation, of some other, and must maintain its ground by a successful struggle against the encroachments of other plants and animals.... The Hessian fly, the locust, and the aphis, caused famines ere now amongst the "lords of creation." The most insignificant and diminutive species, whether in the animal or vegetable kingdom, have each slaughtered their thousands, as they disseminated themselves over the globe, as well as the lion, when first it spread itself over the tropical regions of Africa. (*Principles* 2:156)

Lyell produced a natural history that reflected his times, for he naturalized European colonialism, seeing the spread of species and the life-and-death struggles that it created as a direct expression of the laws of nature. In *Principles of Geology*, the global expansion of Europeans and their domestic biotic allies was simply fulfilling a territorial drive that was present in all species. From this perspective, the European colonization of the earth was no different than the invasion of Hessian flies in the United States or the expansion of lions across Africa. *Principles of Geology* is a book about the natural revolutions that had taken place across the globe and across time. Human agency was speeding up this process, but the process itself, which left a succession of extinctions in its wake, was neither new nor unnatural. Here we have the beginnings of the theorization of the *end of nature*, following a trajectory from Fleming, through Lyell, to Darwin, as well as Alfred Wallace, who envisioned a "time when the earth will produce only cultivated plants and domestic animals; when man's selection shall have supplanted 'natural selection'; and when the ocean will be the only domain in which that power can be exerted, which for countless cycles of ages ruled supreme over the earth" ("Origin of Human Races" 52).

Ecologies in Motion

Volume 2 of *Principles of Geology* could not have appeared at a better time for Darwin, because the ecological changes taking place in America were the showpiece of Lyell's account of the revolutionary capacity of Europeans to transform global habitats to meet their needs:

> The fact of so many millions of wild and tame individuals of our domestic species, almost all of them the largest quadrupeds and birds, having been propagated through the new continent within the short period that has elapsed since the discovery of America, while no appreciable improvement can have been made in the productive powers of that vast continent, affords abundant evidence of the extraordinary changes which accompany the diffusion and progressive advancement of the human race over the globe. That it should have remained for us to witness such mighty revolutions is a proof, even if there was no other evidence, that the entrance of man into the planet is, comparatively speaking, of extremely modern date, and that the effects of his agency are only beginning to be felt. (*Principles* 2:155)

The impact of Lyell on Darwin's understanding of ecology can be seen in his September 1833 description of the changing character of the surface vegetation, near Buenos Aires and Guardia del Monte, from a "coarse herbage" to "a carpet of fine green verdure." At first, he had assumed that this change was being caused by the soil, but he soon discovered that "the whole was to be attributed to the manuring and grazing of the cattle" (*Journal of Researches* 106). Based on Caleb Atwater's observation in the first volume of Silliman's *American Journal of Science* that the high grasslands of the prairies had been changed into pastureland through cattle grazing, Darwin initially speculated that grazing had altered the growth patterns of the native grasses. However, recalling Felix Azara's observation that foreign plants often first appeared along "any track that leads to a newly-constructed hovel," Darwin wondered whether domestic animals and their dung might not also be the means by which European plants had spread into interior regions, as "lines of richly manured land" served "as channels of communication across wide districts" (106). Here manure is as important as rivers in the expansion of Europeans and their biota, and Darwin suggests that a new kind of mapping is necessary to understand the dynamics of ecological change in the colonial world. Such a mapping would no longer be focused on the "rootedness" of species, but their "routedness," as Darwin was now seeing biological distribution in terms of the pathways (often surprising ones) taken by plants and animals as they expanded across the globe.

From thinking about *ecologies in place*, Darwin was now thinking of *ecologies in motion*, and this shift in perspective only deepened over his career. He noted that many European domestic plants were successfully colonizing the New World as weeds. Fennel, he remarks, was prolific in "the ditch-banks in the neighbourhood of Buenos Ayres, Monte Video, and other towns." Even more pernicious was the cardoon or artichoke thistle (*Cynara cardunculus*), which Darwin encountered in "unfrequented spots" in many parts of Chile, Northern Argentina, and Uruguay. Vast areas of the latter were now covered to a depth of "very many (probably several hundred) square miles" with "one mass of these prickly plants ... impenetrable by man or beast" (*Journal of Researches* 107). On 19 September 1833, in his diary, Darwin indicates his amazement that a domestic plant could become so prolific: "The whole country between the Uruguay & M. Video is choked up with it; yet Botanists say it is the common artichoke, run wild" (*Beagle Diary* 190). Commenting on the massive ecological impact that cardoon was having on indigenous plant species, Darwin wrote, "I doubt whether any case is on record of an invasion on so grand a scale of one plant over the aborigines" (*Journal of Researches* 107). His use of the word "invasion" suggests the manner in which Darwin was understanding colonialism as a biological event in which new settler species and their ecological partners seized territory from indigenous populations. In most cases, the outcome of this competition was not the maintenance of an established natural order, but instead the displacement of one nature by another. Darwin writes, "Over the undulating plains, where these great beds occur, nothing else can now live. Before their introduction, however, the surface must have supported, as in other parts, a rank herbage" (107).

At this point, Darwin refers directly to "the principles so well laid down by Mr Lyell," suggesting that

> few countries have undergone more remarkable changes, since the year 1535, when the first colonist of La Plata landed with seventy-two horses. The countless herds of horses, cattle, and sheep, not only have altered the whole aspect of the vegetation, but they have almost banished the guanaco, deer and ostrich. Numberless other changes must likewise have taken place; the wild pig in some parts probably replaces the peccary; packs of wild dogs may be heard howling on the wooded banks of the less-frequented streams; and the common cat, altered into a large and fierce animal, inhabits rocky hills. . . . No doubt many plants, besides the cardoon and fennel, are naturalized; thus the islands near the mouth of the Parana, are thickly clothed with peach and orange trees, springing from seeds carried there by the waters of the river. (*Journal of Researches* 107)

The ecological changes taking place in South America were "numberless" and ongoing, yet they had little to do with soil or climate, arising instead from the competition brought by species coming from elsewhere. The idea that ecological changes were caused by the mobility and resulting competition of species would come to shape Darwin's understanding of the geographical distribution and extinction of species. It would become a foundational element of his understanding of the history of the earth.

Darwin's visit to Saint Helena in July 1836 provided a culminating reflection on the enormous ecological changes that were affecting colonial environments. In a letter to his sister Caroline, he described the island as "a curious little world within itself" (18 July 1836, *Correspondence* 1:502). Prior to European contact, the flora of this isolated island was limited to about eighty species, most of them endemic. At the time of its discovery in 1502, the island had been richly covered with forests and other vegetation, but in less than three centuries most of the island had become a desert. The speed with which Saint Helena had passed from a tropical paradise to a wasteland placed it "at the Romantic centre," as Richard Grove suggests, of "debates and anxieties" about the relationship between human agency, climate change, and species extinction (*Green Imperialism* 343). In *Tracts Relative to the Island of St. Helena; Written During a Residency of Five Years* (1816), Alexander Beatson, its governor from 1808 to 1813, summarized the accepted view that it was now a "rocky and unproductive island; mostly devoid of soil; scantily supplied with water; subject to severe and unusual droughts; abounding with rats, and wholly incapable of extensive cultivation, or improvement" (vii).[5] Following in the footsteps of Sir Joseph Banks, Beatson believed that the original fertility of the island might be recovered and the aridity of the climate reduced by reforesting the island with pine plantations and plants from other parts of the globe. He also introduced European agricultural methods and indentured Chinese laborers.

When Darwin arrived, the impact of this activity on the ecological character of the island was manifest: a degraded tropical island had been transformed into a sad imitation of Wales. "It is surprising," he writes, "to behold a vegetation possessing a character decidedly British. The hills are crowned with irregular plantations of Scotch firs; and the sloping banks are thickly scattered over with thickets of gorse, covered with its bright yellow flowers. Weeping-willows are common on the banks of the rivulets, and the hedges are made of the blackberry, producing its well-known fruit" (*Journal of Researches* 435). Everywhere Darwin looked, he saw dramatic evidence of the ecological changes wrought by colonialism. The extent to which the indigenous flora had been displaced by newcomers was astounding. Using Sir William Roxburgh's annotated list of the flora of Saint Helena, compiled in 1808 and appended to Beatson's *Tracts*, Darwin calculated that of the 476 species that were to be found on the island, only 52, about 11 percent, were native (*Beagle Diary* 428). By 1839, when the *Journal of Researches* was published, the list of introduced species had increased to 694 (435). Much of the responsibility for this new state of affairs lay with Beatson, who, as Joseph Hooker commented in 1866,

> proposed and carried out the introduction of exotic plants on a large scale, and from all parts of the world; these have propagated themselves with such rapidity, and grown with such vigour, that the native plants cannot compete with them. The struggle for existence had no sooner begun, than the issue was pronounced; English Broom, Brambles, Willows and Poplars, Scotch Pines and Gorse bushes, Cape of Good Hope bushes, Australian trees and American weeds, speedily overran the place; and wherever established, they

have virtually extinguished the indigenous Flora, which, as I said before, is now almost confined to the crest of the central ridge. ("Insular Floras" 67)

Hooker's use of the Darwinian phrase "struggle for existence" brings evolutionary language back to the colonial contexts from which it arose. The natural history of Saint Helena, as Hooker further noted, "has undergone such a revolution within the last 400 years, as under the ordinary operations of Nature can only be measured by the geological chronometer" (63). Colonial change, in other words, was geological time moving in fast motion.

Darwin had also recognized the degree to which European plants were colonizing the island at the expense of the native flora, and he concluded that "it is not improbable that even at the present day similar changes may be in progress" (*Beagle Diary* 428). He observed that the remnants of the native vegetation were now restricted to the farthest removes of the island, mostly the "highest and steepest ridges" (*Journal of Researches* 435). Darwin also noted that originally the island had neither mammals nor landbirds (he apparently was unaware of its one endemic species, the wirebird). When he observed its chukar partridges and ring-necked pheasants, originally introduced by the Portuguese, he assumed that they had been brought by the British, concluding that with the birds had also come English game laws. "The island is much too English," he writes, "not to be subject to strict game-laws" (438). The colonial ordinances, however, were more severe than their English counterparts, for they applied not only to the birds but also to the grass that they used for nesting. "The poor people formerly used to burn a plant, which grows on the coast-rocks, and export the soda from its ashes; but a peremptory order came out prohibiting this practice, and giving as a reason that the partridges would have nowhere to build" (439).

In the interior of the island, Darwin discovered eight species of land snail, originally thought to be marine snails, which he quickly recognized as having recently become extinct. He conjectured that their destruction was connected with the disappearance of the Great Wood, composed of gumwood, redwood, and ebony trees, which had once covered the uplands. Now there was only pastureland with "scarcely a tree [to] be found" (*Journal of Researches* 437). For Darwin, the circumstances surrounding the disappearance of the forest and the extinction of land snails "of a very peculiar form" (437) provided a dramatic Lyellian illustration of the intricate interdependencies among species within an environment and of the enormous impact that the appearance of a single plant or animal could have on an entire ecology. Although the forest had disappeared early in the eighteenth century, the cause of its destruction could be traced back to the beginning of the sixteenth century, when the Portuguese introduced goats to the island. The animals reproduced so rapidly that by the time Captain Thomas Cavendish visited the island in 1588, feral goats numbered in the thousands and a single flock could often be a mile long (Hooker, "Insular Floras" 66). The goats fed on the young saplings, so without a means of renewal the entire forest gradually disappeared as the aging trees died or were used for tanning or producing lime. In 1731, an order was finally issued to destroy the feral goats, but by this time, Darwin writes, "the evil was complete and irretrievable." The goats had "change[d] the whole aspect of the island," destroying not only the forests and the land snails that lived there but also "a multitude of insects" (*Journal of Researches* 438).

Darwin's observation of the impact of European colonialism on native peoples provided additional evidence of the violent life-and-death struggles arising from the migration and settlement of newcomers. In South America, he first encountered the brutalities of slavery, and after witnessing firsthand the ruthless "war of extermination" being waged by General Juan Manuel de Rosas against the native peoples, he would write that "not only have whole tribes been exterminated, but the remaining Indians have become more barbarous: instead of living in large villages, and being employed in the arts of fishing, as well as of the chase, they now wander about the open plains, without home or fixed occupation" (*Journal of Researches* 92–93). Near the end of the voyage, he visited Tasmania, where all of the Aboriginal people had just been forcibly removed to Flinders Island. "Thirty years," he would remark, "is a short period, in which to have banished the last aboriginal from his native island,—and that island nearly as large as Ireland" (399). When, in developing the theory of natural selection, Darwin would argue that the struggle for existence is most intense among members of the same or closely allied species rather than between different species, he must have had the violent exterminations affecting members of his own species much in his mind.

Upon returning to England, Darwin did not have a working theory of speciation, but he had fully adopted Lyell's powerful argument that the colonial world could provide a scientific observer with valuable evidence for conceptualizing aspects of the natural history of the earth which were otherwise lost in time. Instead of seeing colonial environments as exceptions to a divinely authorized balance among species operating across the globe, Darwin now saw them as a means for observing the actual laws of nature operating in fast motion. Movement and change now dominated his understanding of nature. With the enormity of the ecological changes wrought by the European colonization of South America clearly in view and simultaneously the changes in species registered by the fossil record, Darwin commented that "it is impossible to reflect on the changed state of this continent without the deepest astonishment" (*Journal of Researches* 154). Yet all of these changes had taken place without any corresponding alteration in either the climate or physical environment of the continent. This situation must have been a factor in Darwin's tendency over the next two decades to move away from seeing geophysical factors as being central to speciation toward understanding the natural world as being the product of intra- and interspecies competition for territory in contexts where ecological change was triggered not only by geological factors but also by the immigration of species from elsewhere. "In considering the distribution of organic beings over the face of the globe," he writes, "neither the similarity nor the dissimilarity of the inhabitants of various regions can be accounted for by their climatal and other physical conditions" (*Origin* 346). In moving toward a model in which human beings were much closer to animals than might otherwise have been thought, Darwin was also moving toward seeing nature as being much more like human life than had been realized. He relinquished a model in which nature was understood as being stationary, held in balance by species that were fitted for the places in which they were found, to one in which balance was at best a temporary state in the life-and-death competition among inherently mobile populations. Instead of seeing nature as naturally seeking to occupy its place, he now understood it in colonial terms, as being caught up in change, structured by the constant pressures of migration and settlement and by the ecological struggles to which this mobility of species gave rise. Darwin was seeing organic nature as being shaped by the same struggles that were at work in the modern world, between the modern and the traditional, newcomers and indigenes, populations that moved and those that did not. He returned to England, in other words, with an idea of nature as being thoroughly *modern*. If human beings were the most successful species on earth, they owed that preeminence to their having somehow obtained advantages over the species with which they competed, the most important among these being their capacity to travel to new places and to adapt to the environments that they encountered there. Mobility had thus become integral to the consideration of biological change.

Evolution and Mobility

With Darwin, the first truly modern conception of nature emerges, one that does not understand the forces shaping the natural world as being fundamentally different from those shaping modern life. Historians of science have long recognized the extent to which Darwin's ideas about the benefits of competition among species and the increasing specialization of organic forms can be seen, as Silvan S. Schweber suggested, as "'biologizing' the explanations political economy gave for the dynamics of the wealth of nations" ("Darwin and the Political Economists" 212).[6] Karl Marx was the first to note the parallel, commenting in an 18 June 1862 letter to Friedrich Engels that "it is remarkable how Darwin recognises among beasts and plants his English society with its division of labour, competition, opening up of new markets, 'inventions,' and the Malthusian 'struggle for existence'" (*Selected Correspondence* 128). In arguing that Darwin found "English society" in the "beasts and plants" that he studied, critics have too narrowly defined the politics of his work, missing the extent to which Darwin had a different objective in view, that of contextualizing British imperial society by writing a global natural history of colonialism, one that would include not only human beings but also plants and animals within its purview.

Although Lyell provided a powerful biogeographical account of the role that mobility and interspecies competition played in the extinction of species, his belief that species were real and could vary only within strictly defined limits prevented him from considering how mobility and interspecies competition might give rise to changes in biological form. This task was left to Darwin. From the beginning, evolutionary theory was as much a theory about the historical movement of biological forms through space as it was an explanation of their descent in time.[7] One of the reasons space played such an important role in Darwin's understanding of the dynamics of speciation was that the problems that most concerned him had less to do with paleontology than with explaining why species were distributed across the earth in the way that they were. As R. Alan Richardson has suggested, biogeography, or "geographical distribution," as it was then called, provided "the data base for Darwin's thought" ("Biogeography and the Genesis" 7) and "the *only* sufficiently elaborated body of knowledge on the relationship of species to their environment, and to changes in that environment, that could constitute a valid test case for Darwin's developing theoretical speculations on transmutation and his search for an efficient cause for change"

(6–7). The relationship between form and geographical movement was consequently essential to Darwin's ideas about evolution.[8] Space and geography, along with biological history, clearly mattered in explaining the forms that species had taken and the places where they were found. As early as 1835, Darwin had recognized that the geographical distribution of mockingbirds and tortoises on the Galapagos Islands and of foxes on the Faulkland Islands was "worth examining; for such facts would undermine the stability of species" ("Ornithological Notes" 262). Three years later, in Notebook C, he was postulating that the "geographi[cal] distribution of animals" provided a "new step in induction" as a "keystone of ancient geography," with "species tell[ing] of Physical relations in time" and "forms and distribution tell[ing] of horizontal barriers" (*Notebooks* 275). By 1845, Darwin dubbed it "that grand subject, that almost key-stone of the laws of creation, Geographical Distribution," and he would hold this viewpoint for the rest of his life (Darwin to J. D. Hooker, 10 Feb. 1845, *Correspondence* 3:140).

Darwin's fascination with biogeography, born from his reading of Lyell, was motivated by his recognition that none of the prevailing theories of creation, with their emphasis on the rootedness and fixity of species, could explain, without major inconsistencies or unjustified claims, the puzzling distribution of plants and animals across the earth. Evolutionary theory emerged as an explanation of these patterns which did not require the appeal to special forms or instances of creation, and it was based on three fundamental commitments: (1) that change was brought about by normal reproduction, (2) that biota were capable of extensive migration, and (3) that their forms had been modified by that travel. Darwin's understanding of the relationship between speciation and mobility, however, underwent extensive revision over the two decades leading up to the publication of the *Origin of Species*. Whereas his early ideas on evolution saw speciation as occurring primarily on islands, where isolation, physical barriers, and other checks on the immigration of foreign species could provide incipient species, adapting to geological changes in their environments, with time to develop into new species, in his later theory Darwin sought to explain how speciation might operate among "the commonest and the most widely-diffused" species (*Origin* 117), those highly mobile, ever-varying, and ever-more-numerous populations whose main evolutionary advantage lay in their capacity to extend their ranges by adapting themselves to new environments and vanquishing competitors. He thus shifted his focus from seeing evolution as a process that was integrally bound up with species occupying boundaries to understanding it as being fundamentally associated with those species most capable of wide-ranging settlement. Underlying this shift was Darwin's recognition that evolution would explain the dominance of certain biological forms in the modern world.

Darwin's commitment to the idea that the distribution of biota was explicable in terms of the different migrational capacities of biological forms distinguishes evolutionary theory from the work of most of his contemporaries. "All the grand leading factors of geographical distribution," he writes in the *Origin of Species*, "are explicable on the theory of migration (generally of the more dominant forms of life), together with subsequent modification and the multiplication of, forms" (408). As Janet Browne explains, "migration and modification were the keys to geographic arrangement.... The organic world thus stood revealed as a product of the unceasing processes of distribution" (*Secular Ark* 219–20). Since Darwin rejected the idea that the same species might have been created in more than one place at different times, arguing instead that every species has come into being "in one area alone" and has "subsequently migrated from that area as far as its powers of migration and subsistence under past and present conditions permitted" (*Origin* 353), migration was a primary component of his theory. In the *Natural Selection* manuscript, written in 1856–58, he declares that if it ever "were proved or rendered probable" "that the same species has ever appeared, independently of migration, on two separate points of the earth's surface... the whole of this volume would be useless" (566). The prevalent theory of multiple centers of creation conformed to the dominant view that nature was structured by boundaries and that plants and animals had limited capacities for dispersal. Thus, the most extreme proponent of this view, Louis Agassiz, would argue that "a Supreme Intelligence... created, at the beginning, each species of animal at the place, and for the place, which it inhabits" (Agassiz and Gould, *Principles of Zoology* 177); "There is only one way to account for the distribution of animals as we find them, namely, to suppose that they are autochthonol, that is to say, that they originated like plants, on the soil where they are found" (179). The *Origin of Species* profoundly undercuts the association between nature and place, the soil–blood–identity triad that shapes much of the racialist and nationalist claims of the nineteenth century, by arguing that species were not created for the places they currently occupy, but instead that they have come to occupy them through a tangled history of migration, settlement, and transformation. Darwin's earth is populated not by Agassiz's native *autochthonoi*, but instead by migrants become settlers. It is the product of extensive biological colonization, as ancient species that had evolved the means or found the opportunity to travel to new places both changed and were changed by the places where they settled. That is why, "on the same continent, under the most diverse conditions, under heat and cold, on mountain and lowland, on deserts and marshes, most of the inhabitants within each great class are plainly related: for they will generally be descendants of the same progenitors and early colonists" (*Origin* 477). Migration and modification are the central components of Darwin's biogeography. Commenting on the character of the fauna and flora of oceanic islands, Darwin argued that they were the product "of colonization from the nearest and readiest source, together with the subsequent modification and better adaptation of the colonists to their new homes" (406).

Colonization plays such an important role in Darwin's evolutionary theory because, within his Malthusian model, the struggle for territory constitutes the primary mechanism by which the relative success of biological forms is registered. "Every single organic being . . . may be said to be striving to the utmost to increase in numbers" (*Origin* 66), writes Darwin, so evolution is fought out most directly in the struggle for new territory. Consequently, when Darwin speaks of the "improvements" that arise from natural selection, these have little to do with notions of beauty, usefulness, or goodness, but instead refer to those aspects of behavior and form which allow an organism to increase the territory that it occupies, usually at the expense of others. The close connection between territorial struggle and Darwin's concept of "adaptation" is clear in his argument that spaces do not need to be "new and unoccupied" for them to be "open" to being "fill[ed] up" by new and "improved" biological forms. "I [do not] believe," he writes, "that any great physical change, as of climate, or any unusual degree of isolation to check immigration, is actually necessary to produce new and unoccupied places for natural selection to fill up by modifying and improving some of the varying inhabitants. For as all the inhabitants of each country are struggling together with nicely balanced forces, extremely slight modifications in the structure or habits of one inhabitant would often give it an advantage over others; and still further modifications of the same kind would often still further increase the advantage" (82). Given the constant variation among competing species, any change in form that gives an organism a new competitive advantage constitutes an "opening"—that is, an opportunity—that will be decided on the ground by its success in acquiring new territory and thus increasing its numbers. Darwin writes, "No country can be named in which all the native inhabitants are now so perfectly adapted to each other and to the physical conditions under which they live, that none of them could anyhow be improved; for in all countries, the natives have been so far conquered by naturalised productions, that they have allowed foreigners to take firm possession of the land. And as foreigners have thus everywhere beaten some of the natives, we may safely conclude that the natives might have been modified with advantage, so as to have better resisted such intruders" (82–83). By using colonial history and the territorial struggles between "natives" and "foreigners" to explain why spaces in the natural world can always be "improved," that is, "filled" by others who use their biological "advantages" to "take firm possession of the land," Darwin makes colonial settlement the basic model for understanding speciation. Drawing upon the same arguments that British colonists used to justify their right to indigenous lands that they believed had not been used to their full potential, Darwin thus makes colonization one of the primary mechanisms of natural selection.

From his knowledge of taxonomy and of contemporary breeding practices, Darwin was fully aware that differences in form regularly appear among the individuals of any species that reproduces sexually. As he commented, "Unless profitable variations do occur, natural selection can do nothing" (*Origin* 82). Yet the continuous generation of difference was not in itself sufficient to produce speciation, for the real question was how these differences were preserved, increased, and given a direction. "How . . . does the lesser difference between varieties become augmented into the greater difference between species?" Darwin asked (111). Darwin understood speciation as a sequence of "very short and slow steps" (471) in which a change in form provides the basis from which subsequent forms could develop: "I look at individual differences, though of small interest to the systematist, as of high importance for us, as being the first step towards such slight varieties as are barely thought worth recording in works on natural history. And I look at varieties which are in any degree more distinct and permanent, as steps leading to more strongly marked and more permanent varieties; and at these latter, as leading to sub-species, and to species" (51–52). Darwin may appear to be using "step" abstractly, but we should never lose sight of the ground on which speciation takes place; it always has a physical dimension, operating in space as well as in time, as the very forms of organisms are themselves the means by which they occupy and resettle space. For Darwin, every new form constitutes a step leading in some direction, so every biological form is not so much a *point of origin* as a *point of departure*, a *branching*, from whence new steps in the evolution of biological forms (from individual differences, to varieties, to subspecies, to species and, then, to new species) can be taken. What appears to be stable is actually "an extremely slow process" (302) of movement and settlement operating across countless generations. Given this highly fluid conception of speciation, the distinction between a variety and a species was, for Darwin, essentially arbitrary: "It will be seen that I look at the term species, as one arbitrarily given for the sake of convenience to a set of individuals closely resembling each other, and that it does not essentially differ from the term variety, which is given to less distinct and more fluctuating forms. The term variety, again, in comparison with mere individual differences, is also applied arbitrarily, and for mere convenience sake" (52). But how do biological "differences" take on the directionality that allows them to be seen as "steps"? Whereas in the case of artificial selection it is the breeder who, by isolating and controlling a breeding population, is able to give reproductive differences a direction, Darwin looked to the struggle for territory as providing the essential mechanism that would differentiate among competing populations. Modifications in form that increased an organic being's capacity to colonize new areas would be preserved by an increase in the population that exhibited this character.

Gillian Beer's observation that Darwin's metaphors often overturn "the bounds of meaning assigned to them" (*Darwin's Plots* 51) certainly applies to his use of the traditional "great Tree of Life" as a metaphor for speciation as a branching of descendants from a common ancestor:

> The affinities of all the beings of the same class have sometimes been represented by a great tree. . . . The green and budding twigs may represent existing species; and those produced during each former year may represent the long succession of extinct species. . . . As buds give rise by growth to fresh buds, and these, if vigorous, branch out and overtop on all sides many a feebler branch, so by generation I believe it has been with the great Tree of Life, which fills with its dead and broken branches the crust of the earth, and covers the surface with its ever branching and beautiful ramifications.
> (*Origin* 129–30)

In adopting this metaphor, Darwin wanted to represent speciation as a process that operated not only across time but also across space—in Lyell's apt phrase, which Darwin cites in the *Essay of 1844*, "as in space, so in time"

(*Foundations* 171). Trees are normally associated with rootedness, not with mobility, so it is understandable why readers have tended to understand speciation as something that occurs "in place" rather than "across space." Since Darwin's "tree of life" looks like a "genealogical tree," it is not surprising that his contemporaries saw it as being only a representation of the phylogenetic descent of species through time, and Ernest Haeckel would adopt this idea in using phylogenetic trees to represent morphological lines of descent. Yet Darwin did not focus exclusively on the growth of the tree in time, for he also stressed that the tree's "ever branching" growth ("ramifications" being the botanical term for that process) was the means by which it spreads its branches over the entire earth, "fill[ing]" and "cover[ing] the surface." For Darwin, speciation was a branching process, and he turned to the concept of "divergence," a word whose etymology is derived from movement, to capture the idea of speciation as a point of departure, a step, or pathway, the "turning apart" by which a species set out on a new path across time and space by slowly stepping apart from itself (in form, space, and time).

For Darwin, the "principle of divergence" explained how organic beings settle the earth by virtue of their unique forms. "Divergence" and "diversity" are etymologically related, and Darwin drew upon that connection to suggest that it was through diversification of form that species were able to seize on and occupy a greater diversity of environments. Divergence, in other words, creates the biological diversity that it exploits, as a species' descendants change in order to settle new ecological environments. In the *Origin of Species*, Darwin placed less emphasis on species adaptation than on its comparative advantage over other competitors in the struggle for existence. As Darwin suggests, if by the slightest changes in form a species can gain "an advantage over some other inhabitant of the country, it will seize on the place of that inhabitant, however different it may be from its own place" (*Origin* 186). Divergences in form increase the opportunities for competitive advantages, which, in turn, lead to the diversification of ecological environments; "the more diversified the descendants from any one species become in structure, constitution, and habits, by so much will they be better enabled to seize on many and widely diversified places in the polity of nature" (112). In his strongest metaphor of the violent interspecies and intraspecies competition for space, Darwin likens "the face of Nature . . . to a yielding surface, with ten thousand sharp wedges packed close together and driven inwards by incessant blows, sometimes one wedge being struck, and then another with greater force" (67). In Notebook D, where Darwin first conceived this metaphor, he further clarifies this process, suggesting that species use their forms to force themselves "into the gaps in the economy of nature, or rather forming gaps by thrusting out weaker ones" (*Notebooks* 375). Instead of simply filling gaps in nature, therefore, divergence creates new biological forms that seize on new places by displacing the prior occupants (which are often their biological antecedents). Through changes in form, organisms are able to establish new biological beachheads that allow them to increase and thus to preserve morphological differences. When Darwin claims that the "modified descendants of any one species will succeed by so much the better as they become more diversified in structure, and are thus enabled to encroach on places occupied by other beings" (*Origin* 116), he is grounding the principles governing speciation in Malthusian battles fought on the ground, driven by every species' goal of settling as much territory as it can. Natural selection and form changing were thus inherently bound up with mobility and settlement in Darwin's thought.

In a world shaped by movement, the distinction between *speciation* and *dispersal*, though useful in practice, is no less arbitrary than the distinction between species and varieties, for it assumes that species come into being in place and then get dispersed, as if the form of a species were not modified by its travels, or its capacity to travel and settle in new places not also linked to its form. In Darwin's biogeography the migration and settlement of species to new regions are important aspects of speciation itself. Migration introduces traveling organisms to new environments and new competitors and native species to foreign ones, which produce the ideal conditions for competition and speciation. Also, the biological advantages that an organism brings to a new environment are bound up with where it has come from and the evolutionary changes produced over the course of a long history of previous ancestral migrations. The similarity between closely allied species thus expresses the branching steps taken by their ancestors; in a broken and often puzzling fashion, these descendants, related by form and travels, indicate the various points where new species have rooted themselves on the dispersal routes of their ancestors. That is why Darwin did not understand the biogeographer's task as being akin to "the grouping of stars in constellations" (*Origin* 411), as if species were simply points fixed in the sky, but instead as a history of *moves*, "a grand game of chess with the world for a Board" (Darwin to C. J. F. Bunbury, 21 Apr. 1856, *Correspondence* 6:80). By examining and comparing the different forms that species had taken as they spread across the earth, he was able to read the history of their movement, particularly in those beings whose taxonomic status was not yet clear: "Wherever many closely-allied species occur, there will be found many forms which some naturalists rank as distinct species, and some as varieties; these doubtful forms showing us the steps in the process of modification" (*Origin* 404). Also, the primary reason, he argued, why one area, such as a continent, was rich in species while another, such as an oceanic island, was not had to do with issues of transport. "Species [are] few in proportion to the difficulty of transport" (*Notebooks* 210), he would write in Notebook B.

Darwin's commitment to understanding evolution in terms of mobility changed how he understood the relationship between organic beings and their habitats. Instead of seeing them as being native to the places in which they were found, as in Agassiz's notion of species as *autochthonoi*, he understood them as being the modified descendants of settlers that had arrived sometime in the distant past. Not surprisingly, some of these organic beings were not entirely suited to the places in which they found themselves. Darwin's interest in plants and animals whose bodies or habits marked them as being "foreign" or "out of place" can be seen in an entry in the first transmutation notebook, where he writes that "it is a point of great interest to prove animals not adopted [sic] to each country" (*Notebooks* 198). He compiled examples of species whose habits were radically at odds with their forms. "He who believes that each being was created as we now see it, must occasionally have felt surprise when he has met with an animal having habits and structure not at all in agreement," he writes (*Origin* 185). Although woodpeckers have bodies that make them adept at climbing trees and searching in their bark for insects, there are North American species that eat fruit and pursue insects on elongated wings. On the treeless plains of La Plata, Darwin even came upon "a woodpecker, which in every essential part of its organization, even in its colouring, in the harsh tone of its voice, and undulatory flight, told me plainly of its close blood-relationship to our common species; yet it is a woodpecker which never climbs a tree!" (184). Darwin commented on upland geese whose webbed feet rarely touched water; on shorebirds that inhabited forests or drylands, such as wood sandpipers and sheathbills; on petrels that instead of excelling at flight had taken to water and now looked and acted like auks or grebes; and on landrails, whose long toes indicated that they were best suited for marshy terrain, yet they inhabited meadows. He remarked that the water ouzel has all the appearance of a terrestrial thrush, and yet it normally finds its food "by diving—grasping the stones with its feet and using its wings under water" (185).

Rather than seeing these species as being maladapted, Darwin saw in them the telltale signs that their progenitors had come from elsewhere. On isolated islands, he found "endemic plants with beautifully hooked seeds" that were clearly adapted "for transportal by the wool and fur of quadrupeds," and yet these islands were uninhabited by mammals. Such a case, wrote Darwin, "presents no difficulty on my view, for a hooked seed might be transported to an island by some other means; and the plant then becoming slightly modified, but still retaining its hooked seeds, would form an endemic species, having as useless an appendage as any rudimentary organ" (*Origin* 392). In occupying territory not normally suited to them, these newcomers were taking advantage of the absence of successful competitors in those habitats. Thus, Darwin concluded that "species in a state of nature are limited in their ranges by the competition of other organic beings quite as much as, or more than, by adaptation to particular climates" (140); and elsewhere, more emphatically, that natural selection adapts "the inhabitants of each country only in relation to the degree of perfection of their associates" (472). In a nature shaped by the historical migration and settlement of species, the character of the organic beings inhabiting a given place was less an expression of geophysical factors such as climate and physical conditions than of which species had originally found the means or the occasion to land on its shores.

Darwin did not simply posit the mobility of species as a central component of evolution; he made it a primary element of his research and sought to develop a theoretical understanding of biological mobility. Underpinning this work, which would culminate in *The Power of Movement in Plants* (1880), was his recognition that plant and animal mobility was not simply a given, but an evolutionary advantage guided by natural selection, as organic beings had developed unique structures that increased their capacity to find resources and disperse themselves across the earth. In an age in which technological advances in transportation had revolutionized British society and increased its capacity to trade with and control other parts of the globe, it would have been difficult for Darwin not to see every new form of mobility as being advantageous, particularly in the global competition with others. But human beings were not the only creatures who had advanced their power by inventing new modes of travel. Whether it was by walking, flying, slithering, burrowing, or swimming on legs, wings, ribs, or fins, or by developing forms that allowed them to climb toward light or spread their seeds—some by hooking rides with passing animals, others by lying hidden inside fruit eaten by birds, or still others, like the dandelion seed, by tethering themselves to flying parasols—all plants and animals had developed forms, through natural selection, that determined how and where they moved and traveled. For Darwin, the movement of organic beings was neither accidental nor unworthy of study, because it was inherently linked to the success of a species and its ability to access new territory.

Since most of Darwin's contemporaries did not believe that plants and mammals could travel extensively without human assistance, much of Darwin's research in the two decades leading up to the publication of the *Origin of Species* was on the different modes of transport and dispersal available to them. When, in 1845, Joseph Hooker declared that under Darwin's tutelage he now considered himself to be "a good migrationist" (late Feb. 1845, *Correspondence* 3:149), Darwin responded that "we cannot pretend, with our present knowledge, to put any limit to the possible & even probable migration of plants" (19 Mar. 1845, 3:159). During the 1850s, he devised a number of ingenious experiments aimed at demonstrating that plants could travel great distances. For instance, to confirm whether they could be transported by ocean currents, he soaked seeds in saltwater tanks for varying periods of time and then tested their viability. "To my surprise," he writes in the *Origin of Species*, "I found that out of 87 kinds, 64 germinated after an immersion of 28 days, and a few survived an immersion of 137 days" (358). To prove that viable seeds could be transported in the gullets of birds, he germinated seeds that he had found in the excrement of small birds, and he also fed dead fish stuffed with seeds to ospreys, storks, and pelicans. Having ascertained that three tablespoons of pond mud contained 537 seeds, Darwin proposed that plants could be transported from one place to another on the dirty feet of migrating birds, and he noted that from just one foot of a partridge, he had removed "twenty-two grains of dry argillaceous earth" (362). He even studied whether freshwater snails could attach themselves to the feet of sleeping ducks by dangling a pair of duck's feet in an aquarium and then counting how many immature snails clung to them. At the same time, he recognized that plants and animals could take advantage of other forms of occasional transport, such as tree roots, flotsam, and icebergs. The experiments were tentative and inconclusive, but they allowed Darwin to counter biologists, such as Edward Forbes, who had explained long-distance plant and animal migration by positing hypothetical land bridges and extended continents. In a 3 January 1860 letter to Hooker, he staunchly insisted that he would "maintain against all the world that no man knows anything about [the] power of transoceanic... migration" (*Correspondence* 8:7). Since increased forms of mobility were the inventions of biological organisms, Darwin was also interested in situations where organisms had gone the other direction. An 1854 paper by T. V. Wollaston noted, for instance, the remarkable fact that of the 550 species of beetle on the island of Madeira, 200 had wings that were so shriveled and atrophied that they could no longer fly. Darwin's answer was that on a wind-swept island, where flying beetles were continually subject to being blown out to sea, flight had proven to be an evolutionary disadvantage and, apparently, the best strategy was to stay grounded (*Natural Selection* 291–93). Flightless birds similarly illustrated that in certain environments a bird might trade flight for a stronger pair of legs and the capacity to grow larger.

Although islands occupy a central place in Darwin's early theorization of evolution, as he initially thought that speciation might operate best in a manner similar to artificial selection through the physical isolation of breeding populations, by the time he drafted the *Origin of Species*, he had shifted away from this largely sedentarist model to one in which speciation functioned best in large open areas, where the occasional isolation of populations through the lowering of continents could combine, once the continents rose again, with the free movement and intense competition of highly mobile populations. This shift rested on Darwin's belief that island species, despite their unique forms, had shown themselves to be at a disadvantage in struggling against species traveling from elsewhere. In a biogeographical model in which colonization was a preeminent value, it was not enough for new species to be created; they also had to be capable "of enduring for a long period, and of spreading widely" (*Origin* 105). In this regard, it was not insular species but continental biota that displayed the competitive advantages that resided in increased forms of mobility and the ready modification of form. Comparing English and New Zealand biota, Darwin would declare, for instance, that "from the extraordinary manner in which European productions have recently spread over New Zealand, and have seized on places which must have been previously occupied, we may believe, if all the animals and plants of Great Britain were set free in New Zealand, that in the course of time a multitude of British forms would become thoroughly naturalized there, and would exterminate many of the natives" (337). Darwin's biogeography thus represents a radical shift in focus, for instead of being fascinated by those rare and exotic endemic species that naturalists had sought out for two hundred years, Darwin's primary interest was in the attributes of common weeds. For Darwin, weeds were evolutionary success stories, whose numbers and widespread ranges marked them as being among the most successful and dominant organisms on earth. Products of large landmasses, the great "manufactories of species" (470), weeds were the species that had proven themselves to be most capable of competing with other biota and spreading widely across the earth, and because they passed on these attributes to their descendants, they gave rise to the greatest number of varieties and species. Unlike the local products of islands, many of which were facing extinction or severe reductions in number, weeds were, for Darwin, the agents of world history, as they played "an important part in the changing history of the organic world" (106).

That is why, in the years immediately preceding the publication of the *Origin of Species*, Darwin turned to the study of weeds in order to learn "why one species ranges widely and is very numerous, and why another allied species has a narrow range and is rare" (*Origin* 6). If all species branched from a single geographical area, why had some shown themselves to be so successful at expanding their ranges, while others had not? Differences in forms of mobility were also expressed in what constituted barriers to immigration, as a barrier to a quadruped would not be the same for a bird, a bat, or a fish. In 1857, Darwin cleared a thirty-two-square-foot patch of ground—his "weed-garden"—at Down and studied the manner in which it was recolonized by weeds. At the same time, he took up "botanical arithmetic" to prove statistically that wide-ranging genera also produced the greatest number of new species and varieties.[9] Speciation seemed to be partly a numbers game. As Darwin remarked, drawing upon the Parable of the Sower, "In the great scheme of nature, to that which has much, much will be given" (*Natural Selection* 248). On the basis of this work, he believed that he had evidence that "plants which are dominant, that is, which are commonest in their own homes, and are most widely diffused" produce "the greatest number of new varieties" (*Origin* 325). Variation and mobility, in other words, were strongly correlated. Together, they were bound up with power, for as Darwin observed, "to range widely implies not only the power of crossing barriers, but the more important power of being victorious in distant lands in the struggle for life with foreign associates" (405).

A pervasive aspect of British representations of the natural world during the nineteenth century was that those that were increasingly being colonized were seen as occupying a different time than the natures that were about to replace them. The same translational gesture that created modern natures also produced a panoply of natures past. Representations of the vast wildernesses of North America, the jungles of Africa and South America, or the ancient landscapes of China and India usually retain a sense that these natures and the cultures that supported them are already outmoded, continuing to exist through their isolation and marginality. "Going up that river was like traveling back to the earliest beginnings of the world, when vegetation rioted on the earth and the big trees were kings," Marlow comments, as he describes a journey through space that is also a journey in time (Conrad, *Heart of Darkness* 35). Darwin's *Origin of Species* produces a similar temporal division between native or indigenous natures and peoples (which are understood as being the biological remnants of insularity) and the "more dominant forms, generated in the larger areas and more efficient workshops of the north" (*Origin* 380), which he believed would soon displace them. Much of the cultural power of the *Origin of Species* comes from the strong homology Darwin draws between colonizing peoples and colonizing natures, which share in the evolutionary capacity to cross borders and seize new territories by virtue of their highly evolved forms and their enhanced mobility. Similarly, he believed that less mobile biological and human societies were fated, through the colonizing mechanisms of natural selection, to be replaced by them. Associated with world history, dominance, change, and modernity, these northern natures and peoples are set against traditional, indigenous societies and biota that are understood as being less adaptive to change and less modern. Viewed as stationary beings in a changing world, the latter's hold on the ground is more precarious, and they occupy territory that they will lose if they do not change. As Darwin wrote in Notebook B, "They die; without they change; like golden Pippens, it is a *generation of species* like generation of *individuals*" (*Notebooks* 187).

Whereas in Darwin's early work organisms change in order to adapt themselves to geophysical changes, in his later work it is traveling natures themselves that introduce the changes that they capitalize on. Change constitutes an opportunity for evolutionary expansion. Darwin's dominant species take advantage of what in the business world is called "creative disruption." Having witnessed on the *Beagle* voyage the destructive dimensions of colonialism, Darwin concluded that a "mysterious agency" is at work: "Wherever the European has trod, death seems to pursue the aboriginal. The varieties of man seem to act on each other in the same way as different species of animals—the strong always extirpating the weaker" (*Origin* 388).[10] In evolutionary theory, Darwin found what he thought was an explanation for this agency, but by attributing the outcome of these struggles to natural selection, rather than to human power and decision making, Darwin made the colonial struggle for territory a natural law and the primary factor in evaluating the value of diversity. Such a viewpoint, which assumes that the history of the earth has been shaped more by competition than by cooperation, an ongoing state of war based on the idea that populations must always push beyond the territories that sustain them, limits the possibilities of thinking about our relationship to nature in alternative ways. A peaceful coexistence with the natures around us is increasingly necessary if we do not want to end up living in a world in which evolutionary diversity has been obliterated and the only creatures that thrive are human beings, weeds, domestic animals, and the creatures that associate with them. A truly postcolonial understanding of the natural world would require human beings to provide space for beings other than themselves. What William Bartram said more than two hundred years ago still applies: "Give place to one another; the Gifts of Providence . . . are abundantly sufficient to satisfy the necessities and even the conveniencies of every creature, without a necessity of contention; & these are equally attainable, as well as the right of every one" ("Genetic Text" bk. 1, 280).

CHAPTER NINE

Frankenstein and the Origin and Extinction of Species

In this book, I have been arguing that during the colonial period indigenous and local natures across the globe, and the peoples and cultures that depended on and supported them, were profoundly changed by the appearance of new natures whose modernity lay in their mobility and the fact that they had been translated from elsewhere. At a time when traveling natures were disrupting long-established associations between natures, places, and peoples, the struggle between European settlers and indigenous peoples came to be powerfully registered in colonial environments and landscapes in the struggle between indigenous plants and animals and introduced or invasive biota. A new history of nature was being written in the biological resettlement of the globe, and Charles Darwin would extend what he saw happening in colonial environments to the idea of nature itself. During the Romantic period, no text conveys more powerfully the contradictions shaping the coming into being of these modern natures than Mary Shelley's extraordinary story about the appearance and destruction of an absolutely new species of being in *Frankenstein*. More than forty years before the publication of Darwin's *On the Origin of Species*, Shelley was already writing a history of the origin and extinction of a species. This radically new form of being, whose patchwork body chronicles its heterogeneous ancestry, is a thoroughly modern being, lacking a fixed species form, without a biological lineage, and unable to claim an identity bound up with time or place. Whereas Darwin in the 1850s remained silent about the implications of evolution for understanding the biological origins and destiny of human beings, Shelley placed this question at the very heart of her understanding of modern natures, suggesting that human beings as a species were not immune to change and that their future was being threatened by their almost pathological narcissism as a species and by their need to control nature. Their best hope, she suggests, lies in adopting a more inclusive idea of what it means to be "human" and in seeking to live in peace with the nonhuman species around them.

With the exception of Maureen McLane, in her chapter on *Frankenstein* entitled "Literate Species," most critics make a categorical mistake in treating the Creature as a human being, when for most of the novel, at best, he has only been trying to pass for one. This tendency is, of course, not without justification. The subtitle of the novel, "The Modern Prometheus," suggests that the story rewrites the ancient Greek myth about the origin of humankind and that Victor Frankenstein is a Promethean scientist who creates a new and improved version of human being. His goal is the technological creation of a new form of human being that, by bypassing women and sexual reproduction, will no longer be subject to the suffering, disease, and death that Victor associates with them. "I began the creation of a human being" (*Frankenstein* 49), he declares near the outset of his narrative. No sooner is the project under way, however, than he runs into difficulties, for in the next sentence, he changes his plans, deciding "contrary to my first intention, to make *the being* of gigantic stature" (49; my emphasis). His understanding of the Creature has changed, for he no longer identifies him as a "human" being, but instead puts "human" under erasure for the remainder of the narrative: the creature becomes "the being I had created" (53, 163), the "being whom I myself had formed" (72), and "the being to whom I gave existence" (210). Of course, Victor is quite capable of using more extreme terms to describe him, such as "the monster whom I had created" (87), a "demoniacal corpse" (53), an "abhorred devil" (96), "a mummy endued with animation" (53), "the filthy daemon" (71), or "my own vampire" (72); but when he uses the taxonomic language of natural philosophy, it is clear that he is no longer certain what kind of being he has created. Thus, what began as a Promethean narrative about the re-creation of human beings ends up being concerned with what it means to be human, with how one becomes human, and with whether "human being" can be claimed by species other than *Homo sapiens*. Like many other Enlightenment promoters of cosmopolitan natures, Victor quickly loses faith in his capacity to control the travels of this new being, and by the final volume of the novel, instead of extolling the huge benefits that will flow from such a creation, he suffers from xenophobia and intense anxiety about the risk posed to human beings by what he now understands to be an "invasive species." The dream of a new world brought into being by science—"A new species would bless me as its creator and source. Many happy and excellent natures would owe their being to me"

(49)—turns into a nightmare struggle between species, with Victor worrying that he has engineered a species that threatens all humankind. "I shuddered to think," he declares, "that future ages might curse me as their pest, whose selfishness had not hesitated to buy its own peace at the price perhaps of the existence of the whole human race" (163).

Mobility and the Displacement of Species

The first moments of the being's life tell us a good deal about why Victor changed his mind about the status of the species that he created:

> How can I describe my emotions at this catastrophe, or how delineate the wretch whom with such infinite pains and care I had endeavoured to form? His limbs were in proportion, and I had selected his features as beautiful. Beautiful!—Great God! His yellow skin scarcely covered the work of muscles and arteries beneath; his hair was of a lustrous black, and flowing; his teeth of a pearly whiteness; but these luxuriances only formed a more horrid contrast with his watery eyes, that seemed almost of the same colour as the dun white sockets in which they were set, his shrivelled complexion and straight black lips. . . . I had worked hard for nearly two years, for the sole purpose of infusing life into an inanimate body . . . but now that I had finished, the beauty of the dream vanished, and breathless horror and disgust filled my heart. Unable to endure the aspect of the being I had created, I rushed out of the room. (*Frankenstein* 54)

Here race appears explicitly in the story as the primary reason why Victor rejects his creation. The "black lips," "lustrous black" hair, and "yellow skin" associated with Ethiopians, Americans, and Mongols (to use eighteenth-century categories) indicate that Victor was planning to improve this human being by creating a cosmopolitan man from the best features drawn from the different races of human beings.[1] Like an animal breeder choosing the best from each variety of man, Victor hoped that this multiracial being, stitched together from the bodily parts taken from many races, would improve on the original species. Such a project, though radical in scope and ambition, was fully in keeping with the kind of research that was being done in German universities at this time. From the 1770s onward, the physiological and taxonomic dimensions of natural philosophy, exemplified at the University of Ingolstadt by Professor Krempe, were being used to develop modern racial science, most notably in physical anthropology, comparative anatomy, and art history. Referred to in the novel as "some of the physiological writers of Germany" (6), this group included Johann Joachim Winckelmann, the Dutch Petrus Camper, and Swiss Johann Caspar Lavater, but the most famous of these men was Johann Friedrich Blumenbach, whose published dissertation *De generis humani varietate nativa* (1775) set this new science in motion. It was based on Blumenbach's having used his international contacts to collect and assemble a famous collection of skulls representing human races across the globe.[2] From this perspective, the being created by Victor might be said to be a kind of walking *Habilitationsschrift*, less a corpse brought back to life than an ingenious animated human comparative anatomy collection, a display of the "variety of mankind" which Victor has curated with "infinite pains and care" after having "spent some months in successfully collecting and arranging [his] materials" (49). The being is, as Anne Mellor suggests, "a racial hybrid" ("*Frankenstein*, Racial Science" 22), but where previous discussions of race and othering in the novel have primarily sought to identify his race, what seems to appall Victor most about his creature is not race itself (for he clearly accepts the idea of human races), nor even the giant size of his mobile display in comparative anatomy, but instead the aesthetic impact of his living diversity. Once again, Victor displays his intellectual debts to German Enlightenment racial theory, which primarily approached racial taxonomies in terms of aesthetics.[3] "I had selected his features as beautiful. Beautiful!—Great God!" he declares, but instead of seeing the harmony wrought by bringing these different races together in one body, all he sees is their "horrid contrast," productive only of feelings of "horror" and "disgust."

Shelley is here providing a critique not only of the taxonomic and aesthetic categories underpinning European racial science but also of the rigid conception of nature on which they were based, because Victor's dread and disgust are less about race than about interracial mixing and the uncertainties about racial boundaries which it produced. At a time when Europeans feared that increased interaction between peoples through mobility, migration, and trade might lead to a dissolution or confusion of the biological differences on which different nations or "races" were built, this being's body, whose scars and heterogeneous parts draw attention to its mixed or hybrid nature, threatens what Victor understands as "human" society. At the same time, it makes the creature a thoroughly modern being, whose nature allows him to speak for all the "varieties of mankind," and particularly for those mixed racial populations—the metis, mestizo, "gypsies," "half-castes," and "half-breeds," to name just a few—that have been excluded and denied status and rights because they cannot claim to be from one race or place.[4] Thus, at the heart of Victor's fears about sexual reproduction are anxieties about miscegenation and the mixing of peoples from different places. As we know from his first words in the novel, being identified by birth with a place matters a great deal to Victor: "I am by birth a Genevese; and my family is one of the most distinguished of that republic" (*Frankenstein* 27).

But the radicalism of *Frankenstein* goes much further than this, for we learn that "the component parts" of this "creature ... manufactured, brought together, and endued with vital warmth" (227) were drawn from both "the dissecting room and the slaughter-house" (50). From a biological perspective, therefore, the creature is not actually a multiracial human being at all, but rather an entirely new biological entity composed of bits and pieces drawn from many races and natures. As McLane astutely writes, "the physiologically indeterminate being he creates brings us to the threshold of species being" (*Romanticism and the Human Sciences* 87). Like the translated environments I have been discussing in this book, the creature is a translated being, a new creation composed of human and animal materials collected from many places and transferred to Victor's laboratory. Lacking a single origin, he cannot trace himself back to any one place or parentage, but instead owes his being to a multitude of places and parents. Also, he could have appeared anywhere, from Ingolstadt or the Orkneys, for that matter; he is a product of mobility, an ontologically deracinated being, not simply uprooted *from* nature or place, but displaced or unplaced in his very being. Unlike other living creatures, which are related to each other by a shared environment or through reproduction, the being created by Victor stands outside of all organic relationships or kinships. Unable to claim a biological or cultural lineage, he troubles eighteenth-century categories of race and species. His body is not so much an origin as a site where many human and nonhuman beings have been brought together, yet remain strangers. In this sense, he can be likened to a moving network or, perhaps, a walking metropolis, maybe even an example of urban development gone awry. Consequently, he cannot make a claim that he has "natural rights," for he cannot claim on the basis of his being or kind that he belongs to the earth. Created by the collection and exchange of specimens from across the earth, he is at once the most ontologically displaced of beings and yet the one being who might reasonably speak for how all human beings and natures might live together. Unable to claim a nature, he is truly a utopian being: at once an ideal and a being that belongs to "no place." Rather than struggling, like most Promethean Romantics, against his roots, he spends most of his life trying to find or create them.

The being's status as a mixture of races and natures, of human beings and animals, disrupts the aesthetic and taxonomic categories with which Victor separates human being from nature, thus returning Victor to the fears of an unlicensed or uncontrollable nature that his race engineering originally sought to replace. Throughout the novel, Victor reacts to this mixing using the language of abjection. He refers, for instance, to the being as "the filthy daemon to whom I had given life" (*Frankenstein* 71), a denomination that evokes, in turn, the "work-shop of filthy creation" (50) where the being was produced. When the being speaks to him, Victor is sympathetic to his words but cannot look at him without seeing a "filthy mass that moved and talked." All he feels is "horror and hatred" (143). Victor will continue to be haunted by this terrible image of the creature as a loathsome and unclean thing that moves and speaks, and it produces in him what he calls a kind of insanity: "I saw continually about me a multitude of filthy animals inflicting on me incessant torture, that often extorted screams and bitter groans" (145). For Victor, the being is the undifferentiated voice of the "multitude of filthy animals" that went into his creation. He is at once multiracial and many natured, and this complexity of being produces in Victor the kind of anxieties about boundaries and their contamination or pollution that Mary Douglas so effectively discusses in *Purity and Danger* (1966). The original Promethean gesture toward a more-than-human being thus leads to a reactive hardening of species categories, an intense anxiety about the capacity of bodies to mix with each other to create new forms, and a rejection of any kind of boundary crossing.

Sexual Reproduction and the Demand to Become a Species

It need not have turned out this way, for the novel presents the reader with alternatives. Shelley's intellectual debt to Erasmus Darwin's theory that all biological beings are connected by a common evolutionary descent from the simplest form of matter endowed with life is well known.[5] What has not been adequately considered is the vital importance that Darwin accorded "sexual reproduction" in his evolutionary theory. In *Phytologia; or, The Philosophy of Agriculture and Gardening*, he distinguishes between two modes of reproduction: "solitary" reproduction, by which he means the asexual reproduction of progeny through buds, leaves, roots, and bulbs; and sexual reproduction, where love leads to a congress of the sexual parts of plants and animals. Darwin considers asexual reproduction to be the first and simplest mode of biological reproduction. Like scientific cloning, it is equivalent to the propagation of the same individual "ad infinitum" without any change in its nature. The "seminal," "amatorial," or sexual mode of reproduction, on the other hand, is, for Darwin, "the chef d'oeuvre, the master-piece of nature" because through it "a countless variety of animals are introduced into the world, and much pleasure is afforded to those, which already exist in it" (*Phytologia* 103). For Darwin, sexual reproduction is nature's greatest invention, and evolutionary diversity is an expression of the constant changes in form wrought by love among biological organisms that reproduce sexually. Sexual reproduction is the great art by which nature brings together the like and the unlike in order to usher new beings, new varieties of species, and even new species into the world. Instead of associating it with all that is dirty or contaminated in nature, therefore, Darwin celebrates the creative mixing, the boundary crossing, and the new creatures that emerge from the sexual interaction of biota and their pleasures.

That Darwin celebrated love, sex, and diversity is not to say that he was against humans benefiting from this power in nature. As an agriculturalist, he was aware of the immense possibilities opened up by the deliberate crossing of selected individuals and species. He too was interested in finding ways to control reproduction in nature.[6] In *Phytologia*, he cites many examples where a species of plant has been crossed with another to create a new variety that combines the valuable characteristics of its parents. Even though most of these hybrids would be sterile, that is, "mules," once produced, they could still be reproduced asexually. Also, Darwin believed, as did Linnaeus, that occasionally hybrids proved fertile, and these cases would result in the appearance of a new species. *Phytologia* envisions a time when human beings will be able to create "new animal combinations" as numerous "as the fabled monsters of antiquity; as between the ram and the female goat; the stag and the cow; the horse and the doe; the bull and the mare; boar and bitch; dog and sow." Furthermore, "new combinations of fish might thus be generated, and people our rivers with aquatic monsters." And, even more marvelously, there may be a time when "some beautiful productions might be generated between the vegetable and animal kingdoms, like the eastern fable of the rose and nightingale" (119). In imagining these extraordinary new creatures—"monsters," he calls them—Darwin presents a very different understanding than Victor's of the extraordinary capacities of creation and change inherent in nature and sexual reproduction.

For Darwin, the future of nature is mixed, and indeed, its past is the same, as he celebrates the biological hybridity produced by love and sexual reproduction, the moving force in nature. That is why, in a note on "Reproduction" in *The Temple of Nature*, which Shelley in all probability read, Darwin writes that "the more perfect orders of animals are propagated by sexual intercourse only" (37) and that sexual reproduction underlies the evolutionary improvement of species. Is it not possible, he asks, that "all vegetables and animals now existing were originally derived from the smallest microscopic ones . . . and that they have by innumerable reproductions, during innumerable centuries of time, gradually acquired the size, strength, and excellence of form and faculties, which they now possess? And that such amazing powers were originally impressed on matter and spirit by the Great Parent of Parents! Cause of Causes. Ens Entium" (38). For Darwin, the improvement of nature was not to be achieved by turning to technology, but through harnessing, through breeding, the tremendous powers of manufacture inherent in sexual reproduction.

Discussing Shelley's reading of Darwin, Anne Mellor comments that "we can see that Mary Shelley directly pitted Victor Frankenstein, that modern Prometheus, against those gradual evolutionary processes of nature described by Darwin" (*Mary Shelley* 100). Certainly, Darwin's discussion of "monsters," vegetable or otherwise, provides the primary context for understanding the urgent request that the being makes of his creator at the end of his account of a life spent in a failed attempt to root himself among human beings: "Man will not associate with me; but one as deformed and horrible as myself would not deny herself to me. My companion must be of the same species, and have the same defects. This being you must create" (*Frankenstein* 140). Prior to making this request, the being has been trying to pass as a human being, as a kind of mimic man. Living in a "hovel" (101) or "kennel" (102) adjoining the De Lacey cottage, he has adopted the humanist belief that it is not biology but education that determines one's place in society. At the same time, this education has made plain the huge social, economic, and physical distances separating him from human beings:

> All my past life was now a blot, a blind vacancy in which I distinguished nothing. From my earliest remembrance I had been as I then was in height and proportion. I had never yet seen a being resembling me, or who claimed any intercourse with me. What was I? The question again recurred, to be answered only with groans. . . . I found myself similar, yet at the same time

strangely unlike the beings concerning whom I read, and to whose conversation I was a listener. I sympathized with, and partly understood them, but I was unformed in mind; I was dependent on none, and related to none.... My person was hideous, and my stature gigantic: what did this mean? Who was I? What was I? Whence did I come? What was my destination? These questions continually recurred, but I was unable to solve them. (117, 124)

The creature's body raises continual questions that he cannot answer. Even as his humanist education has taught him to see himself as a monster, it also provides him with the means of finally discovering his true nature as a multiracial and multi-natural being. Reading through Victor's journal and learning both that he belongs to no race and that his creator disowned him produces tremendous anguish: "I sickened as I read.... Cursed creator! Why did you form a monster so hideous that even you turned from me in disgust?" (126). Confirmed in his greatest fear that he is a "monster," he hastily puts into immediate action the disastrous plan of making himself known to the De Laceys, hoping that they will "overlook my personal deformity" and will recognize him as an admirer of *"their* virtues" (my emphasis) and as one who solicits "their compassion and friendship" (126). When he is rejected by them, he loses "the only link that held me to the world" (134), that is, to the world of human beings, and he admits that for the "first time the feelings of revenge and hatred filled my bosom" (134).

Operating within the animal/human binaries that shape the novel, the being initially responds to this rejection by identifying with his nonhuman nature. "I was like a wild beast that had broken the toils," and "I gave vent to my anguish in fearful howlings" (132). Shelley is probably referencing the ideas of Buffon, whom she was reading at this time and cites in the novel. In the essay "On Domestic Animals" in his *Natural History*, Buffon argues that the first empire established by human beings was the "empire over animals." Justified by God and by human sagacity, this empire, he argues, was based on human beings' success either (1) in domesticating wild animals as "slaves" to provide labor, clothing, and food; or (2) in destroying those that competed with human beings, often by employing the assistance of other domestic animals, such as dogs and horses, for that purpose. This empire, he writes,

> like every other empire, was not founded till after society was instituted.
> It is from society that man derives his power; from that he perfects his reason, exercises his genius, and unites his strength.... When the human

race multiplied, and spread over the earth, and when, by the aid of the arts and society, man was able to conquer the universe, he by degrees lessened the number of ferocious beasts; he purged the earth of those gigantic animals of which we sometimes still find the enormous bones; he destroyed, or reduced to a small number, every hurtful and voracious species; he opposed one animal to another, and conquered some by fraud, others by force; and attacking them by every rational method he arrived at the means of safety, and has established an empire which is only bounded by inaccessible solitudes, burning sands, frozen mountains, and obscure caverns, which now serve as retreats for the small number of species of ferocious animals that remains. (*Natural History* 5:92–93)

Both John Fleming and Shelley radically questioned the idea that human beings have an imperial right to tame or destroy any nonhuman animal that competes with them for space on earth. Mary also recognized, as did her husband, that the first empire over animals had led, in turn, to other empires over human beings. That is why, for political and ecological reasons, they both promoted vegetarianism, as well as why the being manufactured by Victor is a vegetarian, though, of course, his way of life also represents a deep acknowledgement of the many animal natures that make up his being.

To be a "wild animal" in "everlasting war against the species" (*Frankenstein* 133) is a choice not unlike Prometheus's curse at the opening of Percy Shelley's "Prometheus Unbound," but here "wildness" has been forced on the being. For the remainder of the novel, he will inhabit the "inaccessible solitudes" on the fringes of mankind's first empire, and particularly the "frozen mountains" of the Alps, which Victor sees as being "the habitations of another race of beings" (90). But it is also here, in the borderlands, that the being conceives a radically new alternative that reconceptualizes the relationship between human beings and animals within his nature. "I demand a creature of another sex, but as hideous as myself," he declares. "Our lives will not be happy, but they will be harmless. Let me see that I excite the sympathy of some existing thing; do not deny me my request" (142). The being makes his request in full knowledge of who he is. Furthermore, his request can be seen, as McLane suggests, as marking "his formal and explicit renunciation of human being" (*Romanticism and the Human Sciences* 101). A perhaps less radical claim might be that this request constitutes a fundamental acceptance of who he is, without requiring that he choose between his human and animal natures, and an appeal for the continuance of his unique kind of mixed being.

The creature is clearly asking for a partner with whom he can have sex, and to Victor, who is unable to admit his own desires, the idea is both threatening and repulsive, made more so by the fact that, initially, the creature seemed to be referring to Victor when he declared that he was "consumed by a burning passion which you alone can gratify. . . . O my creator, make me happy" (*Frankenstein* 140, 142). But sex, pleasure, and companionship are only part of this request, for it represents a much deeper and encompassing appeal for recognition. The being knows that he is a creature that reproduces through sexual reproduction, so his appeal for "a creature of another sex" is also for control over his own biological destiny. Having recognized the limits of humanism, and no longer believing, as McLane remarks, that literature and education will ever allow him to claim fellowship with human kind, the being has seen a new way to rid himself of his status as a monster: by asking for a female partner, he is asking to become a species.

As I have already suggested, for eighteenth-century biologists and breeders, biological monstrosity was not inherently linked to appearance, but instead had to do with whether a plant or animal had the capacity to reproduce itself by sexual reproduction. Buffon shared this view, arguing that the essence of a species does not reside in the individuals of a species, but instead in their capacity to produce "beings similar to themselves, this successive chain of individuals . . . constitutes the real existence of the species" (*Natural History* 2:16). Reproduction was thus understood as being the primary means of distinguishing species. "Every species of animal is distinguishable from another by copulation," he writes. "Those may be regarded as of the same species which, by copulation, uniformly produce and perpetuate beings every way similar to their parents; and those which, by the same means, either produce nothing, or dissimilar beings, may be considered as being of different species." Lacking the ability to reproduce his likeness, the being created by Victor is not a species; thus, he only holds a foothold in the present. Having no biological ancestors and unable to produce like progeny, he is truly a biological monster. His plan, however, is to gain a foothold in time and across space through sexual reproduction. Having recognized that he is "not . . . the same nature as man" (*Frankenstein* 115–16), he seeks a partner "of the same nature as myself" (142), one whom he specifically insists should not be a "human being": "My companion must be of the same species, and have the same defects. This being you must create" (140).

Although the being will continue to use demeaning language in referring to himself (for this is what Victor expects), he already sees himself in a new way. Having accepted his complex hybridity, he now asks his creator to do the same:

> If you consent, neither you nor any other human being shall ever see us again: I will go to the vast wilds of South America. My food is not that of man; I do not destroy the lamb and the kid, to glut my appetite; acorns and berries afford me sufficient nourishment. My companion will be of the same nature as myself, and will be content with the same fare. We shall make our bed of dried leaves; the sun will shine on us as on man, and will ripen our food. The picture I present to you is peaceful and human, and you must feel that you could deny it only in the wantonness of power and cruelty. (*Frankenstein* 142)

This moment represents the high point of the novel, for here Shelley articulates a vision of a "peaceful and human" mode of being that does not see itself as being separate from nature, but instead acknowledges the right of all natures to exist, no matter what their appearance or being, and this is the ethic that underlies the being's request for recognition of his rights as an organic being. The Promethean gesture of the novel is thus toward a posthuman ecology for it allows readers, along with Victor, to entertain the possibility of conceiving of a different relationship to the natures around them, one that adopts a humanism that is based not on human beings separating themselves from their animal nature and animal natures, but instead on their seeking to preserve the integrity of all that lives. Indeed, Victor is given the opportunity to use his knowledge to usher in a world that recognizes love, pleasure, and community as being basic to all being, not power, control, and solitary reproduction.

The being had some basis for his hope that Victor would support his request, that he would step beyond a narrow European humanism to one that was inclusive of subaltern and nonhuman others. Having read the journal of his creation, he must have known that it was always Victor's plan to engineer a new species, not simply an isolated being that would eventually disappear from the earth with his death. Like other Enlightenment scientists, Victor wanted to unlock the secrets of nature in order to change it. However, the being's request to become a species only intensifies the conflict between the being and his creator, giving rise, as McLane puts it, "to a further and more horrifying problem, that of species competition" (*Romanticism and the Human Sciences* 103). Victor's fears center on the female being, whose appearance would make his creature a reproductive being, and all of his anxieties about women, sex, dirt, and death are replayed almost pathologically. He worries that "she might become ten thousand times more malignant than her mate" (*Frankenstein* 163). Given his belief in his own biological and intellectual superiority over his creation, Victor speculates that even if he were to create a female companion, she would turn from the being in disgust and quit him for the "superior beauty of man," and this would only produce "the fresh provocation of being deserted by one of his own species" (163). Xenophobia and fears of miscegenation haunt Victor, but even more, he recognizes that with the creation of a female the being will gain control over his biological destiny: "The first results of those sympathies for which the daemon thirsted would be children, and a race of devils would be propagated upon the earth, who might make the very existence of the species of man a condition precarious and full of terror. Had I a right, for my own benefit, to inflict this curse upon everlasting generations?" (163). Victor now portrays himself as a heroic guardian of human beings against the threat posed by other races and natures. The novel thus echoes the fears about invasive species which I discussed in chapter 4, even as it suggests their refraction into anxieties about racial interaction. Faced with the question of whether other species have a right to control their own reproduction, Victor concludes, "My duties towards my fellow-creatures had greater claims to my attention because they included a greater proportion of happiness or misery" (215). In the 1831 edition, Shelley revised this passage to stress interspecies struggle by changing "my fellow-creatures" to "the beings of my own species" (Appendix B, *Frankenstein* 259).

Elegy for an Unborn Species

The scene in which Victor destroys and scatters the remains of the female being while her male companion looks on in torment is one of the most terrible scenes in Romantic literature. The sheer violence directed toward her (and through her against the creature) is difficult to read: it is a murder, an abortion, a gender-driven hate crime, a wholesale indignity to a dead body, it is all these things, and even worse, it was all unnecessary. When he declares that "I *almost felt* as if I had mangled the living flesh of a human being" (*Frankenstein* 167; my emphasis), the boundary between human ethics and ecological thinking is pretty much erased for any reader. The destruction of the female nevertheless provides the imaginative core for the despair-ridden final movement of the novel. Since this was not a hypothetical female being, but one that Victor had the capacity to bring to life, his decision to destroy her is tantamount to condemning a new species to extinction. From this moment on, the being recognizes that he is the last of his race, and he speaks of his species in the past tense, rather than in the future conditional, for he has no future. His anger in the final part of the novel is thus deeply linked to an intense sorrow, which speaks for many natures existing before and after him, in knowing that he is the last of his species, one that was destroyed before it even had the chance to be born. "Shall each man," he expostulates, "find a wife for his bosom, and each beast have his mate, and I be alone. . . . Man, you shall repent of the injuries you inflict" (165). Despite the language of heroic sacrifice that concludes *Frankenstein*, the novel provides a deep ecological appeal for the necessity of rethinking the category of the human in relation to nonhuman animals if we are to preserve those that we still have left, even as it explores the contradictions that shaped the global translation of natures that I have examined in this book. The novel ends with a bleak apocalyptic ecological vision of a world in which human beings and nature are engaged in perennial self-destruction. In the final lines of the novel, the being justifies himself, declaring in his farewell words to Victor "thou didst seek my extinction" (220), and stating to Walton that nothing more is needed "to consummate *the series* of my being" (220; my emphasis) than death. All that is left him are symbolic, commemorative gestures, and in his last act, he performs, as others have noted, an act of *sati*. Although this act does express the bond that he shares with Victor, it also memorializes the female companion and biological relationship that he was denied. Instead of ushering in a new world, the novel ends with a single individual, lost in the expanse of an arctic waste and borne into "darkness and distance" (221), a living reminder (or ghost) of a species that was never given a chance.

This image of a lonely, solitary figure inhabiting an inhospitable nature haunted Shelley, as it should haunt us, for she returned to it, in 1824, in *The Last Man*. There, however, her ecological reflection on the self-destructive violence with which human beings seek to control the natural world and deny space for the nonhuman natures that they nevertheless depend on had given way to an extended reflection on the possibility that the human species itself might disappear as a consequence of a new nature that it brought into being, the agent this time being another traveling nature, a global pathogen.[7] *The Last Man* rewrites *Frankenstein* through the eyes of Lionel Verney, reminding us both of the extraordinary achievements of human beings and of why it is important to listen to the appeals, inherently forms of grievance, silently being made to us by present-day natures and the ghosts of natures past.

NOTES

Preface

1. This kind of reading, though applied to landscape, has much in common with what Cannon Schmitt, in "Tidal Conrad (Literally)," has usefully called "denotative or literal reading," which engages in sustained attention to the meaning conveyed by specialized denotative language present in texts.

Introduction · Natures in Translation

1. For a discussion of the history of the American wilderness movement, see Cronon, "Trouble with Wilderness."

2. For recent work on Romanticism and ecology, see K. Thomas, *Man and the Natural World*; Worster, *Nature's Economy*; Bate, *Romantic Ecology* and *Song of the Earth*; Kroeber, *Ecological Literary Criticism*; Morton, *Shelley and the Revolution in Taste* and *Ecology without Nature*; Buell, *Environmental Imagination*; McKusick, *Green Writing*; Oerlemans, *Romanticism and the Materiality of Nature*; Hutchings, *Imagining Nature* and *Romantic Ecologies and Colonial Cultures*; Hess, *William Wordsworth*; and Hall, *Romantic Naturalists*.

3. For work on the relationship between natural history and empire, see Arnold, *Tropics and the Traveling Gaze*; Schiebinger, *Plants and Empire*; Endersby, *Imperial Nature*; Drayton, *Nature's Government* and "Knowledge and Empire"; Gascoigne, *Science in the Service of Empire*; J. Browne, "Biogeography and Empire"; Stafford, *Scientist of Empire*; Headrick, *Tentacles of Progress*; McClellan, *Colonialism and Science*; and Brockway, *Science and Colonial Expansion*.

4. For discussions of British imperialism as part of world history, see Bayly, *Imperial Meridian*; Colley, *Britons*; Cain and Hopkins, *British Imperialism*; Hancock, *Citizens of the World*; and K. Wilson, *Sense of the People*. D. E. White shows how this approach can illuminate our understanding of British Romanticism in *From London to Little Bengal*.

5. For a recent critique of the separation of natural and human history, see Chakrabarty, "Climate of History."

6. For the strongest assertion of this viewpoint, see Bate, *Romantic Ecology*.

7. See also Agrawal, *Environmentality*.

8. See DeLoughrey and Handley, *Postcolonial Ecologies*; V. D. Anderson, *Creatures of Empire*; Arnold, *Problem of Nature*; Grove, *Green Imperialism*; Crosby, *Ecological Imperialism*; Melville, *Plague of Sheep*; and Cronon, *Changes in the Land*.

9. For natural history and urban spectacle, see Altick, *Shows of London* 22–33, 235–52.

10. See Crosby, *Ecological Imperialism*; and Grove, *Green Imperialism*.

11. Even the more traditional naturalist Gilbert White argued in his advertisement to *Selborne* that *"parochial histories"* should be used to create "the most complete county-histories," and his own work emerged from his participation in a correspondence network centered around Thomas Pennant (*Natural History of Selborne*, ed. Foster, 7; future references are to this edition unless otherwise noted).

12. For Darwin's use of this idea, see *Descent of Man* 239-40.

13. See Cheyfitz, *Poetics of Imperialism*; Niranjana, *Siting Translation*; and Rafael, *Contracting Colonialism*.

14. See also Miller, "Joseph Banks"; and Spary, *Utopia's Garden*.

15. See Latour, "Visualization and Cognition."

16. Mungo Park's interest in botany came to Banks's attention through his brother-in-law John Dickson, who owned a seed business in Covent Garden. Banks arranged for Park to serve as ship's surgeon on the East Indiaman *Worcester*, which sailed to Bencoolan, Sumatra, in 1793. On 4 November 1794, Park read a paper before the Linnean Society entitled "Descriptions of Eight New Fishes from Sumatra," subsequently published in the *Transactions of the Linnean Society*, vol. 3 (1797).

17. For an argument that the eighteenth century saw the beginning of an information culture, see Headrick, *When Information Came of Age*.

18. See, e.g., Anthony Pagden's *European Encounters with the New World*, which discusses how early explorers used the "principle of attachment" to find ways of reducing the "incommensurability" between the New World and the Old (17-49); or Bernard Smith's account in *European Vision and the South Pacific* of the manner in which eighteenth-century representations of the Pacific were made to conform to European ideas of classical beauty. Elizabeth Bohls has demonstrated that Enlightenment travelers to the Caribbean applied the conventions of Enlightenment aesthetics to unfamiliar places and thus reinstituted traditional hierarchies centered on the educated, property-owning gentleman (*Women Travel Writers*).

19. Darwin also recounts this anecdote in *Descent of Man* 1:236.

Chapter 1 · *Erasmus Darwin's Cosmopolitan Nature*

1. See also Walpole to Mary Berry, 23 Aug. 1791, *Correspondence* 11:339.

2. For the influence of Darwin on other Romantic writers, see King-Hele, *Erasmus Darwin and the Romantic Poets*; and Fulford, "Coleridge, Darwin, Linnaeus."

3. For general studies of consumer culture during the Romantic period, see McKendrick, Brewer, and Plumb, *Birth of a Consumer Society*; Brewer and Porter, *Consumption and the World of Goods*; Campbell, *Romantic Ethic*; Douglas and Isherwood, *World of Goods*; and Walvin, *Fruits of Empire*.

4. See also E. Robinson, "Eighteenth Century Commerce and Fashion" 39-60.

5. Harvey, *Early Gardening Catalogues* 28.

6. *The New Flora of the British Isles* (1991) by Clive Stace records that 2,990 species currently exist naturally in Britain. Of these, 81 are ferns, and these are mostly native. Of the remaining 2,909 species (of flowering plants), 1,737, or nearly 60% of them, are naturalized exotic species. Among the thirty-five conifers now found on the two islands, only three are native; see Ingoville, *Historical Ecology of the British Flora* 297.

7. For the "United Catalogue," see Weston, *Universal Botanist and Nurseryman* 4:96-130.

8. See Richard French's letter to Walpole, 14 Feb. 1790, Walpole, *Correspondence* 42:268-69.

9. For early plant catalogs and nursery activities, see Harvey, *Early Gardening Catalogues* and *Early Nurserymen*; and Henrey, *British Botanical and Horticultural Literature* 2:317-410.

10. This viewpoint did not stop him, however, from seeking shortly thereafter the position of poet laureate. In 1790, Darwin entered into correspondence with Hon. Dudley Ryder about the laureateship, which went instead to Henry Pye.

11. Charles Darwin's father told him that Erasmus was paid "a thousand guineas for the part which was published last" (*Life of Erasmus Darwin* 92). In his journal, William Bagshaw Stevens, who wrote a prefatory poem for *The Botanic Garden*, complains about not being presented with a copy of the poem, despite the fact that "the Copyright was sold to the Bookseller for £800, an astonishing Price for a Poem at the end of the eighteenth century" (*Journal* 31).

12. Elsewhere, Darwin likens the poem to "an unskilful exhibition in some village-barn" (132), requiring good humor on the part of its audience.

13. I am here indebted to King-Hele's "Erasmus Darwin," see esp. 166–67.

14. See also Pope's "Windsor-Forest," where "Unbounded *Thames* shall flow for all Mankind, / Whole Nations enter with each swelling Tyde" (*Poems* 1:144–94, lines 398–99).

15. See also Joseph Banks to Henry Dundas, 1st Viscount Melville, 15 June 1787, Banks, *Indian and Pacific Correspondence* 2:205.

16. Deirdre Coleman discusses the relationship between women, luxury, and tea drinking in "Conspicuous Consumption."

17. For an overview of Jones's work, see Christmas, *Laboring Muses* 130–56.

18. For a suggestive discussion of the politics of landscape gardening, see Bending, "Natural Revolution?"

19. In *The Task*, William Cowper brings together the foreign natures of empire in a more domestic setting—the greenhouse—where the task of establishing a cosmopolitan order is enacted by the middle-class gardener poet:

> All plants, of ev'ry leaf, that can endure
> The winter's frown, if screen'd from his shrewd bite,
> Live there, and prosper. Those Ausonia claims,
> Levantine regions these; th' Azores send
> Their jessamine, her jessamine remote
> Caffraia; foreigners from many lands
> They form one social shade, as if conven'd
> By magic summons of th' Orphean lyre. (Poetical Works 3:580–87)

For a discussion of Romanticism and greenhouse gardening, see Lynch, "Young ladies are delicate plants."

20. For imperial plant transfer, see Brockway, *Science and Colonial Expansion*; MacKay, *In the Wake of Cook* 123–43; McCracken, *Gardens of Empire*; Gascoigne, *Science in the Service of Empire*; and Drayton, *Nature's Government*.

21. Jackson further notes that occasionally William and Dorothy also sought to make up for deficiencies in Withering's botanical descriptions, as in Dorothy's marginal comment beside Withering's description of the common butterwort (*Pinguicula vulgaris*): "In great abundance all round the Grasmere fells. This plant is here very ill described, a remarkable circumstance belonging to it is the manner in which its leaves grow, lying close to the ground, and diverging from the stalk so as exactly to resemble a starfish, the tall slender stalk surrounded by the blue flower and rising from the middle of the starfish renders the appearance of this plant very beautiful especially as it is always found in the most comfortless and barren places, upon moist rocks for instance" (*Romantic Readers* 78).

Chapter 2 · Traveling Natures

1. For discussion and relevant documents, see Lai, *Chinese in the West Indies* 22–46.

2. See Van den Boogaart and Emmer, "Colonialism and Migration" 3.

3. See Glissant, "Distancing, Determining" 146.

4. For Kincaid's use of botany to understand Caribbean colonial history, see Braziel, "Caribbean Genesis"; Tiffin, "Man Fitting the Landscape" and "Flowers of Evil"; and O'Brien, "Garden and the World."

5. For the introduction of bamboo into Jamaica, see Long, *History of Jamaica* 3:903. Since Patrick Browne does not mention bamboo in his *Civil and Natural History of Jamaica* (1756), the plant must have arrived sometime in the 1760s.

6. See D. Powell, *Botanic Garden, Liguanea* 6.

7. See Mackay, *In the Wake of Cook* 123–43. Also relevant are his "A Presiding Genius of Exploration"; "Myth, Science, and Experience"; and "Agents of Empire." For an interpretation of this voyage from the perspective of the history of globalization, see DeLoughrey, "Globalizing the Routes of Breadfruit."

8. Bligh is quoting from Hawkesworth's *Account of the Voyages* 2:197.

9. See also Fulford, Lee, and Kitson, *Literature, Science and Exploration* 108–26; and D. Powell, *Voyage of the Plant Nursery*. For discussions of British plant transfer, see Mackay, *In the Wake of Cook*; Frost, *Sir Joseph Banks* and "Antipodean Exchange"; Gascoigne, *Science in the Service of Empire*; Brockway, *Science and Colonial Expansion*; Drayton, *Nature's Government*; and Rigby, "Politics and Pragmatics."

10. For Linnaeus's acclimatization schemes for Lapland, see Koerner, *Linnaeus* 113–39.

11. For eighteenth-century maritime specimen transport, see Parsons and Murphy, "Ecosystems under Sail."

12. The essay, its authorship mistakenly ascribed to a "D. Walker," is in *Papers of Sir Joseph Banks*, sec. 5, ser. 21.02, at the Library of New South Wales. The earliest date of composition for the essay would be 1777, because Walker refers on the third page to the "late" John Ellis, who died in 1776. Since the title page indicates that Walker was still living at Moffat, the latest date of composition is probably 1779, when Walker took up his duties at the University of Edinburgh, though he still retained his pastoral duties there until 1783.

13. See also Burgess, "Transport" 237–41; Baker, *Written on the Water*; K. Wilson, *Island Race*; Gillis, "Taking History Offshore" and *Islands of the Mind*; Beer, "Writing Darwin's Islands"; and Armitage, "Elizabethan Idea of Empire."

14. Pineapples first arrived in England in the 1720s and became a popular hothouse plant grown in pineries. Pineapple prices are indicated in the documents collected by Harvey, *Early Nurserymen* 206–7. William Speechly published a treatise on their cultivation in 1779.

15. See also Wiles and Smith's letter to Banks, 17 Dec. 1792, *Papers of Sir Joseph Banks*, sec. 9, ser. 52.09.

16. For a discussion of colonial ideas about "indigenes" and the manner in which they expressed a specifically European conception of how human beings should relate to the land, see Pocock, "Tangata Whenua."

17. For Cook and gardening, see Tobin, *Colonizing Nature* 2–9.

18. Johann Reinhold Forster remarked that "the only disagreeable animal in O-Taheitee, is the common black rat, which is very numerous there, and often does mischief by its voracity" (*Observations* 128). This situation did not prevent Cook, on the third voyage, from seeking to rid the *Resolution* of its own rat problem by encouraging them to take permanent shore leave: "The Ship being a good [deal] pestered with rats, I hauled her within thirty yards of the Shore, being as near as the depth of water would allow, and made conveniences for them to go a Shore, being in hopes some would be induced to it, but I beleive [sic] we got clear of very few if any" (*Journals* 3:226).

19. See James Wiles and Christopher Smith to Banks, 26 Jan. 1793, *Papers of Sir Joseph Banks*, sec. 9, ser. 52.12.

Chapter 3 · Translating Early Australian Natural History

1. Bligh's drawing of an echidna, done in 1792, was included in Everard Home's "Description of the Anatomy of the *Ornithorhynchus Hystrix*."
2. See P. Carter, *Road to Botany Bay*; and N. Thomas, *Entangled Objects*.
3. For plant and animal introductions to Australia, see Low, *Feral Future*. For early plant invasions, see Black, "Naturalised Flora of South Australia"; Michael, "Weeds Themselves"; Swarbrick, "Weeds of Sydney Town"; and Parsons, "Introduced Weeds." For the introduction of animals to Australia, see Rolls, *They All Ran Wild*; Jenkins, *Noah's Ark Syndrome*; E. Wilson, "On the Introduction of the British Song Bird"; and Parsonson, *Australian Ark*.
4. See also Burt and Williams, "Plant Introduction in Australia" 254; and Low, *Feral Future* 25.
5. At the same time, France sent out a similarly equipped natural history expedition under the command of Nicolas Baudin in the ship *La Géographie*.
6. For discussions of the printing history of this book, see Pigott, "John White's *Journal*"; and Nelson, "John White's *Journal*."
7. See, e.g., McCarthy, "Notes on the Cave Paintings."

Chapter 4 · An England of the Mind: Gilbert White and the Black-bobs of Selborne

1. *Natural History of Selborne*, ed. Foster, 112. Unless otherwise noted, all subsequent references are to this edition, cited as *Selborne*.
2. Wordsworth mistakenly believed that he had read the book "with great pleasure when a Boy at school" (*Letters* 2:270).
3. For accounts of Linnaeus's traveling "disciples," see Koerner, "Purposes of Linnaean Travel" and *Linnaeus* 113–15; and Stafleu, *Linnaeus and the Linnaeans*. For Banks, see Mackay, "Agents of Empire."
4. See Mabey, *Gilbert White*; Foster, *Gilbert White*; Maddox, "Gilbert White"; Menely, "Traveling in Place"; and Bellanca, *Daybooks of Discovery* 47–50.
5. In addition to *Selborne*, White published, for example, the work of Mark Catesby, Pehr Osbeck, George Forster, Rousseau, John Ellis, John Matthews, Thomas Martyn, Thomas Pennant, William Curtis, Philip Miller, and John Aiken, along with the second edition of Patrick Browne's *Natural History of Jamaica* and the *Transactions of the Linnaean Society*; see Noblett, "Pennant and His Publisher."
6. For White's reading in natural history, see *Selborne* 183n11.
7. Influenced by Bruno Latour's *Science in Action*, extensive work has been done on the role of correspondence networks in the production of scientific knowledge. See, e.g., Secord, "Corresponding Interests"; and Spary, *Utopia's Garden*, particularly chap. 2.
8. I am drawing on James Buzard's use of the term "autoethnography" in *Disorienting Fiction*; see also his article "On Auto-ethnographic Authority."
9. See Hogden, *Early Anthropology*; and Stocking, *Race, Culture and Evolution*.
10. Although not included with the first edition, this poem has normally been included in subsequent editions of *Selborne* by its editors.
11. *Selborne* included on its title page an epigraph from Scopoli: "To attempt a description of everything there is in the world, whether of God's work or of the forces of natural creation, is a fruitless endeavour far beyond the skills, or even the lifetime, of one man. For this reason the most useful task is the preparation of books on the fauna and flora of a particular region: it is monographs that are of the greatest value" (262n5).
12. White kept track of the appearance and departure of migratory birds by listening for their distinctive notes. That he used language in order to distinguish species is most evident in the fact that he was the first naturalist to distinguish between the chiffchaff and the willow

warbler (two birds that are almost identical in physical appearance) on the basis of their songs and arrival dates. He writes that the smaller of these, the chiffchaff, usually arrives around 23 March and "chirps till September," while the larger willow warbler arrives in the middle of April and has a "sweet plaintive note" (*Selborne* 102).

13. Milton, *Paradise Lost* 8.395.

14. See also Festa, "Humanity without Feathers."

15. A similar absence of expressed concern for the poor of Selborne has also been noted. Virginia Woolf, for instance, observed that White "is far less tender to the poor—'We abound with poor,' he writes, as if the vermin were beneath his notice—than to the grasshopper whom he lifts out of its hole so carefully and once inadvertently squeezed to death" (*Collected Essays* 3:125).

16. For natural theology and ecology, see Ospovat, *Development of Darwin's Theory* 23-38.

17. For eighteenth-century ideas about the economy of nature, see Spary, "Political, Natural, and Bodily Economies"; Worster, *Nature's Economy* 31-55; Limoges, introduction to *Carl Linné*; and Egerton, "Changing Concepts of the Balance of Nature."

18. See Mellor's "Baffling Swallow" for an in-depth study of eighteenth-century natural history and the controversy surrounding the idea of swallows and underwater hibernation in the work of White and Charlotte Smith.

19. White dismissed Michel Adanson's testimony that European swallows overwintered in Senegal as being based on "very poor evidence" (*Selborne* 85), yet it is now known that British swallows overwinter as far south as South Africa.

20. W. J. Keith argues that the absence of political references in the natural history was more a matter of "literary decorum" (*Rural Tradition* 57).

21. For accounts of what Richard Hamblyn calls the "European Weather Panic 1783," see his *Terra* 65-121; and Menely, "Present Obfuscation."

22. E. M. Nicholson provides a good account of the changes in species populations at Selborne up to the publication of his edition of the *Natural History of Selborne* (58-74).

23. The report, which published Joseph Banks's memoranda concerning the insect, was reprinted in *Annals of Agriculture* (A. Young, "Proceedings"). Banks's notes and correspondence are collected in the Joseph Banks Papers in the Sutro Branch, California State Library, referred to hereafter as the Joseph Banks Manuscript Collection.

24. For a recent study of this event, see Pauly, *Fruits and Plains* 33-50. Additional material can be found in Fitch, *Hessian Fly* 4-18; Hindle, *Pursuit of Science* 364-65; Ritcheson, *Aftermath of Revolution* 199-202; Jefferson, "Editorial Note III," *Papers* 20:445-49; Gascoigne, *Science in the Service of Empire* 114-15; and Todd, *Tinkering with Eden* 39-46.

25. Others, such as Banks and Thomas Jefferson, who chaired a committee that investigated the matter, were of the opinion that the fly was native to America. "I suspect [the Hessian fly] is aboriginal here," wrote Jefferson (Letter to Thomas Mann Randolph Jr., 15 June 1792, *Papers* 23:14). In 1817, Thomas Say reached a similar opinion, arguing that the insect was "absolutely unknown in Europe" and "must be a species entirely new to the systems" ("Observations on the Hessian Fly" 46).

26. James Bateman, the author of the massive and luxurious *Orchidaceae of Mexico and Guatemala* (1837-43), also seems to have enjoyed a strange love-hate relationship with cockroaches, for at the end of the section devoted to the Spotted Catasetum (*Catasetum maculatum*) he provides "a full-length portrait of Blatta gigantean," a specimen of which arrived in a shipment of orchids, "upon which, judging from the condition of the plants, he must have made many a hearty meal." A plate later, Bateman provided another illustration of the same cockroach now flipped over to display "a hero" with "literally two faces." In a volume filled with glorious illustrations of fantastic orchids, the repeated appearance of this insect bal-

ances beauty with disgust and constitutes a kind of memento mori to the joys of collecting and consuming beautiful plants.

27. La Pérouse, who left France in 1785, writes of a similar experience with cockroaches, which destroyed most of the provisions on his ship shortly after he visited the Isle de France. Jacques-Julien Labillardière, who set out in search of La Pérouse, never found the explorer, but he did find cockroaches. In *An Account of a Voyage in Search of La Pérouse*, he writes that *B. orientalis* appeared on the ship shortly after it left Brest; these, however, were replaced by an infestation of *B. germanica* that appeared when they passed the tropics. The "insects, not contented with our biscuits, gnawed our linen, paper, &c. Nothing came amiss to them. Their taste for vegetable acids was surprising; no sooner was a citron opened than they attacked it; but it was still more astonishing to observe the rapidity with which they emptied my ink-glass, when I happened to leave it open. The causticity of the vitriol which they swallowed, appeared to have no bad effect on them" (431).

28. Ebenezer Sibly, in his fourteen-volume *Magazine of Natural History*, writes that the cockroach's love of ink is so great that they often drown in inkstands: "In this case they soon turn most offensively putrid, so that a man might as well sit over the cadaverous body of a large animal as write with the ink in which they have died" (13:171).

Chapter 5 · William Bartram's Travels *and the Contested Natures of Southeast America*

1. See, e.g., Hallock, *From the Fallen Tree* 158.
2. For Bartram's scientific reception, see Ray, "Podophyllum peltatum."
3. See his comment, in a letter to Benjamin Smith Barton, that the *Travels* would include a "Catalogue of the most curious Vegetables discovered in these Travels" (Ewan, *William Bartram* 12).
4. See Laird, *Flowering of the Landscape Garden* 70. For the correspondence between Bartram and Collinson, see Irmscher, *Poetics of Natural History* 13–32; and A. W. Armstrong, "John Bartram and Peter Collinson."
5. While apprenticing as a merchant in Philadelphia over the winter of 1756–57, Bartram was hard at work drawing plants and animals. George Edwards made use of many of his bird illustrations for his *Gleanings of Natural History* (1758–64); Philip Miller included his illustration of green *Veratrum* in his *Figures of the most Beautiful, Useful and Uncommon Plants described in the Gardener's Dictionary* (1755–56); and two of his turtle drawings appeared in the *Gentleman's Magazine* in 1758.
6. For the view that Bartram is a straightforward spokesperson for the impersonal, objectifying dimensions of Enlightenment science, see Regis, *Describing Early America* 40–78; and Looby, "Construction of Nature."
7. See D. Anderson, "Bartram's *Travels* and the Politics of Nature"; C. H. Adams, "William Bartram's *Travels*"; Bellin, *Demon of the Continent* 123–30; Cashin, *William Bartram*; and Hallock, *From the Fallen Tree* 149–73.
8. See Walters, "'Peaceable Disposition' of Animals" and "Creator's Boundless Palace."
9. This association was first drawn by Lowes in *Road to Xanadu* 332–40.
10. For John Bartram's description of the settlement, see his *Diary of a Journey* 37, 46. For information on Rolle's plantation, see Schafer, *William Bartram* 82–86.
11. Emily W. B. Russell observes that in North America, the fruits of the Kentucky coffee tree (*Gymnocladus dioicus*), American chestnut (*Castanea dentate*), and butternut (*Juglans cinerea*) were traded by First Nations people, and that they were often propagated near their settlements. In Wisconsin, Kentucky coffee trees are rare but can be found near Native American village sites (*People and the Land through Time* 94).

12. On the subsequent American antiquarian mythologization of the "Mound Builders," see Sayre, "Mound Builders"; and Silverberg, *Mound Builders of Ancient America*.

13. See "Observations on the Creek and Cherokee Indians," in Waselkov and Braund, *William Bartram on the Southeastern Indians* 139–43.

Chapter 6 · "I see around me things which you cannot see": William Wordsworth and the Historical Ecology of Human Passion

1. For the dating of Coleridge's reading of Bartram, see Lowes, *Road to Xanadu* 468–71.

2. See also *"Ruined Cottage"* MS. D 68^r, 373.

3. Most of the criticism has focused on "Ruth," where Wordsworth's debt to Bartram is explicit. For a suggestive study of the poem as a narrative about the perils of "ethnographic seduction," see McLane, *Romanticism and the Human Sciences* 69–78.

4. For the relationship of the poem to tales of abandonment, see J. Wordsworth, *Music of Humanity* 50–67, 264–68; see also Jacobus, *Tradition and Experiment* 143–48.

5. Ian H. Thompson considers Wordsworth as a "landscape architect" in "William Wordsworth, Landscape Architect."

6. In the manuscript monograph "Decline and the Depths of Time: Historicity and the Forms of Ruin in British Romanticism," Jonathan Sachs offers a salient study of the multiple temporalities at work in "The Ruined Cottage," suggesting that Margaret, like Robert, has lost her capacity to be in time and no longer knows "what time it is."

7. For material on early Florists' Feasts, see Duthie, "English Florists' Societies and Feasts" and "Florists' Societies and Feasts after 1750." For working-class botanical societies, see Secord, "Science in the Pub."

8. See also Hadfield, *History of British Gardening* 262.

9. For the poem's connection to contemporary ideas about colonial disease, see Bewell, *Romanticism and Colonial Disease* 119–26.

10. Hudson, *Complete Guide to the Lakes*. For Wordsworth and geology, see Bewell, *Wordsworth and the Enlightenment* 237–79; see also Wyatt, *Wordsworth and the Geologists*.

11. For the documentary record relating to the actual site, see Chandler, "Hart-Leap Well."

12. For a discussion of the ways in which place-names not only point to the early history and settlement of England but also embody early interactions with animals and nature, see Reaney, *Origin of English Place-Names* 82, 146–58.

13. See Yalden, *History of British Mammals* 131–36; and Aybes and Yalden, "Place-Name Evidence."

14. See Yalden, *History of British Mammals* 171–73.

15. For a good account of the landscape history of the Lake District, see Denyer, "Lake District Landscape" 3–29.

16. See Binns, *Notes on the Agriculture of Lancashire* 41; and Murdoch, "Landscape of Labour" 178.

17. For Wordsworth's use of the picturesque in the *Guide*, see Nabholtz, "Wordsworth's 'Guide to the Lakes'"; and Noyes, *Wordsworth and the Art of Landscape*.

Chapter 7 · John Clare and the Ghosts of Natures Past

1. For a chronological presentation of Clare's work during this year, see Chilcott, *Living Year 1841*.

2. Johanne Clare presents a similar argument in *John Clare and the Bounds of Circumstance* 120–31.

3. For further discussion of the circumstances surrounding work on this publication, see Vardy, *Politics and Poetry* 135–66; and Clare, *Natural History Prose Writings*.

4. For an insightful account of Clare's environmentalism, see McKusick, *Green Writing* 77–94.

5. For a valuable assessment of the political dimensions of "The Deserted Village," see Lutz, "Politics of Reception."

6. For a suggestive examination of Clare's idea of the lyric, which sees it as a form of resistance to the ways in which enclosure imprisoned the landscape and the laboring class, see Nicholson, "Itinerant 'I.'"

7. Helsinger discusses the contradictions of the stereotype of the "peasant poet" in *Rural Scenes and National Representation* 141–61.

8. On the ambiguity of ownership and tenantry in Clare, see Helsinger, *Rural Scenes and National Representation* 149–54.

9. For an excellent discussion of these poems, see Johanne Clare, *John Clare and the Bounds of Circumstance* 164–88.

10. Keegan situates Clare's writing within a literature of wetlands in *British Labouring-Class Nature Poetry* 148–71.

Chapter 8 · *Of Weeds and Men: Evolution and the Science of Modern Natures*

1. See also Ospovat, *Development of Darwin's Theory*; and Kohn, "Theories to Work By."

2. See Egerton, "Studies of Animal Populations" 225–59; and J. Browne, *Secular Ark* 182–83.

3. In understanding habitats as being the locus of interspecies struggle, Lyell was anticipated by the Swiss botanist Augustin-Pyramus de Candolle, whose *Dictionnaire des science naturelles* (1817) he cites, noting that "all the plants of a given country . . . are at war one with another. The first which establish themselves by chance in a particular spot, tend, by the mere occupancy of space, to exclude other species—the greater choke the smaller, the longest livers replace those which last for a shorter period, the more prolific gradually make themselves masters of the ground, which species multiplying more slowly would otherwise fill" (*Principles* 2:131). R. Alan Richardson rightly suggests that when Darwin read Malthus in 1838, he was not so much discovering "a new principle from Malthus, but rather amplification and further application of ideas he already held about overpopulation and struggle in plant and animal populations, to man himself" ("Biogeography and the Genesis" 9n). Reading Malthus allowed Darwin to extend the Malthusian model beyond Lyell's concern with extinctions to the species question.

4. For a valuable discussion of the use of colonial metaphors in nineteenth-century biogeography, see J. Browne, "Biogeography and Empire."

5. For current evaluations of the natural history of the island, see Cronk, *Endemic Flora of St Helena*; and Ashmole and Ashmole, *St Helena and Ascension Island*.

6. See also R. Young, *Darwin's Metaphor*.

7. For Darwin and biogeography, see J. Browne, *Secular Ark* 182–220 and her "Biogeography and Empire"; R. A. Richardson, "Biogeography and the Genesis"; Mayr, *Growth of Biological Thought* 439–55; Egerton, "Studies of Animal Populations" and "Humboldt, Darwin, and Population"; P. J. Darlington, "Darwin and Zoography"; Greene, *Death of Adam*; and Ghiselin, *Triumph of the Darwinian Method*. For broader discussions of eighteenth- and nineteenth-century biogeography, see Kinch, "Geographical Distribution"; and Larson, "Not without a Plan."

8. For the role of mobility in the history of biogeography, see J. Browne, *Secular Ark*.

9. See J. Browne, "Darwin's Botanical Arithmetic."

10. Lyell made a similar observation about indigenous societies, remarking that "a faint image of the certain doom of a species less fitted to struggle with some new condition in a region which it previously inhabited, and where it has to contend with a more vigorous

species, is presented by the extirpation of savage tribes of men by the advancing colony of some civilized nation.... Yet few future events are more certain than the speedy extermination of the Indians of North America and the savages of New Holland in the course of a few centuries, when these tribes will be remembered only in poetry and tradition" (*Principles* 2:175). R. Alan Richardson notes that Darwin wrote "Good" ("Biogeography and the Genesis" 35) beside this passage in his copy of *Principles of Geology.*

Chapter 9 · Frankenstein *and the Origin and Extinction of Species*

1. For the racial aspects of this description, see, particularly, Malchow, "Frankenstein's Monster"; Mellor, "*Frankenstein*, Racial Science"; and Bohls, *Romantic Literature* 166–78.

2. For the relationship between skull collecting and race theory, see Fulford, Lee, and Kitson, *Literature, Science and Exploration*, chap. 6. For the role of race theory in the novel, see Mellor, "*Frankenstein*, Racial Science."

3. See, e.g., Bindman, *Ape to Apollo.*

4. For a more extended discussion of nineteenth-century racial hybridity, see R. Young, *Colonial Desire.*

5. See, e.g., King-Hele, *Erasmus Darwin* 260.

6. It should be stressed that Darwin was not appealing to a highly speculative group of Enlightenment philosophers in *Phytologia*, but instead to a more practically oriented readership of farmers, animal breeders, and scientists. A comparison of Darwin's ideas with the claims made by Thomas Andrew Knight in a letter he wrote to Sir Joseph Banks on 7 January 1802 suggests the degree to which plant and animal breeding was seen as one of the important "modern" sciences. After discussing experiments that he has done in cross-breeding turnips and cabbage, Knight goes on to write,

> I have cultivated the Primrose into the Polyanthus, & into a Flower similar to the Cowslip; but red; & I have in my Garden every gradation from the Filbert to the common nut, obtained from the seeds of the former.... If we are to admit that productive mules can be obtained between each Kindred Genus, I see no impossibility of alternately converting a Potatoe into a Peach or Pine Apple. It appears to me to be a most important Question, whether new Plants can be obtained between distinct Species & Genera; for, if they can, there is no end of the Labours & study of the Botanist. (Banks, *Scientific Correspondence* 5:123)

7. See Bewell, *Romanticism and Colonial Disease* 296–314.

Abrams, M. H. "Structure and Style in the Greater Romantic Lyric." In *From Sensibility to Romanticism*, edited by Frederick W. Hilles and Harold Bloom, 527–60. Oxford: Oxford University Press, 1965.
Adams, Charles H. "William Bartram's *Travels*: A Natural History of the South." In *Rewriting the South: History and Fiction*, edited by Lothar Honnighausen and Valeria Gennaro Lerda, 112–20. Tubingen: Francke, 1993.
Adams, John. *The Works of John Adams, Second President of the United States*. 10 vols. Boston: Little, Brown, 1850–56.
Agassiz, Louis, and Augustus A. Gould. *Principles of Zoology: Touching the Structure, Development, Distribution, and Natural Arrangement of the Races of Animals, Living and Extinct; with Numerous Illustrations for the Use of Schools and Colleges. Part I. Comparative Physiology*. Boston: Gould, Kendall & Lincoln, 1848.
Agrawal, Arun. *Environmentality*. Durham, NC: Duke University Press, 2005.
Aiken, John, ed. "Biographical Memoirs of the Late Dr. Darwin." *Monthly Magazine* 13, no. 87 (June 1802): 457–63.
Allen, David Elliston. *The Naturalist in Britain: A Social History*. London: Allen Lane, 1976.
Altick, Richard D. *Shows of London*. Cambridge, MA: Belknap, 1978.
Anderson, Benedict. *Imagined Communities: Reflections on the Origin and Spread of Nationalism*. London: Verso, 1983.
———. *The Spectre of Comparisons: Nationalism, Southeast Asia, and the World*. London: Verso, 1998.
Anderson, Douglas. "Bartram's *Travels* and the Politics of Nature." *Early American Literature* 25 (1990): 3–17.
Anderson, Edgar. *Plants, Man and Life*. Berkeley: University of California Press, 1971.
Anonymous. *Archaeologia Americana: Transactions and Collections of the American Antiquarian Society*. Worcester, MA: American Antiquarian Society, 1820.
———. *Jesuite's Reasons Unreasonable*. London: s.n., 1662.

———. "Remarks on Some of the Advantages and Disadvantages of Periodical Works on Natural History." *Magazine of Natural History* 3 (1830): 296–308.

———. Review of *History and Antiquities of the County of Somerset*, by John Collinson. *Critical Review*. London: A. Hamilton, 1793.

———. "Selborne." *New Monthly Magazine* 30, no. 2 (1830): 565–66.

Appadurai, Arjun. "Disjuncture and Difference in the Global Cultural Economy." *Public Culture* 2 (1990): 1–23.

Apter, Emily. *The Translation Zone: A New Comparative Literature*. Princeton, NJ: Princeton University Press, 2011.

Armitage, David. "The Elizabethan Idea of Empire." *Transactions of the Royal Historical Society* 14 (2004): 269–77.

Armstrong, Alan W. "John Bartram and Peter Collinson: A Correspondence of Science and Friendship." In *America's Curious Botanist: A Tercentennial Reappraisal of John Bartram, 1699–1777*, edited by Nancy E. Hoffmann and John C. Van Horne, 23–42. Philadelphia: American Philosophical Society, 2004.

Armstrong, Douglas V. "The Afro-Jamaican Community at Drash Hall." *Jamaica Journal* 24 (1991): 1–2.

Arnold, David. *The Problem of Nature: Environment, Culture, and European Expansion*. Oxford: Blackwell, 1996.

———. *The Tropics and the Traveling Gaze: India, Landscape, and Science, 1800–1856*. Seattle: University of Washington Press, 2006.

Ashmole, Philip, and Myrtle Ashmole. *St Helena and Ascension Island: A Natural History*. Oswestry: Anthony Nelson, 2000.

Austen, Jane. *Pride and Prejudice*. Edited by R. W. Chapman. 3rd ed. Oxford: Oxford University Press, 1959.

Aybes, C., and Derek W. Yalden. "Place-Name Evidence for the Former Distribution and Status of Wolves and Beavers in Britain." *Mammal Review* 25 (1995): 201–27.

Bacon, Francis. *Works of Francis Bacon*. 4 vols. London: Walthoe, Midwinder, Innys, et al., 1740.

Baker, Samuel. *Written on the Water: British Romanticism and the Maritime Empire of Culture*. Charlottesville: University of Virginia Press, 2009.

Banks, Joseph. *The Endeavour Journal of Joseph Banks, 1768–1771*. Edited by J. C. Beaglehole. 2 vols. Sydney: Angus & Robertson, 1962.

———. "Extract of a Letter from Sir Joseph Banks, President of the Royal Society to an Hon. Member of the Assembly of this Island." *Royal Gazette*, 25 July 1791. The Papers of Sir Joseph Banks, sec. 9, ser. 50.09.

———. "General Report of Sir Joseph Banks, Respecting the Hessian Fly, and Flying Wevil, 24th July, 1788." In A. Young, "Proceedings of His Majesty's Most Honourable Privy Council," 437–47.

———. *The Indian and Pacific Correspondence of Sir Joseph Banks, 1768–1820*. Edited by Neil Chambers. 8 vols. London: Pickering & Chatto, 2008–14.

———. "Instructions for Mr. James Wiles appointd to Proceed with Capt Bligh in his majesties Ship the Providence to the Society Islands in the Pacific Ocean for the Purpose of Collecting Bread fruits trees & other usefull Productions of the Islands in these Seas." *The Papers of Sir Joseph Banks*, sec. 9, ser. 49.09.

———. Joseph Banks Manuscript Collection. Sutro Branch, California State Library, San Francisco.

———. "Memorandum Concerning the Purchase of Plants and the Route of the Voyage, 1791." *The Papers of Sir Joseph Banks*, sec. 9, ser. 49.11.

———. *The Papers of Sir Joseph Banks*. State Library of New South Wales. http://www.sl.nsw.gov.au/banks.

———. *Scientific Correspondence of Sir Joseph Banks, 1765–1820*. Edited by Neil Chambers. 6 vols. London: Pickering & Chatto, 2007.

Barbauld, Anna Letitia. *Poems of Anna Letitia Barbauld*. Edited by William McCarthy and Elizabeth Kraft. Athens: University of Georgia Press, 1994.

Barrell, John. *The Dark Side of the Landscape*. Cambridge: Cambridge University Press, 1980.

———. *The Idea of Landscape and the Sense of Place, 1730–1840: An Approach to the Poetry of John Clare*. Cambridge: Cambridge University Press, 1972.

Barrington, Daines. *Naturalist's Journal*. London: W. Sandey, 1767.

Bartram, John. *Correspondence of John Bartram, 1734–1777*. Edited by Edmund Berkeley and Dorothy Smith Berkeley. Gainesville: University Press of Florida, 1992.

———. *Diary of a Journey through the Carolinas, Georgia, and Florida from July 1, 1765, to April 10, 1766*. Edited by Francis Harper. *Transactions of the American Philosophical Society* 33, part 1. Philadelphia: American Philosophical Society, 1942.

Bartram, William. "The Construction of William Bartram's Narrative Natural History: A Genetic Text of the Draft Manuscript for *Travels Through North and South Carolina, Georgia, East and West Florida*." Edited by Nancy Everill Hoffman. *Dissertation Abstracts International* (1996).

———. "[The Dignity of Human Nature]." In *William Bartram: The Search for Nature's Design*, edited by Thomas Hallock and Nancy E. Hoffman, 348–58. Athens: University of Georgia Press, 2010.

———. "Observations on the Creek and Cherokee Indians." In Waselkov and Braund, *William Bartram on the Southeastern Indians*, 139–43.

———. "Preface to a Catalogue of Trees, Shrubs, and Herbaceous Plants, Indigenous to the United States of America." In Slaughter, *William Bartram*, 585–87.

———. "Report to Dr. Fothergill." In Slaughter, *William Bartram*, 429–522.

———. "Some Hints & Observations, Concerning the Civilization, of the Indians, or Aborigines of America." In Waselkov and Braund, *William Bartram on the Southeastern Indians*, 193–98.

———. *The Travels of William Bartram: Naturalist Edition*. Edited by Francis Harper. Athens: University of Georgia Press, 1998.

Baskin, Yvonne. *A Plague of Rats and Rubbervines: The Growing Threat of Species Invasions*. Washington, DC: Shearwater, 2002.

Bassnet, Susan. "The Translation Turn in Cultural Studies." In *Constructing Cultures: Essays on Literary Translation*, edited by Susan Bassnett and André Lefevere, 123–40. Cleveland: Multilingual Matters, 1998.

Bate, Jonathan. *Romantic Ecology: Wordsworth and the Environmental Tradition*. London: Routledge, 1991.

———. *The Song of the Earth*. London: Picador, 2000.

Bayly, Christopher A. *Imperial Meridian: The British Empire and the World 1780–1830*. London: Longman, 1989.

Beaglehole, J. C. *The Life of Captain James Cook*. Stanford: Stanford University Press, 1974.

Beatson, Alexander. *Tracts Relative to the Island of St Helena; Written during a Residence of Five Years*. London: G & W. Nicol & J. Booth, 1816.

Beer, Gillian. *Darwin's Plots: Evolutionary Narrative in Darwin, George Eliot and Nineteenth-Century Fiction*. 2nd ed. Cambridge: Cambridge University Press, 2000.

———. "Writing Darwin's Islands: England and the Insular Condition." In *Inscribing Science: Scientific Texts and the Materiality of Communication*, edited by Timothy Lenoir, 119–39. Stanford: Stanford University Press, 1998.

Belich, James. *Replenishing the Earth: The Settler Revolution and the Rise of the Anglo-World, 1783–1939*. Oxford: Oxford University Press, 2009.

Bellanca, Mary Ellen. *Daybooks of Discovery: Nature Diaries in Britain 1770–1870*. Charlottesville: University of Virginia Press, 2007.

Bellin, Joshua. *The Demon of the Continent: Indians and the Shaping of American Literature*. Philadelphia: University of Pennsylvania Press, 2001.

Bending, Stephen. "A Natural Revolution? Garden Politics in Eighteenth-Century England." In *Refiguring Revolutions: Aesthetics and Politics from the English Revolution to the Romantic Revolution*, edited by Kevin Sharpe and Steven N. Zwicker, 241–66. Berkeley: University of California Press, 1998.

Benjamin, Walter. "The Task of the Translator." In *Walter Benjamin: Selected Writings*, vol. 1, *1913–1926*, edited by Marcus Bullock and Michael W. Jennings, 253–63. Cambridge, MA: Belknap, 1996.

Berkenhout, John. *Synopsis of the Natural History of Great-Britain and Ireland*. London: T. Cadell, 1789.

Beston, Henry. *The Outermost House: A Year of Life on the Great Beach of Cape Cod*. New York: Doubleday, Doran, 1928.

Bewell, Alan. *Romanticism and Colonial Disease*. Baltimore: Johns Hopkins University Press, 1999.

———. *Wordsworth and the Enlightenment: Nature, Man, and Society in the Experimental Poetry*. New Haven, CT: Yale University Press, 1989.

Bewick, Thomas. *A General History of Quadrupeds: The Figures Engraved on Wood*. Newcastle upon Tyne: S. Hodgson, R. Beilby, & T. Bewick, 1790.

Bhabha, Homi. *The Location of Culture*. London: Routledge, 1994.

Bibery, Isaac J. "Oeconomia Naturae." In *Miscellaneous Tracts Relating to Natural History, Husbandry, and Physic*, translated by Benjamin Stillingfleet, 39–109. 2nd ed. London: R. J. Dodsley, 1762.

Bight, Chris. *Life Out of Bounds: Bioinvasion in a Borderless World.* New York: Norton, 1998.
Bindman, David. *Ape to Apollo: Aesthetics and the Idea of Race in the 18th Century.* Ithaca, NY: Cornell University Press, 2002.
Binns, Jonathan. *Notes on the Agriculture of Lancashire, with Suggestions for Its Improvement.* Preston, UK: Dobson & Son, 1851.
Black, John MacConnell. "The Naturalised Flora of South Australia." Adelaide: J. M. Black, 1909.
Blackstone, William. *Commentaries on the Laws of England.* 4 vols. Oxford: Clarendon, 1765–69.
Blake, William. *The Complete Poetry and Prose of William Blake.* Edited by David V. Erdman. Rev. ed. Berkeley: University of California Press, 2008.
Bligh, William. *A Voyage to the South Sea.* London: George Nicol, 1792.
Blomley, Nicolas. *Law, Space and the Geographies of Power.* New York: Guilford, 1994.
Bloom, Harold. *The Visionary Company: A Reading of English Romantic Poetry.* Rev. ed. Ithaca, NY: Cornell University Press, 1971.
Bohls, Elizabeth A. *Romantic Literature and Postcolonial Studies.* Edinburgh: Edinburgh University Press, 2013.
———. *Slavery and the Politics of Place: Representing the Colonial Caribbean, 1770–1833.* Cambridge: Cambridge University Press, 2014.
———. *Women Travel Writers and the Language of Aesthetics, 1716–1818.* Cambridge: Cambridge University Press, 1995.
Bohrer, Martha Adams. "Tales of Locale: *The Natural History of Selborne* and *Castle Rackrent.*" *Modern Philology* 100 (2003): 393–416.
Bolton, Geoffrey. *Spoils and Spoilers: Australians Make Their Environment, 1788–1980.* Sydney: Allen & Unwin, 1981.
Boorstin, Daniel. *The Lost World of Thomas Jefferson.* New York: Henry Holt, 1948.
Boswell, Thomas D. Review of *The West Indies: Patterns of Development, Culture and Environmental Change since 1492,* by David Watts. *Geographical Review* 79, no. 2 (July 1989): 362–64.
Bourdieu, Pierre. *The Field of Cultural Production: Essays on Art and Literature.* Edited by Randal Johnson. Cambridge: Polity, 1993.
Braudel, Fernand. *Capitalism and Material Life.* London: Weidenfeld & Nicolson, 1973.
Braund, Kathryn E. Holland. "William Bartram's Gustatory Tour." In Braund and Porter, *Fields of Vision,* 33–53.
Braund, Kathryn E. Holland, and Charlotte M. Porter, eds. *Fields of Vision: Essays on the "Travels" of William Bartram.* Tuscaloosa: University of Alabama Press, 2010.
Braziel, Jana Evans. "'Caribbean Genesis': Language, Gardens, Worlds (Jamaica Kincaid, Derek Walcott, Édouard Glissant)." In DeLoughrey, Gosson, and Handley, *Caribbean Literature and the Environment,* 110–26.
Brewer, John, and Roy Porter, eds. *Consumption and the World of Goods.* New York: Routledge, 1993.

Britton, Alexander. *Historical Records of New South Wales.* 7 vols. in 8. Sydney: Government Printer, 1892.
Brockway, Lucile H. *Science and Colonial Expansion: The Role of the British Royal Botanic Gardens.* New York: Academic Press, 1979.
Brooke, Frances. *The History of Emily Montague.* 4 vols. London: J. Dodsley, 1769.
Browne, Janet. "Biogeography and Empire." In *Cultures of Natural History*, edited by Nicholas Jardine, Emma Spary, and James A. Secord, 305–21. Cambridge: Cambridge University Press, 1996.
———. "Darwin's Botanical Arithmetic and the 'Principle of Divergence,' 1854–58." *Journal of the History of Biology* 13 (1980): 53–89.
———. *The Secular Ark: Studies in the History of Biogeography.* New Haven, CT: Yale University Press, 1983.
Browne, Patrick. *The Civil and Natural History of Jamaica.* London: B. White & Son, 1789.
Bryant, William Cullen. *Letters of William Cullen Bryant.* Edited by William Cullen Bryant II and Thomas G. Voss. 6 vols. New York: Fordham University Press, 1975–92.
———. *Poetical Works of William Cullen Bryant.* Edited by Parke Godwin. 2 vols. New York: Russell & Russell, 1883.
Buell, Lawrence. *The Environmental Imagination: Thoreau, Nature Writing, and the Formation of American Culture.* Cambridge, MA: Belknap, 1995.
Buffon, comte de. *Buffon's Natural History: Containing a Theory of the Earth. . . .* 10 vols. London: H. D. Symonds, 1797.
Burgess, Miranda. "Transport: Mobility, Anxiety, and the Romantic Poetics of Feeling." *Studies in Romanticism* 49 (2010): 229–55.
Burt, R. L., and W. T. Williams. "Plant Introduction in Australia." In *Australian Science in the Making*, edited by R. W. Home, 252–76. Cambridge: Cambridge University Press, 1988.
Buzard, James. *Disorienting Fiction: The Autoethnographic Work of Nineteenth-Century British Novels.* Princeton, NJ: Princeton University Press, 2005.
———. "On Auto-ethnographic Authority." *Yale Journal of Criticism* 16, no. 1 (2003): 61–91.
Byron. *Lord Byron: The Complete Poetical Works.* Edited by Jerome J. McGann. 7 vols. Oxford: Clarendon, 1980–93.
Cain, P. J., and A. G. Hopkins. *British Imperialism.* London: Longman, 1993.
Caley, George. *Reflections on the Colony of New South Wales.* Edited by John E. B. Currey. Melbourne: Lansdowne, 1966.
Callon, Michel, and Bruno Latour. "Unscrewing the Big Leviathon; Or How Actors Macrostructure Reality, and How Sociologists Help Them to Do So." In *Advances in Social Theory and Methodology: Toward an Integration of Micro and Macro Sociologies*, edited by K. Knorr-Cetina and A. V. Cicourel, 277–303. London: Routledge & Kegan Paul, 1981.
Campbell, Colin. *The Romantic Ethic and the Spirit of Modern Consumerism.* Oxford: Basil Blackwell, 1989.

Carey, Matthew. "On the Hessian Fly." *American Museum* 1 (February 1787): 133–35.
Carlyle, Thomas, and Jane Welsh Carlyle. *Collected Letters of Thomas and Jane Welsh Carlyle*. Edited by Charles Richard Sanders. Durham, NC: Duke University Press, 1970.
Carter, Landon. "Observations Concerning the Fly-Weevil, that destroys the wheat, with some useful discoveries and conclusions, concerning the propagation and progress of that pernicious insect, and the methods to be used to prevent the destruction of grain by it." *Transactions of the American Philosophical Society* 1 (1769): 205–17.
Carter, Paul. *The Road to Botany Bay: An Explanation of Landscape and History*. Chicago: University of Chicago Press, 1987.
Cashin, Edward J. "The Real World of Bartram's *Travels*." In Braund and Porter, *Fields of Vision*, 3–14.
———. *William Bartram and the American Revolution on the Southern Frontier*. Columbia: University of South Carolina Press, 2000.
Chakrabarty, Dipesh. "The Climate of History: Four Theses." *Critical Inquiry* 35, no. 2 (Winter 2009): 197–222.
Chambers, Douglas. "'A love for every simple weed': Clare, Botany and the Poetic Language of Lost Eden." In *John Clare in Context*, edited by Hugh Haughton, Adam Phillips, and Geoffrey Summerfield, 238–58. Cambridge: Cambridge University Press, 1994.
———. *The Planters of the English Landscape Garden: Botany, Trees, and the "Georgics."* New Haven, CT: Yale University Press, 1993.
Chambers, William. *Plans, Elevations, Sections, and Perspective Views of the Gardens and Buildings at Kew in Surrey*. London: J. Haberkorn, 1763.
Chandler, David. "Hart-Leap Well: A History of the Site of Wordsworth's Poem." *Notes & Queries* 49, no. 1 (March 2002): 19–25.
Cheyfitz, Eric. *The Poetics of Imperialism: Translation and Colonization from the Tempest to Tarzan*. Philadelphia: University of Pennsylvania Press, 1997.
Chilcott, Tim. *John Clare: The Living Year 1841*. Nottingham: Trent, 1999.
———. *"A Real World & Doubting Mind": A Critical Study of the Poetry of John Clare*. Hull: University of Hull Press, 1985.
Christmas, William J. *The Lab'ring Muses: Work, Writing, and the Social Order in English Plebeian Poetry, 1730–1830*. Newark: University of Delaware Press, 2001.
Clare, Johanne. *John Clare and the Bounds of Circumstance*. Kingston, ON: McGill-Queen's University Press, 1987.
Clare, John. *Early Poems*. Edited by Eric Robinson and David Powell. 2 vols. Oxford: Clarendon, 1989.
———. *Later Poems*. Edited by Eric Robinson and David Powell. 2 vols. Oxford: Clarendon, 1984–85.
———. *Letters of John Clare*. Edited by Mark Storey. Oxford: Clarendon, 1985.
———. *Natural History Prose Writings of John Clare*. Edited by Margaret Grainger. Oxford: Clarendon, 1983.

———. *Poems of the Middle Period, 1822–1837.* Edited by Eric Robinson, David Powell, and P. M. S. Dawson. 5 vols. Oxford: Clarendon, 1996.

———. *Prose of John Clare.* Edited by J. W. and Anne Tibble. New York: Barnes & Noble, 1951.

Clark, William, Jan Golinski, and Simon Schaffer, eds. *The Sciences in Enlightened Europe.* Chicago: University of Chicago Press, 1999.

Colbey, John. *Sydney Cove 1788.* London: Hodder, 1962.

Coleman, Deirdre. "Conspicuous Consumption: White Abolitionism and English Women's Protest Writing in the 1790s." *ELH* 61 (1994): 341–62.

———. *Romantic Colonization and British Anti-Slavery.* Cambridge: Cambridge University Press, 2005.

Coleridge, Samuel Taylor. *Biographia Literaria.* Edited by James Engell and W. Jackson Bate. 2 vols. In *Collected Works of Samuel Taylor Coleridge.*

———. *Collected Letters.* Edited by Earl Leslie Griggs. 6 vols. Oxford: Clarendon, 1959–71.

———. *Collected Works of Samuel Taylor Coleridge.* Edited by Kathleen Coburn. Princeton, NJ: Princeton University Press, 1969–.

———. *Marginalia.* Edited by George Whalley and Heather Jackson. 6 vols. In *Collected Works of Samuel Taylor Coleridge.*

———. *Notebooks of Samuel Taylor Coleridge.* Edited by Kathleen Coburn et al. 5 vols. Princeton, NJ: Princeton University Press, 1957–2002.

———. *Poetical Works.* Edited by J. C. C. Mays. 6 vols. In *Collected Works of Samuel Taylor Coleridge.*

———. *Shakespearean Criticism.* Edited by T. M. Raysor. 2 vols. London: Constable, 1930.

Colley, Linda. *Britons: Forging the Nation, 1707–1820.* New Haven, CT: Yale University Press, 1992.

Collier, Price. *England and the English from an American Point of View.* London: Duckworth, 1909.

Columbus, Christopher. *Journal of First Voyage to America.* Freeport, NY: Books for Libraries Press, 1924.

Conrad, Joseph. *Heart of Darkness.* 3rd ed. New York: Norton, 1988.

Cook, James. *Journals of Captain James Cook on His Voyages of Discovery.* Edited by J. C. Beaglehole. 4 vols. Cambridge: Hakluyt Society, 1967.

———. *Voyage Towards the South Pole, and Round the World.* 2 vols. London: W. Strahan, 1777.

Cook, James, and James King. *Voyage to the Pacific Ocean.* 3 vols. London: G. Nicol, 1784.

Cooper, Lane. "Bartram redivivus?" *Nation* 80 (1905): 152.

Cooper, Susan Fenimore. *Rural Hours.* Edited by Rochelle Johnson and Daniel Patterson. Athens: University of Georgia Press, 1998.

Corner, Betsy C., and Christopher C. Booth, eds. *Chain of Friendship: Selected Letters of Dr. John Fothergill of London, 1735–1780.* Cambridge, MA: Belknap, 1971.

Corson, Richard. *Fashions in Hair.* London: Peter Owen, 1965.

Cowper, William. *The Poetical Works of William Cowper.* Edited by H. S. Milford. 4th ed. London: Oxford University Press, 1934.

Crabbe, George. *Complete Poetical Works.* Edited by Norma Dalrymple-Champneys and Arthur Pollard. 3 vols. Oxford: Clarendon, 1988.

Cresswell, Tim. "Mobilities—An Introduction." *new formations* 43 (Spring 2001): 9–10.

Cronk, Q. C. B. *The Endemic Flora of St Helena.* Oswestry: Anthony Nelson, 2000.

Cronon, William. *Changes in the Land: Indians, Colonists, and the Ecology of New England.* New York: Hill & Wang, 1983.

———. *Nature's Metropolis: Chicago and the Great West.* New York: Norton, 1991.

———. "The Trouble with Wilderness, or, Getting Back to the Wrong Nature." In *Uncommon Ground: Rethinking the Human Place in Nature*, 69–90. New York: Norton, 1995.

Crosby, Alfred W. *Ecological Imperialism: The Biological Expansion of Europe, 900–1900.* Cambridge: Cambridge University Press, 1986.

Crumley, Carole L. "Historical Ecology: A Multidimensional Ecological Orientation." In *Historical Ecology: Cultural Knowledge and Changing Landscapes*, edited by C. Crumley, 1–16. Sante Fe, NM: School of American Research Press, 1994.

Curtis, William. *The Botanical Garden; or, Flower-Garden Displayed.* 14 vols. Lambeth Marsh: W. Curtis, 1787–1800.

Dadswell, Ted. *The Selborne Pioneer: Gilbert White as Naturalist and Scientist: A Re-Examination.* Aldershot: Ashgate, 2003.

Darlington, Philip J., Jr. "Darwin and Zoography." *Proceedings of the American Philosophical Society* 103 (1959): 307–19.

Darlington, William, ed. *Memorials of John Bartram and Humphry Marshall; with notices of their Botanical Contemporaries.* [s.l.]: [s.n.], 1849.

Darwin, Charles. *Autobiography of Charles Darwin 1809–1882.* Edited by Nora Barlow. London: Collins, 1958.

———. *Charles Darwin's Beagle Diary.* Edited by Richard Darwin Keynes. Cambridge: Cambridge University Press, 1988.

———. *Charles Darwin's Natural Selection Being the Second Part of His Big Species Book Written from 1856 to 1858.* Edited by R. C. Stauffer. Cambridge: Cambridge University Press, 1975.

———. *Charles Darwin's Notebooks, 1836–1844: Geology, Transmutation of Species, Metaphysical Enquiries.* Edited by Paul H. Barrett, Peter J. Gautrey, Sandra Herbert, David Kohn, and Sydney Smith. Ithaca, NY: Cornell University Press, 1987.

———. *Charles Darwin's Zoology Notes and Specimen Lists from H.M.S. Beagle.* Edited by Richard Darwin Keynes. Cambridge: Cambridge University Press, 2000.

———. *Correspondence of Charles Darwin.* Edited by Frederick Burkhardt and Sydney Smith. 17 vols. Cambridge: Cambridge University Press, 1985–2008.

———. "Darwin's Ornithological Notes." Edited by Nora Barlow. *Bulletin of the British Museum (Natural History) Historical Series* 2, no. 7 (1963): 201–78.

———. *The Descent of Man, and Selection in Relation to Sex.* Introduction by John Tyler Bonner and Robert M. May. Princeton, NJ: Princeton University Press. 1981.

———. *The Foundations of the Origin of Species: Two Essays Written in 1842 and 1844 by Charles Darwin.* Edited by Francis Darwin. Cambridge: Cambridge University Press, 1909.

———. *Life of Erasmus Darwin.* Edited by Ernst Krause. London: John Murray, 1887.

———. *On the Origin of Species: A Facsimile of the First Edition.* Introduction by Ernst Mayr. Cambridge, MA: Harvard University Press, 1964.

———. *The Voyage of the Beagle: Journal of Researches into the Natural History and Geology of the Countries Visited during the Voyage of H.M.S. "Beagle" Round the World.* Edited by Steve Jones. New York: Modern Library, 2001.

Darwin, Erasmus. *The Botanic Garden. Part 1: The Economy of Vegetation. Part 2: The Loves of the Plants.* London: J. Johnson, 1791.

———. *Phytologia; or, The Philosophy of Agriculture and Gardening.* London: J. Johnson, 1800.

———. *The Temple of Nature; or, The Origin of Society.* London: J. Johnson, 1803.

Darwish, Mahmoud. *La Palestine comme métaphore: Entretiens.* Paris: Sindhad, Actes Sud, 1997.

———. *Unfortunately, It Was Paradise: Selected Poems.* Edited by Munir Akash and Carolyn Forché. Berkeley: University of California Press, 2003.

Darwish, Mahmoud, Samih al-Qasim, and Adonis. *Victims of a Map.* Translated by Abdullah al-Udhari. London: Al Saqi Books, 1984.

de Galaup, Jean François, comte de La Pérouse. *Voyage de Lapérouse autor du monde pendant des années 1785, 1786, 1787 et 1788.* Edited by Pierre Sabbagh. Geneva: Cercle du bibliophile, 1970.

DeJohn Anderson, Virginia. *Creatures of Empire: How Domestic Animals Transformed Early America.* Oxford: Oxford University Press, 2004.

DeLoughrey, Elizabeth M. "Globalizing the Routes of Breadfruit and Other Bounties." *Journal of Colonialism and Colonial History* 8 (2007). doi:10.1353/cch.2008.0003.

———. "Island Ecologies." *Tijdschrift voor economische en sociale geografie / Journal for Economic and Social Geography* 93, no. 3 (1994): 298–310.

DeLoughrey, Elizabeth M., René Gosson, and George B. Handley, eds. *Caribbean Literature and the Environment: Between Nature and Culture.* Charlottesville: University of Virginia Press, 2005.

DeLoughrey, Elizabeth M., and George B. Handley, eds. *Postcolonial Ecologies: Literatures of the Environment.* Oxford: Oxford University Press, 2011.

Denham, John. *Poetical Works of Sir John Denham.* Edited by Theodore Howard Banks. 2nd ed. Hamden, CT: Archon Books, 1969.

Dening, Greg. *Islands and Beaches: Discourse on a Silent Land, Marquesas, 1774–1880.* Melbourne: Melbourne University Press, 1980.

Denyer, Susan. "The Lake District Landscape: Cultural or Natural?" In *The Making of a Cultural Landscape: The English Lake District as Tourist Destination, 1750–2010,* edited by John K. Walton and Jason Wood, 3–29. Farnham: Ashgate, 2013.

De Quincey, Thomas. *Confessions of an English Opium-Eater and Other Writings.* Edited by Barry Milligan. London: Penguin, 2003.

Derham, William. *Physico-Theology; or, A Demonstration of the Being and Attributes of God, from his Work of Creation.* 3rd ed. London: W. Innys, 1714.

Descartes, René. *Discourse on Method and the Meditations.* Translated by F. E. Sutcliffe. London: Penguin, 1968.

Desmond, Ray. *Great Natural History Books and Their Creators.* London: British Library, 2003.

———. *Kew: The History of the Royal Botanic Gardens.* Kew: Harvill Press with the Royal Botanic Gardens, Kew, 1995.

Dickens, Charles. *The Mystery of Edwin Drood.* 1870. Reprint, Harmondsworth: Penguin, 1974.

Dilke, Charles Wentworth. *Greater Britain: A Record of Travel in English-Speaking Countries during 1866 and 1867.* 2nd ed. 2 vols. London: Macmillan, 1869.

Dillard, Annie. *Teaching a Stone to Talk: Expeditions and Encounters.* New York: Harper Colophon, 1982.

Douglas, Mary. *Purity and Danger: An Analysis of Concepts of Pollution and Taboo.* New York: Frederick A. Praeger, 1966.

Douglas, Mary, and Baron Isherwood. *The World of Goods: Towards an Anthropology of Consumption.* Harmondsworth: Penguin, 1979.

Drayton, Richard. "Knowledge and Empire." In *The Oxford History of the British Empire: The Eighteenth Century,* edited by P. J. Marshall, 231–52. Oxford: Oxford University Press, 1998.

———. *Nature's Government: Science, Imperial Britain, and the "Improvement" of the World.* New Haven, CT: Yale University Press, 2000.

Dunlap, Thomas R. *Nature and the English Diaspora.* Cambridge: Cambridge University Press, 1999.

Durling, Dwight L. *Georgic Tradition in English Poetry.* New York: Columbia University Press, 1935.

Duthie, Ruth E. "English Florists' Societies and Feasts in the Seventeenth and First Half of the Eighteenth Centuries." *Garden History* 10, no. 1 (Spring 1982): 17–35.

———. "Florists' Societies and Feasts after 1750." *Garden History* 12, no. 1 (Spring 1984): 8–38.

Edwards, Bryan. *History, Civil and Commercial, of the British West Indies.* 5 vols. London: T. Miller, 1819.

Edwards, Phyllis I. "Botany of the Flinders Voyage." In *People and Plants in Australia,* edited by D. J. Carr and S. G. M. Carr, 139–66. Sydney: Academic Press, 1981.

Egan, Dave, and Evelyn Howell, eds. *The Historical Ecology Handbook: A Restorationist's Guide to Reference Ecosystems.* Washington, DC: Island Press, 2005.

Egerton, Frank N. "Changing Concepts of the Balance of Nature." *Quarterly Review of Biology* 48 (1973): 322–50.

———. "Humboldt, Darwin, and Population." *Journal of the History of Biology* 3 (1970): 325–60.

---. "Studies of Animal Populations from Lamarck to Darwin." *Journal of the History of Biology* 1 (1968): 225–59.
Ellis, John. *Directions for Bringing Over Seeds and Plants, from the East-Indies and Other Distant Countries, in a State of Vegetation*. . . . London: L. Davis, 1770.
Endersby, Jim. *Imperial Nature*. Chicago: University of Chicago Press, 2008.
Evelyn, John. *Sylva; or, A Discourse of Forest-Trees*. 2nd ed. London: Martyn & Allestry, 1670.
Ewan, Joseph. *William Bartram: Botanical and Zoological Drawings, 1756–1788*. Philadelphia: American Philosophical Society, 1968.
Fabian, Johannes. *Time and the Other*. New York: Columbia University Press, 1983.
Faulkner, William. *Requiem for a Nun*. New York: Random House, 1951.
Festa, Lynn. "Humanity without Feathers." *Humanity: An International Journal of Human Rights, Humanitarianism, and Development* 1, no. 1 (2010): 3–27.
---. *Sentimental Figures of Empire in Eighteenth-Century Britain and France*. Baltimore: Johns Hopkins University Press, 2006.
Fitch, Asa. *The Hessian Fly, Its History, Character, Transformations, and Habits*. Albany, NY: Joel Munsell, 1846.
Fleming, John. "Remarks Illustrative of the Influence of Society on the Distribution of British Animals." *Edinburgh Philosophical Journal* 11 (1824): 287–305.
Forster, Johann Reinhold. *Catalogue of British Insects*. Warrington: William Eyres, 1770.
---. *Catalogue of the Animals of North America*. London: B. White, 1771.
---. *Observations Made During a Voyage Round the World*. Edited by Nicholas Thomas, Harriet Guest, and Michael Dettelbach. Honolulu: University of Hawaii Press, 1996.
Foster, Paul G. M. "The Gibraltar Correspondence of Gilbert White." *Notes and Queries* 32 (1985): 227–36, 315–28, 489–500.
---. *Gilbert White and His Records: A Scientific Biography*. London: Christopher Helm, 1988.
Franklin, Benjamin. *The Papers of Benjamin Franklin*. Edited by Leonard W. Labaree. 40 vols. New Haven, CT: Yale University Press, 1960–.
Fraser, Nancy, and Linda Gordon. "Contract versus Charity: Why Is There No Social Citizenship in the United States?" In *The Citizenship Debates: A Reader*, edited by Gershon Shafir, 113–30. Minneapolis: University of Minnesota Press, 1998.
Friel, Brian. *Translations*. London: Faber & Faber, 1981.
Frost, Alan. "Antipodean Exchange: European Horticulture and Imperial Designs." In Miller and Reill, *Visions of Empire*, 58–79.
---. "Science for Political Purposes: European Explorations of the Pacific Ocean, 1764–1806." In *Nature in Its Greatest Extent: Western Science in the Pacific*, edited by Roy MacLeod and Philip F. Rehbock, 27–44. Honolulu: University of Hawaii Press, 1988.
---. *Sir Joseph Banks and the Transfer of Plants to and from the South Pacific, 1786–1798*. Melbourne: Colony Press, 1993.

Fulford, Tim. "Coleridge, Darwin, Linnaeus: The Sexual Politics of Botany." *Wordsworth Circle* 28 (1997): 124–30.

———. *Romantic Indians: Native Americans, British Literature, and Transatlantic Culture 1756–1830*. Oxford: Oxford University Press, 1989.

Fulford, Tim, Debbie Lee, and Peter J. Kitson. *Literature, Science and Exploration in the Romantic Era: Bodies of Knowledge*. Cambridge: Cambridge University Press, 2004.

Galloway, Andrew. "William Cullen Bryant's American Antiquities: Medievalism, Miscegenation, and Race in *The Prairies*." *American Literary History* 22 (2010): 724–51.

Garrard, Greg. "An Absence of Azaleas: Imperialism, Exoticism and Nativity in Romantic Biogeographical Ideology." *Wordsworth Circle* 28 (1997): 148–55.

Gascoigne, John. *Science in the Service of Empire: Joseph Banks, the British State and the Uses of Science in the Age of Revolution*. Cambridge: Cambridge University Press, 1998.

Gentzler, Edwin. *Translation and Identity in the Americas: New Directions in Translation Theory*. New York: Routledge, 2008.

Ghiselin, Michael T. *The Triumph of the Darwinian Method*. Berkeley: University of California Press, 1969.

Giddens, Anthony. *Consequences of Modernity*. Stanford: Stanford University Press, 1990.

———. *In Defence of Sociology: Essays, Interpretations and Rejoinders*. Cambridge: Polity, 1996.

———. *Modernity and Self-Identity: Self and Society in the Late Modern Age*. Cambridge: Polity, 1991.

Gillis, John R. *Islands of the Mind: How the Human Imagination Created the Atlantic World*. New York: Palgrave Macmillan, 2004.

———. "Taking History Offshore: Atlantic Islands in European Minds 1400–1800." In *Islands in History and Representation*, edited by Rod Edmond and Vanessa Smith, 19–31. London: Routledge, 2003.

Glissant, Edouard. *Caribbean Discourse*. Charlottesville: University of Virginia Press, 1989.

———. "Distancing, Determining." In *Poetics of Relation*, translated by Betsy Wing, 141–58. Ann Arbor: University of Michigan Press, 2006.

Goethe, Johann Wolfgang. *West-Easterly Divan*. In *Translating Literature: The German Tradition from Luther to Rosenzweig*, edited by André Lefevere, 35–37. Assan: Van Gorcam, 1977.

Goldsmith, Oliver. *Collected Works*. Edited by Arthur Friedman. 5 vols. Oxford: Clarendon, 1966.

Goodman, Kevis. *Georgic Modernity and British Romanticism: Poetry and the Mediation of History*. Cambridge: Cambridge University Press, 2004.

Gosse, Edmund. *Gossip in a Library*. London: Heineman, 1913.

Graham, Gerald S. *Tides of Empire: Discussions on the Expansion of Britain Overseas*. Kingston, ON: McGill-Queen's University Press, 1972.

Grainger, James. *The Sugar-Cane: A Poem in Four Books.* London: R. & J. Dodsley, 1764.
Greene, John C. *The Death of Adam: Evolution and Its Impact on Western Thought.* Ames: Iowa State University Press, 1959.
Griffin, Harry. "If they go, it is the end of the Lakeland." Country Diary. *Guardian*, 11 April 2001. http://www.theguardian.com/uk/2001/apr/11/footandmouth.features11.
Grove, Richard H. *Green Imperialism: Colonial Expansion, Tropical Island Edens and the Origins of Environmentalism, 1600–1860.* Cambridge: Cambridge University Press, 1995.
Hadfield, Miles. *A History of British Gardening.* 3rd ed. Chatham: John Murray, 1979.
Hall, Dewey W. *Romantic Naturalists, Early Environmentalists: An Ecocritical Study, 1780-1912.* Farnham: Ashgate, 2014.
Hallock, Thomas. *From the Fallen Tree: Frontier Narratives, Environmental Politics, and the Roots of a National Pastoral, 1749–1826.* Chapel Hill: University of North Carolina Press, 2003.
Hallock, Thomas, and Nancy E. Hoffman, eds. *William Bartram: The Search for Nature's Design.* Athens: University of Georgia Press, 2010.
Hamblyn, Richard. *Terra: Tales of the Earth: Four Events That Changed the World.* London: Picador, 2009.
Hanbury, William. *A Complete Body of Planting and Gardening.* 2 vols. London: Edward & Charles Dilly, 1770.
Hancock, David. *Citizens of the World: London Merchants and the Integration of the British Atlantic Community.* New York: Cambridge University Press, 1995.
Harper, Francis. "William Bartram and the American Revolution." *Proceedings of the American Philosophical Society* 97 (1953): 571–77.
Harris, Wilson. *Whole Armour.* London: Faber, 1962.
Harrison, Gary. "Hybridity, Mimicry and John Clare's 'Child Harold.'" *Wordsworth Circle* 34, no. 3 (Summer 2003): 149–55.
Hartman, Geoffrey. *The Unremarkable Wordsworth.* Minneapolis: University of Minnesota Press, 1987.
Harvey, John H. *Early Gardening Catalogues.* London: Phillimore, 1972.
———. *Early Nurserymen with Reprints of Documents and Lists.* London: Phillimore, 1974.
Hawkesworth, John. *An Account of the Voyages Undertaken by the Order of His Present Majesty for Making Discoveries in the Southern Hemisphere.* 3 vols. London: Strahan & Cadell, 1773.
Hazucha, Andrew. "Neither Deep nor Shallow but National: Eco-Nationalism in Wordsworth's *Guide to the Lakes.*" *Interdisciplinary Studies in Literature and Environment* 9, no. 2 (2002): 61–73.
Headrick, Daniel R. *Tentacles of Progress: Technology Transfer in the Age of Imperialism, 1850–1940.* New York: Oxford University Press, 1988.

———. *When Information Came of Age: Technologies of Knowledge in the Age of Reason and Revolution, 1700–1850*. New York: Oxford University Press, 2000.

Hedges, William. "Toward a National Literature." In *Columbia Literary History of the United States*, edited by Emory Elliott, 187–202. New York: Columbia University Press, 1988.

Helsinger, Elizabeth. *Rural Scenes and National Representation: Britain, 1815–1850*. Princeton, NJ: Princeton University Press, 1997.

Henrey, Blanche. *British Botanical and Horticultural Literature before 1800*. 3 vols. London: Oxford, 1975.

Herder, Johann Gottfried von. *Outlines of a Philosophy of the History of Man*. Translated by T. Churchill. London: J. Johnson, 1800.

Hess, Scott. *William Wordsworth and the Ecology of Authorship: The Roots of Environmentalism in Nineteenth-Century Culture*. Charlottesville: University of Virginia Press, 2012.

Hindle, Brook. *The Pursuit of Science in Revolutionary America, 1735–1789*. Chapel Hill: University of North Carolina Press, 1956.

Hogden, Margaret T. *Early Anthropology in the Sixteenth and Seventeenth Century*. Philadelphia: University of Pennsylvania Press, 1964.

Holt-White, Rashleigh, ed. *The Life and Letters of Gilbert White of Selborne*. 2 vols. London: John Murray, 1901.

Home, Everard. "Description of the Anatomy of the *Ornithorhynchus Hystrix*." *Philosophical Transactions of the Royal Society* 92 (1802): 348–64.

Hooker, Joseph Dalton. *Botany of the Antarctic Voyage of the H.M. Discovery Ships Erebus and Terror, in the Years 1839–1843*. 3 vols. London: Reeve, Brothers, 1844.

———. "The Distribution of the North American Flora." *American Naturalist* 13 (1879): 155–70.

———. *Flora Novae-Zelandiae*. Part 2 of *The Botany of the Antarctic Voyage*. London: Lovell Reeve, 1853.

———. "Insular Floras." Edited by M. Williamson. *Biological Journal of the Linnean Society* 22, no. 1 (1984): 59–77.

Hooker, William. *The British Flora*. 4th ed. London: Longman, Orme, Brown, Green & Longmans, 1838.

Houghton, John. *A Collection for Improvement of Husbandry and Trade*. London: Woodman & Lyon, 1727.

Hudson, J., ed. *A Complete Guide to the Lakes, Comprising Minute Directions for the Tourist, with Mr. Wordsworth's Description of the Scenery of the Country, &c. and Three Letters on the Geology of the Lake District, by the Rev. Professor Sedgwick*. Kendal: Hudson & Nicholson, 1842.

Hughes, Robert. *Australia: Beyond the Fatal Shore*. DVD Video. Directed by Christopher Spencer. 2010.

Humboldt, Alexander von. *Views of Nature*. Translated by Mark W. Person. Chicago: University of Chicago Press, 2014.

Humboldt, Alexander von, and Aimé Bonpland. *Essai sur la Géographie des plantes; accompané d'un Tableau Physique des Régions Équinoxiales*. Paris: Levrault, Schoell, 1805.
Hutchings, Kevin. *Imagining Nature: Blake's Environmental Poetics*. Kingston, ON: McGill-Queen's University Press, 2002.
———. *Romantic Ecologies and Colonial Cultures in the British-Atlantic World, 1770–1850*. Kingston, ON: McGill-Queen's University Press, 2009.
Ingold, Tim. "The Temporality of the Landscape." *World Archaeology* 25, no. 2 (October 1993): 152–74.
Ingoville, Martin. *The Historical Ecology of the British Flora*. London: Chapman & Hall, 1995.
Irmscher, Christoph. *Poetics of Natural History: From John Bartram to William James*. New Brunswick, NJ: Rutgers University Press, 1999.
Jackson, Christine E. *Sarah Stone: Natural Curiosities from the New Worlds*. London: Merrell Holberton and the Natural History Museum, London, 1998.
Jackson, Heather J. *Romantic Readers: The Evidence of Marginalia*. New Haven, CT: Yale University Press, 2005.
Jacobus, Mary. *Tradition and Experiment in Wordsworth's "Lyrical Ballads" (1798)*. Oxford: Oxford University Press, 1976.
Janowitz, Anne. *England's Ruins: Poetic Purpose and the National Landscape*. Oxford: Basil Blackwell, 1990.
Jefferson, Thomas. *The Papers of Thomas Jefferson*. Edited by Julian P. Boyd. Princeton, NJ: Princeton University Press, 1950.
———. *Thomas Jefferson: Writings*. Edited by Merrill D. Peterson. New York: Library of America, 1984.
Jenkins, C. F. H. *The Noah's Ark Syndrome*. Perth: Zoological Gardens Board, 1977.
Johnson, Samuel. *A Dictionary of the English Language: in which the words are deduced from their originals, and illustrated in their different significations by examples from the best writers. To which are prefixed, A history of the language, and An English grammar*. 2 vols. London: W. Strahan, 1765. Eighteenth Century Collections Online. http://find.galegroup.com.myaccess.library.utoronto.ca/ecco/infomark.do?&source=gale&prodId=ECCO&userGroupName=utoronto_main&tabID=T001&docId=CB131376708&type=multipage&contentSet=ECCOArticles&version=1.0&docLevel=FASCIMILE.
Johnston, Kenneth R. "The Romantic Idea-Elegy: The Nature of Politics and the Politics of Nature." *South Central Review* 9 (Spring 1992): 24–43.
Jones, Henry. *Kew Garden*. 2nd ed. London: J. Brown, 1767.
———. *The Royal Vision: In an Ode to Peace*. Dublin: William Watson, 1763.
Jones, William. *The Collected Works of Sir William Jones*. 13 vols. London: Stockdale, Walker, 1807.
Josselyn, John. *New-England's Rarities, Discovered in Birds, Beasts, Fishes, Serpents, and Plants of that Country*. Edited by Edward Tuckerman. Boston: William Veazie, 1865.

Kalm, Peter. *Travels into North America: Containing its Natural History, and a Circumstantial Account of its Plantations and Agriculture in General, with the Civil, Ecclesiastical and Commercial State of the Country, the Manners of the Inhabitants*. Translated by John Reinhold Forster. 2nd ed. 2 vols. London: T. Lowndes, 1772.
Keats, John. *The Letters of John Keats 1814–1821*. Edited by Hyder E. Rollins. 2 vols. Cambridge, MA: Harvard University Press, 1958.
Keegan, Bridget. *British Labouring-Class Nature Poetry, 1730–1837*. Houndmills: Palgrave Macmillan, 2008.
———. "'Camelion' Clare." *European Romantic Review* 18 (2007): 445–57.
Keith, W. J. *The Rural Tradition: William Cobbett, Gilbert White, and Other Non-Fiction Prose Writers*. Hassocks: Harvester, 1975.
Kelley, Theresa M. *Clandestine Marriage: Botany and Romantic Culture*. Baltimore: Johns Hopkins University Press, 2012.
Kimber, Clarissa Thérèse. *Martinique Revisited: The Changing Plant Geographies of a West Indian Island*. College Station: Texas A&M University Press, 1988.
Kincaid, Jamaica. "Alien Soil." In *Nature Writing: The Tradition in English*, edited by Robert Finch and John Elder, 1015–22. New York: Norton, 2002.
———. *My Garden (Book):*. Illustrated by Jill Pox. New York: Farrar, Straus & Giroux, 1999.
Kinch, Michael Paul. "Geographical Distribution and the Origin of Life: The Development of Early Nineteenth Century British Explanations." *Journal of the History of Biology* 13 (1980): 91–119.
King-Hele, Desmond. *Erasmus Darwin and the Romantic Poets*. London: Macmillan, 1986.
———. "Erasmus Darwin, Man of Ideas and Inventor of Words." *Notes and Records of the Royal Society of London* 42 (1988): 149–80.
Kipling, Rudyard. *Cambridge Edition of the Poems of Rudyard Kipling*. Edited by Thomas Pinney. 3 vols. Cambridge: Cambridge University Press, 2013.
———. *Jungle Books*. Toronto: Penguin, 1987.
Kirby, William, and William Spence. *An Introduction to Entomology; or, Elements of the Natural History of Insects*. 2nd ed. London: Longman, Hurst, Rees, Orme, & Brown, 1816.
Klingender, Francis D. *Art and the Industrial Revolution*. London: Noel Carrington, 1947.
Koerner, Lisbet. "Linnaeus' Floral Transplantation." *Representations* 47 (Summer 1994): 144–69.
———. *Linnaeus: Nature and Nation*. Cambridge, MA: Harvard University Press, 1999.
———. "Purposes of Linnaean Travel." In Miller and Reill, *Visions of Empire*, 117–52.
Kohn, David. "Theories to Work By: Rejected Theories, Reproduction, and Darwin's Path to Natural Selection." *Studies in History of Biology* 4 (1980): 67–170.
Kosek, Jake. *Understories*. Durham, NC: Duke University Press, 2006.

Kraak, Deborah. "Eighteenth-Century English Floral Silks." *Magazine Antiques* 153, no. 6 (June 1998): 843–49.
Kroeber, Karl. *Ecological Literary Criticism: Romantic Imagining and the Biology of Mind.* New York: Columbia University Press, 1994.
Kumar, Deepak. "The Evolution of Colonial Science in India: Natural History and the East India Company." In *Imperialism and the Natural World*, edited by John M. MacKenzie, 51–66. Manchester: Manchester University Press, 1990.
Labillardière, Jacques-Julien. *An Account of a Voyage in Search of La Pérouse, undertaken by order of the Constituent Assembly of France, and Performed in the Years 1791, 1792, and 1793.* London: J. Debrett, 1800.
Lai, Walton Look. *The Chinese in the West Indies, 1806–1995.* Kingston, Jamaica: University of the West Indies, 1998.
Laird, Mark. *The Flowering of the Landscape Garden: English Pleasure Grounds, 1720–1800.* Philadelphia: University of Pennsylvania Press, 1999.
Lambert, José, and Clem Robyns. "Translation." In *Semiotics: A Handbook on the Sign-Theoretic Foundations of Nature and Culture*, edited by Roland Posner, Klaus Robering, and Thomas A. Sebeok, 3:3594–614. Berlin: W. de Gruyter, 1997.
Larson, James. "Not without a Plan: Geography and Natural History in the Late Eighteenth Century." *Journal of the History of Biology* 19 (1986): 447–88.
Latour, Bruno. *Politics of Nature: How to Bring the Sciences into Democracy.* Cambridge, MA: Harvard University Press, 2004.
———. *Science in Action: How to Follow Scientists and Engineers through Society.* Cambridge, MA: Harvard University Press, 1999.
———. "Visualization and Cognition: Thinking with Eyes and Hands." *Knowledge and Society: Studies in the Sociology of Culture Past and Present* 6 (1986): 1–40.
Layman, William. "Hints for the Cultivation of Trinidad" (1802). In *The Chinese in the West Indies, 1806–1995*, edited by Walton Look Lai, 23–27. Kingston, Jamaica: University of the West Indies, 1998.
Lear, Jonathan. *Radical Hope: Ethics in the Face of Cultural Devastation.* Cambridge, MA: Harvard University Press, 2006.
Leckie, Gould Francis. *An Historical Survey of the Foreign Affairs of Great Britain for the Years 1808, 1809, 1810: with a View to Explain the Causes of the Disasters of the Late and Present Wars.* London: D. N. Shury, 1810.
Lee, Ida. *William Bligh's Second Voyage to the South Sea.* London: Longmans, Green, 1920.
Lefebvre, Henri. *The Production of Space.* Translated by Donald Nicholson-Smith. Oxford: Blackwell, 1991.
Leopold, Aldo. *A Sand County Almanac and Sketches Here and There.* New York: Oxford University Press, 1974.
Lévi-Strauss, Claude. *Tristes Tropiques.* New York: Atheneum, 1978.
Limoges, Camille. *Carl Linné, L'Equilibre de la nature.* Paris: Vrin, 1972.
Linnaeus, Carl. *Linnaeus' Philosophia Botanica.* Translated by Stephen Freer. Oxford: Oxford University Press, 2003.

Liu, Alan. *Wordsworth: The Sense of History*. Stanford: Stanford University Press, 1989.
Locke, John. *An Essay Concerning Human Understanding*. Edited by Alexander Campbell Fraser. 2 vols. New York: Dover, 1959.
Logan, James Venable. *The Poetry and Aesthetics of Erasmus Darwin*. Princeton, NJ: Princeton University Press, 1936.
Long, Edward. *History of Jamaica*. 3 vols. London: T. Lowndes, 1774.
Longfellow, Henry Wadsworth. *Complete Poetical Works of Henry Wadsworth Longfellow*. Edited by Horace Elisha Scudder. Boston: Houghton, Mifflin, 1893.
Looby, Christopher. "The Construction of Nature: Taxonomy as Politics in Jefferson, Peale, and Bartram." *Early American Literature* 22 (1987): 252–73.
Loudon, John Claudius. *An Encyclopaedia of Gardening, Comprising the Theory and Practice of Horticulture, Floriculture, Arboriculture, and Landscape-Gardening*. 5th ed. London: Longman, Rees, Orme, Brown, Green, & Longman, 1828.
Louter, David. *Windshield Wilderness: Cars, Roads, and Nature in Washington's National Parks*. Seattle: University of Washington Press, 2006.
Lovejoy, Arthur O. "Some Meanings of 'Nature.'" In *Primitivism and Related Ideas in Antiquity*, edited by Arthur O. Lovejoy and George Boas, 447–56. Baltimore: Johns Hopkins University Press, 1948.
Low, Tim. *Feral Future: The Untold Story of Australia's Exotic Invaders*. 2nd ed. Chicago: University of Chicago Press, 2001.
Lowell, James Russell. *My Study Windows*. Boston: J. R. Osgood, 1874.
Lowes, John Livingston. *The Road to Xanadu: A Study in the Ways of the Imagination*. New York: Vintage, 1959.
Lucas, John. "Places and Dwellings: Wordsworth, Clare and the Anti-Picturesque." In *The Iconography of Landscape: Essays on the Symbolic Representation, Design and Use of Past Environments*, edited by Denis Cosgrove and Stephen Daniel, 83–97. Cambridge: Cambridge University Press, 1989.
Lutz, Alfred. "The Politics of Reception: The Case of Goldsmith's 'The Deserted Village.'" *Studies in Philology* 95 (1998): 174–96.
Lyell, Charles. *Life, Letters, and Journals of Sir Charles Lyell, Bart*. Edited by K. Lyell. 2 vols. London: Murray, 1881.
———. *Principles of Geology, Being an Attempt to Explain the Former Changes of the Earth's Surface, by Reference to Causes Now in Operation*. 3 vols. Chicago: University of Chicago Press, 1990.
Lyly, John. *Euphues*. London: Gabriel Cawood, 1578.
Lynch, Deidre Shauna. "'Young ladies are delicate plants': Jane Austen and Greenhouse Romanticism." *ELH* 77 (2010): 689–729.
Mabey, Richard. *Gilbert White: A Biography of the Author of "The Natural History of Selborne."* London: Century Hutchinson, 1986.
Macaulay, Thomas. *History of England*. London: Macmillan, 1913.
Mackaness, George. *The Life of Vice-Admiral William Bligh*. 2 vols. New York: Farrar & Rinehart, 1936.

Mackay, David. "Agents of Empire." In Miller and Reill, *Visions of Empire*, 38–57.
———. *In the Wake of Cook: Exploration, Science and Empire, 1780–1801*. London: Croom Helm, 1985.
———. "Myth, Science, and Experience in the British Construction of the Pacific." In *Voyage and Beaches: Pacific Encounters, 1769–1840*, edited by Alex Calder, Jonathan Lamb, and Bridget Orr, 100–113. Honolulu: University of Hawaii Press, 1999.
———. "A Presiding Genius of Exploration: Banks, Cook, and Empire, 1767–1805." In *Captain James Cook and His Times*, edited by Robin Fisher and Hugh Johnston, 21–39. Vancouver: Douglas & McIntyre, 1979.
Macnaghton, Phil, and John Urry. *Contested Natures*. London: Sage, 1998.
Maddox, Lucy B. "Gilbert White and the Politics of Natural History." *Eighteenth-Century Life* 10 (1986): 45–57.
Mahood, M. M. *The Poet as Botanist*. Cambridge: Cambridge University Press, 2008.
Makdisi, Saree. *Romantic Imperialism: Universal Empire and the Culture of Modernity*. Cambridge: Cambridge University Press, 1998.
Malchow, H. L. "Frankenstein's Monster and Images of Race in Nineteenth-Century Britain." *Past and Present* 139 (1993): 90–130.
Malkki, Lisa H. "National Geographic: The Rooting of Peoples and the Territorialization of National Identity among Scholars and Refugees." *Cultural Anthropology* 7 (1992): 24–44.
Marsh, George Perkins. *Man and Nature; or, Physical Geography as Modified by Human Action*. New York: C. Scribner, 1864.
Marshall, Ian. *Story Line: Exploring the Literature of the Appalachian Trail*. Charlottesville: University Press of Virginia, 1998.
Marx, Karl. *Capital: A Critique of Political Economy*. New York: International, 1967.
———. *Grundrisse*. Harmondsworth: Penguin, 1973.
Marx, Karl, and Friedrich Engels. *Selected Correspondence*. 2nd ed. Moscow: Progress, 1965.
Mathias, Thomas James. *The Pursuits of Literature. A Satirical Poem in Four Dialogues*. 5th ed. London: T. Becket, 1798.
Mayr, Ernst. *The Growth of Biological Thought: Diversity, Evolution, and Inheritance*. Cambridge, MA: Harvard University Press, 1982.
McCarthy, F. D. "Notes on the Cave Paintings of Groote and Chasm Islands in the Gulf of Carpentaria." *Mankind* 5, no. 2 (September 1955): 68–75.
McClellan, James E. *Colonialism and Science: Saint Domingue in the Old Regime*. Baltimore: Johns Hopkins University Press, 1992.
McCracken, Donald. *Gardens of Empire*. Washington, DC: Leicester University Press, 1997.
McGann, Jerome K. *The Romantic Ideology: A Critical Investigation*. Chicago: University of Chicago Press, 1983.
McGregor, Deborah. "Coming Full Circle: Indigenous Knowledge, Environment, and Our Future." *American Indian Quarterly* 28 (2004): 385–410.

McKendrick, Neil. "Commercialization of Fashion." In McKendrick, Brewer, and Plumb, *Birth of a Consumer Society*, 35–99.

———. "Josiah Wedgwood and the Commercialization of Potteries." In McKendrick, Brewer, and Plumb, *Birth of a Consumer Society*, 100–145.

McKendrick, Neil, John Brewer, and J. H. Plumb, eds. *The Birth of a Consumer Society: The Commercialization of Eighteenth-Century England*. London: Hutchinson, 1983.

McKibben, Bill. *The End of Nature*. New York: Random House, 1989.

———. "A Special Moment in History." *Atlantic Monthly*, May 1998, 54–78.

McKusick, James C. *Green Writing: Romanticism and Ecology*. New York: St. Martin's, 2000.

McLane, Maureen N. *Romanticism and the Human Sciences: Poetry, Population, and the Discourse of the Species*. Cambridge: Cambridge University Press, 2000.

McLeod, Donald. *Donald McLeod's Gloomy Memories in the Highlands of Scotland: versus Mrs. Harriet Beecher Stowe's Sunny Memories in (England) a Foreign Land: or a Faithful Picture of the Extirpation of the Celtic Race from the Highlands of Scotland*. Glasgow: A Sinclair, 1892.

Mellor, Anne K. "The Baffling Swallow: Gilbert White, Charlotte Smith and the Limits of Natural History." *Nineteenth-Century Contexts: An Interdisciplinary Journal* 31 (2009): 299–309.

———. "*Frankenstein*, Racial Science, and the Yellow Peril." *Nineteenth-Century Contexts* 23, no. 1 (2008): 1–28.

———. *Mary Shelley: Her Life, Her Fiction, Her Monsters*. New York: Methuen, 1988.

Melville, Elinor. *A Plague of Sheep: Environmental Consequences of the Conquest of Mexico*. Cambridge: Cambridge University Press, 1994.

Menely, Tobias. "'The Present Obfuscation': Cowper's 'Task' and the Time of Climate." *PMLA* 127 (2012): 477–92.

———. "Traveling in Place: Gilbert White's Cosmopolitan Parochialism." *Eighteenth-Century Life* 28, no. 3 (2004): 46–65.

Miall, L. C., and Alfred Denny. *The Structure and Life-History of the Cockroach (Periplaneta Orientalis): An Introduction to the Study of Insects*. London: Lovell Reeve, 1886.

Michael, P. W. "The Weeds Themselves—Early History and Identification." *Proceedings of the Weed Society, New South Wales* 5 (1972): 3–18.

Miller, David Philip. "Joseph Banks, Empire, and 'Centers of Calculation' in Late Hanoverian London." In Miller and Reill, *Visions of Empire*, 21–37.

Miller, David Philip, and Peter Hanns Reill, eds. *Visions of Empire: Voyages, Botany, and Representations of Nature*. Cambridge: Cambridge University Press, 1996.

Milton, John. *Paradise Lost*. Oxford: Oxford University Press, 2005.

Morton, Timothy. *Ecology without Nature: Rethinking Environmental Aesthetics*. Cambridge, MA: Harvard University Press, 2007.

———. *Shelley and the Revolution in Taste: The Body and the Natural World.* Cambridge: Cambridge University Press, 1994.

Moyal, Ann Mozley, ed. *Scientists in Nineteenth Century Australia: A Documentary History.* Melbourne: Cassell Australia, 1975.

Muirhead, James Patrick, ed. *The Origin and Progress of the Mechanical Inventions of James Watt.* 3 vols. London: John Murray, 1854.

Muller, Gilbert H. *William Cullen Bryant: Author of America.* Albany: State University of New York Press, 2008.

Mulligan, Martin, and Stuart Hill. *Ecological Pioneers: A Social History of Australian Ecological Thought and Action.* Cambridge: Cambridge University Press, 2001.

Murdoch, John. "The Landscape of Labour: Transformations of the Georgic." In *Romantic Revolutions: Criticism and Theory*, edited by Kenneth Johnston, Gilbert Chaitin, Karen Hanson, and Herbert Marks, 176–93. Bloomington: Indiana University Press, 1990.

Murray, John. *Account of the Larch Plantations on the Estates of Atholl and Dunkeld.* Edinborough: William Blackwood, 1832.

Nabholtz, John R. "Wordsworth's 'Guide to the Lakes' and the Picturesque Tradition." *Modern Philology* 61 (1964): 288–97.

Nelson, E. Charles. "John White's *Journal of a Voyage to New South Wales* (London 1790): Bibliographic Notes." *Archives of Natural History* 25 (1998): 109–30.

Nicholson, Michael. "The Itinerant 'I': John Clare's Lyric Defiance." *ELH* 82 (2015): 637–69.

Nietzsche, Friedrich. *The Gay Science: With a Prelude in Rhymes and an Appendix of Songs.* Translated by Walter Kaufmann. New York: Vintage Books, 1974.

Niranjana, Tejaswini. *Siting Translation: History, Post-Structuralism, and the Colonial Context.* Berkeley: University of California Press, 1992.

Noblett, William. "Pennant and His Publisher; Benjamin White, Thomas Pennant, and *Of London*." *Archives of Natural History* 11 (1982): 61–68.

North, Marianne. *A Vision of Eden: The Life and Work of Marianne North.* Edited by Graham Bateman. Exeter: Webb & Bower, 1980.

Noyes, Russell. *Wordsworth and the Art of Landscape.* New York: Haskell House, 1973.

Nugent, Maria Skinner. *Lady Nugent's Journal of Her Residence in Jamaica from 1801 to 1805.* Edited by Philip Wright. Kingston: Institute of Jamaica, 1966.

O'Brien, Susie. "The Garden and the World: Jamaica Kincaid and the Cultural Borders of Ecocriticism." *Mosaic* 35 (2002): 167–84.

Oerlemans, Onno. *Romanticism and the Materiality of Nature.* Toronto: University of Toronto Press, 2002.

Ospovat, Dov. *The Development of Darwin's Theory: Natural History, Natural Theology, and Natural Selection, 1838–1859.* Cambridge: Cambridge University Press, 1981.

Pagden, Anthony. *European Encounters with the New World: From Renaissance to Romanticism.* New Haven, CT: Yale University Press, 1993.

Park, Mungo. *Travels in the Interior Districts of Africa*. Edited by Kate Ferguson Marsters. Durham, NC: Duke University Press, 2000.

Parry, John Horace, and Philip Manderson Sherlock. *A Short History of the West Indies*. 2nd ed. London: Macmillan, 1960.

Parsons, Christopher M., and Kathleen S. Murphy. "Ecosystems under Sail: Specimen Transport in the Eighteenth-Century French and British Atlantics." *Early American Studies* 10 (2012): 503–29. doi:10.1353/eam.2012.0022.

Parsons, W. J. "Introduced Weeds." In *Plants and Man in Australia*, edited by D. J. and S. G. M. Carr, 179–93. Sydney: Academic Press, 1981.

Parsonson, Ian. *The Australian Ark: A History of Domesticated Animals in Australia*. Collingwood, Victoria: CSIRO, 1998.

Patterson, Annabel. *Pastoral and Ideology: Virgil to Valery*. Berkeley: University of California Press, 1987.

Pauly, Philip J. "Fighting the Hessian Fly: American and British Responses to Insect Invasion, 1776–1779." *Environmental History* 7 (2002): 485–507.

———. *Fruits and Plains: The Horticultural Transformation of America*. Cambridge, MA: Harvard University Press, 2007.

Pennant, Thomas. *British Zoology*. 4 vols. London: Benjamin White, 1768–70.

Perkins, David. *Romanticism and Animal Rights*. Cambridge: Cambridge University Press, 2003.

Peterfreund, Stuart. "Great Frosts and... Some Very Hot Summers: Strange Weather, the Last Letters, and the Last Days in White's *The Natural History of Selborne*." In *Romantic Science: The Literary Forms of Natural History*, edited by Noah Heringman, 85–108. Albany: State University of New York Press, 2003.

Pfau, Thomas. *Wordsworth's Profession: Form, Class, and the Logic of Early Romantic Cultural Production*. Stanford: Stanford University Press, 1997.

Phillip, Arthur. *The Voyage of Governor Phillip to Botany Bay; with an account of the Establishment of the Colonies of Port Jackson and Norfolk Island; compiled from Authentic papers... Embellished with Fifty Five Copper Plates*. London: J. Stockdale, 1789.

Pigott, Louis J. "John White's *Journal of a Voyage to New South Wales* (1790): Comments on the Natural History and the Artistic Origins of the Plates." *Archives of Natural History* 27 (2000): 157–74.

Plotz, John. *Portable Property: Victorian Culture on the Move*. Princeton, NJ: Princeton University Press, 2008.

Pocock, J. G. A. "The Mobility of Property." In *Virtue, Commerce, and History*. Cambridge: Cambridge University Press, 1985.

———. "Tangata Whenua and Enlightenment Anthropology." *New Zealand Journal of History* 26 (April 1992): 28–53.

Poirier, Richard. *A World Elsewhere: The Place of Style in American Literature*. New York: Oxford University Press, 1966.

Pope, Alexander. *The Poems of Alexander Pope*. Edited by John Butt. 6 vols. London: Methuen, 1961.

Powell, Dulcie. *Botanic Garden, Liguanea*. Kingston: Institute of Jamaica, 1972.

———. *Voyage of the Plant Nursery, H. M. S. Providence, 1791–1793*. Kingston: Institute of Jamaica, 1973.

Powell, J. M. *Environmental Management in Australia, 1788–1914: Guardians, Improvers and Profit: An Introductory Survey*. Melbourne: Oxford University Press, 1976.

Pratt, Mary Louise. *Imperial Eyes: Travel Writing and Transculturation*. London: Routledge, 1992.

Rackham, Oliver. *Trees and Woodland in the British Landscape*. London: J. M. Dent, 1976.

Rafael, Vicente L. *Contracting Colonialism: Translation and Christian Conversion in Tagalog Society under Early Spanish Rule*. Ithaca, NY: Cornell University Press, 1988.

Rauschenberg, Roy A. "John Ellis, F.R.S.: Eighteenth Century Naturalist and Royal Agent to West Florida." *Notes and Records of the Royal Society of London* 32 (1978): 149–64.

Ray, Laura E. "Podophyllum peltatum and Observations on the Creek and Cherokee Indians: William Bartram's Preservation of Native American Pharmacology." *Yale Journal of Biology and Medicine* 82 (2009): 25–36.

Reaney, P. H. *Origin of English Place-Names*. London: Routledge, 1960.

Regier, Alexander. *Fracture and Fragmentation in British Romanticism*. Cambridge: Cambridge University Press, 2010.

Regis, Pamela. *Describing Early America: Bartram, Jefferson, Crevecoeur, and the Rhetoric of Natural History*. DeKalb: Northern Illinois University Press, 1992.

Richardson, John. *Fauna boreali-americana; or, The Zoology of the Northern Parts of British America: containing descriptions of the objects of natural history collected on the late northern land expeditions, under command of Captain Sir John Franklin, R.N*. With the assistance of William Swainson and William Kirby. London: J. Murray, 1829–37.

Richardson, R. Alan. "Biogeography and the Genesis of Darwin's Ideas on Transmutation." *Journal of the History of Biology* 14 (1981): 1–41.

Rigby, Nigel. "The Politics and Pragmatics of Seaborne Plant Transportation." In *Science and Exploration in the Pacific*, edited by Margarette Lincoln, 81–100. Woodbridge: Boydell & Brewer, 2001.

Ritcheson, Charles R. *Aftermath of Revolution: British Policy toward the United States 1783–1795*. Dallas: Southern Methodist University Press, 1969.

Roach, Joseph. *Cities of the Dead: Circum-Atlantic Performance*. New York: Columbia University Press, 1996.

Robbins, Bruce. "Actually Existing Cosmopolitanisms." In *Cosmopolitics: Thinking and Leaving beyond the Nation*, edited by Pheng Cheah and Bruce Robbins, 1–19. Minneapolis: University of Minnesota Press, 1998.

Robinson, Eric. "Eighteenth Century Commerce and Fashion: Matthew Boulton's Marketing Techniques." *Economic History Review*, 2nd ser., 14 (1963): 39–60.

Robinson, Henry Crabb. *Correspondence of Henry Crabb Robinson with the Wordsworth Circle*. 2 vols. Oxford: Clarendon, 1927.

Rolls, E. C. *They All Ran Wild*. Sydney: Angus & Robertson, 1969.
Rowley, Trevor. *The English Landscape in the Twentieth Century*. London: Hambledon Continuum, 2006.
Rushdie, Salmon. "Imaginary Homelands." *Imaginary Homelands: Essays and Criticism 1981–1991*. London: Granta, 1991.
Ruskin, John. *The Works of Ruskin*. Edited by E. T. Cook and A Wedderbun. 39 vols. London: George Allen, 1906.
Russell, Emily W. B. *People and the Land through Time: Linking Ecology and History*. New Haven, CT: Yale University Press, 1997.
Rzepka, Charles J. "Sacrificial Sites, Place-Keeping, and 'Pre-History' in Wordsworth's 'Michael.'" *European Romantic Review* 15, no. 2 (2004): 205–13.
Sachs, Jonathan. "Decline and the Depths of Time: Historicity and the Forms of Ruin in British Romanticism." Unpublished manuscript.
Said, Edward W. "Reflections on Exile." *Reflections on Exile and Other Essays*. Cambridge, MA: Harvard University Press, 2000.
Salisbury, William. *Hints addressed to proprietors of orchards, and to growers of fruit in general, comprising observations on the present state of the apple trees, in the cider countries, made in a tour during the last summer*. London: Longman, 1816.
Say, Thomas. "Observations on the Hessian Fly." *Journal of the Academy of Natural Sciences of Philadelphia* 1 (1817): 46.
Sayre, Gordon. "The Mound Builders and the Imagination of American Antiquity in Jefferson, Bartram, and Chateaubriand." *Early American Literature* 33 (1998): 225–49.
Schafer, Daniel L. *William Bartram and the Ghost Plantations of British East Florida*. Gainesville: University Press of Florida, 2010.
Schiebinger, Londa L. *Plants and Empire: Colonial Bioprospecting in the Atlantic World*. Cambridge, MA: Harvard University Press, 2002.
Schimmelpenninck, Mary Anne Galton. *Life of Mary Anne Schimmelpenninck*. Edited by Christiana C. Hankin. London: Longman, Green, Longman, & Roberts, 1860.
Schleiermacher, Friedrich. "On the Different Methods of Translating." In *The Translation Studies Reader*, edited by Lawrence Venuti, 43–63. 3rd ed. London: Routledge, 2012.
Schmitt, Cannon. *Darwin and the Memory of the Human: Evolution, Savages, and South America*. Cambridge: Cambridge University Press, 2009.
———. "Tidal Conrad (Literally)." *Victorian Studies* 55 (2012): 7–29.
Schofield, Robert. *The Lunar Society of Birmingham: A Social History of Provincial Science and Industry in Eighteenth Century England*. Oxford: Clarendon, 1963.
Schweber, Silvan S. "Darwin and the Political Economists: Divergence of Character." *Journal of the History of Biology* 13 (1980): 195–289.
Secord, Anne. "Corresponding Interests: Artisans and Gentlemen in Nineteenth-Century Natural History." *British Journal for the History of Science* 27 (1994): 383–408.

———. "Science in the Pub: Artisan Botanists in Early Nineteenth-Century Lancashire." *History of Science* 32 (1994): 269–315.

Seward, Anna. *Memoirs of the Life of Dr. Darwin*. London: J. Johnson, 1804.

———. *Poetical Works of Anna Seward: With Extracts from Her Literary Correspondence*. Edited by Walter Scott. 3 vols. Edinburgh: James Ballantyne, 1810.

Shakespeare, William. *The Norton Shakespeare*. Edited by Stephen Greenblatt. New York: Norton, 1997.

Shelley, Mary Wollstonecraft. *Frankenstein or, The Modern Prometheus. The 1818 Text*. Chicago: University of Chicago Press, 1974.

Shelley, Mary Wollstonecraft, and Percy Bysshe Shelley. *History of a Six Weeks' Tour*. London: T. Hookham & C. & J. Ollier, 1817.

Shelley, Percy Bysshe. *Shelley's Poetry and Prose*. Edited by Donald H. Reiman and Neil Fraistat. 2nd ed. New York: Norton, 2002.

Short, Thomas. *Medicina Britannica*, 3rd ed. with introduction, notes, and appendix by John Bartram. Philadelphia: B. Franklin & D. Hall, 1751.

Sibly, Ebenezer. *Magazine of Natural History*. 14 vols. London: Sibly, 1794–1808.

Silver, Bruce. "William Bartram's and Other Eighteenth-Century Accounts of Nature." *Journal of the History of Ideas* 39 (1978): 597–614.

Silverberg, Robert. *Mound Builders of Ancient America: The Archaeology of a Myth*. Athens: Ohio University Press, 1968.

Simpson, David. *Wordsworth, Commodification and Social Concern*. Cambridge: Cambridge University Press, 2009.

Slaughter, Thomas P. *The Natures of John and William Bartram*. New York: Knopf, 1996.

———, ed. *William Bartram: "Travels" and Other Writings*. New York: Library of America, 1996.

Smith, Adam. *Theory of Moral Sentiments*. Edinburgh: Kincaid & Bell, 1759.

Smith, Bernard. *European Vision and the South Pacific, 1768–1850: A Study in the History of Art and Ideas*. Oxford: Clarendon, 1960.

Smith, James Edward. "Biographical Memoirs of Several Norwich Botanists, in a Letter to Alexander MacLeay." *Transactions of the Linnean Society* 7 (1804): 295–301.

———. *English Botany; or, Coloured Figures of British Plants, with their Essential Characters, Synonyms, and Places of Growth*. Illustrated by James Sowerby. London: James Sowerby, 1790–1813.

———. *Selection of the Correspondence of Linnaeus and Other Naturalists*. 2 vols. London: Longman, Hurst Rees, Orme, & Brown, 1821.

Smith, John. *A Generall History of Virginia, New England, and the Summer Isles*. 5 vols. London: John Dawson, 1624.

Smith, Robert. *The Universal Directory for Taking Alive and Destroying Rats*. London: n.p., 1768.

Spary, Emma C. "Political, Natural, and Bodily Economies." In *Cultures of Natural History*, edited by N. Jardine, J. A. Secord, and E. C. Spary, 178–96. Cambridge: Cambridge University Press, 1995.

———. *Utopia's Garden: French Natural History from Old Regime to Revolution.* Chicago: University of Chicago Press, 2000.
Spary, Emma C., and Paul White. "Food of Paradise: Tahitian Breadfruit and the Autocritique of European Consumption." *Endeavour* 28, no. 2 (June 2004): 75–80.
Speight, Harry. *Romantic Richmondshire.* London: Elliot Stock, 1897.
Sprat, Thomas. *The History of the Royal-Society of London for the Improving of Natural Knowledge by Tho. Sprat.* London: T.R., 1667.
Stafford, Robert A. *Scientist of Empire: Roderick Murchison, Scientific Exploration and Victorian Imperialism.* Cambridge: Cambridge University Press, 1989.
Stafleu, Frans A. *Linnaeus and the Linnaeans: The Spreading of Their Ideas in Systematic Botany, 1735–1789.* Utrecht: International Association for Plant Taxonomy, 1971.
Stedman, John Gabriel. *Narrative of a Five Years' Expedition.* 2 vols. London: Joseph Johnson, 1796.
Steiner, George. *After Babel: Aspects of Language and Translation.* 2nd ed. Oxford: Oxford University Press, 1992.
Stevens, William Bagshaw. *The Journal of the Rev. William Bagshaw Stevens.* Edited by Georgina Galbraith. Great Britain: Oxford University Press, 1965.
Stocking, George W. *Race, Culture and Evolution: Essays in the History of Anthropology.* Chicago: University of Chicago Press, 1982.
Stork, William. *A Description of East-Florida, with a journal, kept by John Bartram of Philadelphia, botanist to His Majesty for the Floridas. . . .* 3rd ed. 2 vols. London: W. Nicholl, 1769.
Sturtevant, E. Lewis. "Kitchen Garden Esculents of American Origin." *American Naturalist* 5 (1885): 452–54.
Sulloway, Frank J. "Darwin's Conversion: The Beagle Voyage and Its Aftermath." *Journal of the History of Biology* 5 (1982): 325–96.
Swarbrick, J. T. "Weeds of Sydney Town, 1802–4." *Australian Weeds* 3, no. 1 (1984): 42.
Taylor, George. "John Walker, D.D., F.R.S.E., a Notable Scottish Naturalist." *Transactions of the Botanical Society of Edinburgh* 38 (1959): 180–203.
Tennyson, Alfred. *In Memoriam.* New York: Norton, 1973.
Thelwall, John. "On the Causes of the Late Disturbances." *Tribune* 30 (23 September 1795): 305–20.
Thomas, Keith. *Man and the Natural World: Changing Attitudes in England 1500–1800.* London: Allen Lane, 1983.
Thomas, Nicholas. *Entangled Objects: Exchange, Material Culture, and Colonialism in the Pacific.* Cambridge, MA: Harvard University Press, 1991.
Thompson, E. P. *The Making of the English Working Class.* New York: Vintage Books, 1966.
———. *Whigs and Hunters: The Origin of the Black Act.* New York: Pantheon, 1975.
Thompson, Ian H. "William Wordsworth, Landscape Architect." *Wordsworth Circle* 38 (Autumn 2007): 196–203.

Thornton, Richard H. "Letter from Erasmus Darwin to Joseph Johnson, 1784." *Notes and Queries*, 12th ser., 12 (1923): 449.
Tibble, J. W., and Anne Tibble. *John Clare: His Life and Poetry*. London: William Heineman, 1956.
Tiffin, Helen. "'Flowers of Evil,' Flowers of Empire: Roses and Daffodils in the Work of Jamaica Kincaid, Olive Senior, and Lorna Goodison." *SPAN* 46 (1998): 58–71.
———. "'Man Fitting the Landscape': Nature, Culture, and Colonialism." In DeLoughrey, Gosson, and Handley, *Caribbean Literature and the Environment*, 198–212.
Tobin, Beth Fowkes. *Colonizing Nature: The Tropics in British Arts and Letters, 1760–1820*. Philadelphia: University of Pennsylvania Press, 2005.
Todd, Kim. *Tinkering with Eden*. New York: Norton, 2001.
Van den Boogaart, Ernst, and P. C. Emmer. "Colonialism and Migration: An Overview." In *Colonialism and Migration: Indentured Labor before and after Slavery*. Edited by P. C. Emmer. Boston: Martinus Nijhoff, 1986.
Vardy, Alan D. *John Clare, Politics and Poetry*. New York: Palgrave Macmillan, 2003.
Vaux, James Hardy. *Memoirs of James Hardy Vaux, Including His Vocabulary of the Flash Language*. Edited by Noel McLachlan. London: Heinemann, 1964.
Veder, Robin, "Flowers in the Slums: Weavers' Floristry in the Age of Spitalfields' Decline." *Journal of Victorian Culture* 14, no. 2 (2009): 261–81.
Venuti, Lawrence. "Translation, Community, Utopia." In *The Translation Studies Reader*, edited by Lawrence Venuti, 468–88. London: Routledge, 2000.
Walker, George. *The Vagabond*. 2 vols. London: Lee & Hurst, 1799.
Walker, John. *Essay . . . of the Translation of Plants from the East to the West Indies*. *The Papers of Sir Joseph Banks*, sec. 5, ser. 21.02.
———. *Essays on Natural History and Rural Economy*. London: Longman, Hurst, Rees, & Orme, 1812.
———. *Rev. Dr. John Walker's Report of the Hebrides of 1764 and 1771*. Edited by Margaret M. McKay. Edinburgh: J. Donald, 1980.
Wallace, Alfred Russel. "Origin of Human Races and the Antiquity of Man Deduced from the Theory of 'Natural Selection.'" In *Images of Race*, edited by Michael D. Biddiss, 37–54. Surrey: Leicester University Press, 1979.
Walpole, Horace. *Yale Edition of Horace Walpole's Correspondence*. Edited by W. S. Lewis. 48 vols. New Haven, CT: Yale University Press, 1937–83.
Walsh, Benjamin D., trans. *The Comedies of Aristophanes*. 3 vols. London: A. H. Bailey, 1837.
Walters, Kerry S. "The Creator's Boundless Palace: William Bartram's Philosophy of Nature." *Transactions of the Charles S. Peirce Society: A Quarterly Journal in American Philosophy* 25 (1989): 309–32.
———. "The 'Peaceable Disposition' of Animals: William Bartram on the Moral Sensibility of Brute Creation." *Pennsylvania History* 56 (1989): 157–76.
Walvin, James. *Fruits of Empire: Exotic Produce and British Taste, 1660–1800*. Houndmills: Macmillan, 1997.

Waselkov, Gregory A., and Kathryn H. Braund, eds. *William Bartram on the Southeastern Indians*. Lincoln: University of Nebraska Press, 1995.

Weil, Simone. *The Need for Roots: Prelude to a Declaration of Duties toward Mankind*. Translated by Arthur Wills. New York: Harper & Row, 1952.

Weinstock, Jeffrey Andrew. *Spectral America*. Madison: University of Wisconsin Press, 2004.

Wells, Thomas E. *Michael Howe: The Last and Worst of the Bushrangers of van Dieman's Land*. Sydney: Angus & Robertson, 1926.

Welsh, Stanley L. *Flora Societensis: A Summary Revision of the Flowering Plants of the Society Islands*. Orem, UT: E.P.S., 1998.

Weston, Richard. *The Universal Botanist and Nurseryman: Containing Descriptions of the Species and Varieties of all the Trees, Shrubs, Herbs, Flowers, and Fruits*. 4 vols. London: J. Bell, 1770–77.

White, Daniel E. *From London to Little Bengal: Religion, Print, and Modernity in Early British India 1793–1835*. Baltimore: Johns Hopkins University Press, 2013.

White, Gilbert. *Gilbert White's Journals*. Edited by Walter Johnson. New York: Taplinger, 1970.

———. *Natural History and Antiquities of Selborne*. Edited by Thomas Bell. 2 vols. London: John van Voorst, 1877.

———. *Natural History of Selborne*. Edited by E. M. Nicholson. London: Thornton Butterworth, 1929.

———. *Natural History of Selborne*. Edited by James Fisher. Harmondsworth: A. Lane, 1941.

———. *Natural History of Selborne*. Edited by Paul Foster. Oxford: Oxford University Press, 1993.

White, John. *Journal of a Voyage to New South Wales with Sixty-Five Plates of Non-Descript Animals, Birds, Lizards, Serpents, Curious Cones of Trees and Other Natural Productions*. London: Debrett, 1790.

Whittaker, Robert J., and José María Fernández-Palacios, eds. *Island Biogeography: Ecology, Evolution and Conservation*. New York: Oxford University Press, 1998.

Whyte, Ian. "William Wordsworth's *Guide to the Lakes* and the Geographical Tradition." *Area* 32, no. 1 (2000): 101–6.

Willdenow, D. C. *Principles of Botany and of Vegetable Physiology*. Edinburgh: William Blackwood & T. Cadell & W. Davies, 1811.

Williams, Raymond. *Keywords: A Vocabulary of Culture and Society*. 2nd ed. New York: Oxford University Press, 1985.

Wilson, E. "On the Introduction of the British Song Bird." *Transactions of the Philosophical Institute of Victoria* 11 (1858): 77–88.

Wilson, Kathleen. *Island Race: Englishness, Empire and Gender in the Eighteenth Century*. London: Routledge, 2003.

———. *Sense of the People: Politics, Culture, and Imperialism in England, 1715–1785*. New York: Cambridge University Press, 1995.

Withering, William. *A Botanical Arrangement of All the Vegetables Naturally Growing in Great Britain: with Descriptions of the Genera and Species, According* to the System of the Celebrated Linnaeus. 2 vols. London: Cadel, Elmsley, & Robinson, 1776.

Withers, Charles W. J. "A Neglected Scottish Agriculturalist: The Georgical Lectures and Agricultural Writings of the Rev. Dr. John Walker (1731–1803)." *Agricultural History Review* 33 (1985): 132–43.

Woolf, Virginia. *Collected Essays*. Edited by Leonard Woolf. 4 vols. London: Chatto & Windus, 1966–69.

Wordsworth, Dorothy. *Journals of Dorothy Wordsworth*. Edited by Mary Moorman. 2nd ed. Oxford: Oxford University Press, 1971.

Wordsworth, Jonathan. *The Music of Humanity: A Critical Study of Wordsworth's "Ruined Cottage."* London: Thomas Nelson, 1969.

Wordsworth, William. *A Guide through the District of the Lakes*. In *The Prose Works of William Wordsworth*, edited by W. J. B. Owen and J. W. Smyser, 2:123–253. Oxford: Clarendon, 1974.

———. *Home at Grasmere: Part First, Book First, of "The Recluse."* Edited by Beth Darlington. Ithaca, NY: Cornell University Press, 1977.

———. *The Letters of William and Dorothy Wordsworth: The Early Years, 1787–1805*. Edited by Ernest de Selincourt. Revised by Chester L. Shaver. Oxford: Clarendon, 1967.

———. *The Letters of William and Dorothy Wordsworth: The Middle Years*. 2 vols. Oxford: Clarendon, 1969.

———. *Lyrical Ballads and Other Poems, 1797–1800*. Edited by James Butler and Karen Green. Ithaca, NY: Cornell University Press, 1992.

———. *Poetical Works*. Edited by Ernest de Selincourt and Helen Darbishire. 5 vols. Oxford: Clarendon, 1940–49.

———. *The Prelude 1799, 1805, 1850*. Edited by Jonathan Wordsworth. New York: Norton, 1979.

———. *The Prose Works of William Wordsworth*. Edited by W. J. B. Owen and J. W. Smyser. 3 vols. Oxford: Clarendon, 1974.

———. *"The Ruined Cottage" and "The Pedlar."* Edited by James Butler. Ithaca, NY: Cornell University Press, 1979.

Worster, Donald. "Doing Environmental History." In *The Ends of the Earth: Perspectives on Modern Environmental History*, edited by Donald Worster, 289–307. Cambridge: Cambridge University Press, 1988.

———. *Nature's Economy: A History of Ecological Ideas*. Cambridge: Cambridge University Press, 1985.

Wyatt, John. *Wordsworth and the Geologists*. Cambridge: Cambridge University Press, 1995.

Yalden, Derek W. *The History of British Mammals*. London: A. D. Poyser, 1999.

Young, Arthur. "On the Musca Pumilionis." *Annals of Agriculture* 16 (1791): 176–77.

———. "Proceedings of His Majesty's Most Honourable Privy Council, and Information Received, Respecting an Insect, Supposed to Infest the Wheat of the Territories of the United States of America." *Annals of Agriculture* 11 (1789): 406–613.

Young, Robert. *Colonial Desire: Hybridity in Theory, Culture, and Race.* London: Routledge, 1995.

———. *Darwin's Metaphor: Nature's Place in Victorian Culture.* Cambridge: Cambridge University Press 1985.

Zanger, Jules. "The Premature Elegy: Bryant's 'The Prairies' as Political Poem." In *Interface: Essays on History, Myth and Art in American Literature,* edited by Daniel Royot, 13–20. Montpellier: Université de Montpellier, 1984.

INDEX

Aboriginal people, 126, 127, 128–30, 133, 138, 140–41, 150–52, 310, 325–26; effect of changes on, 136; and Gosse, 128, 130; and Howe, 123; and J. White, 140–41, 142, 150–51. *See also* Australia; indigenous peoples; Native Americans
Agassiz, Louis, 314–15, 320
agriculture, 6, 23, 29, 224, 227, 228, 261, 264, 270, 274, 282; and W. Bartram, 215, 217, 218, 219, 220, 222, 224; and Clare, 272, 273
Ahaye, 202
Alexander, James, 58
Allen, Matthew, 270
Alva, Duke of, 241
American Antiquarian Society, 228
American Journal of Science, 306
Anderson, Alexander, 121
Anderson, Benedict, 38–39, 159, 162
Anderson, William, 171, 190, 191–92
Andrews, Henry C., 64
animals, 26–29, 37, 41, 150–52; and W. Bartram, 204–13; and Buffon, 335–37; and language, 39–40, 44–46, 168–78, 182, 204; as moral and intellectual beings, 166, 169, 170, 173, 175–76, 204–5, 276–77; and M. Shelley, 331, 332, 335, 336, 338, 340; transfer of, 23–25, 26, 134, 283; and G. White, 165–67, 168, 170; and W. Wordsworth, 256, 259–62, 282. *See also* species
Argyll, 3rd Duke of, 74
Aristotle, 22, 34
Artis, Edmund Tyrell, 275
Atholl, John Murray, Duke of, 83
Atwater, Caleb, 227, 228, 306

Australia, 16, 25, 88, 111, 123–52; as collaborative translation, 126–27, 137–52; and E. Darwin, 130–33, 137; as empty land, 135. *See also* Aboriginal people
Ayres, William, 146
Azara, Felix, 306

Bacon, Francis, 42
Bacon, Roger, 68
Balmoral Estate, 262
Banks, Sir Joseph, xiv, 22, 25, 27, 71, 77, 78, 87, 103, 119–20, 134, 346n23, 346n25; and Beatson, 308; and Bligh, 109, 122, 125; and Botany Bay, 124, 128–30; and breadfruit expedition, 98–101, 109–13; and Cook, 116, 156, 157–58; and E. Darwin, 65; *Endeavour Journal,* 98; and Flinders, 156; and Hessian fly, 185–88; and Labillardière, 116; letter to Vaughan, 187; letter to Yonge, 109–10; "Memorandum," 111; and natural history, 153–57; and Park, 39, 342n16; and Walker, 104, 105; and Wallen, 101; and G. White, 153, 154, 155–56, 161; and woolly aphid, 188
Barbauld, Anna Laetitia, 2, 16
Barrington, Daines, 154, 165, 168–69, 170, 176, 180
Barton, Benjamin Smith, 204, 227
Bartram, John, 30, 71, 197, 198, 201, 224, 347n5; *Correspondence,* 57, 58, 74, 199–200, 241
Bartram, William, xiv; and American Southeast, 196, 197–98, 199, 200, 201–3, 208, 213, 214, 215, 218–19, 222, 223, 228, 231; and botany, 196, 198, 201, 214, 218, 223, 225; and Bryant, 228; and Coleridge, 196, 198–99, 208–9, 211, 231, 232, 233; and P. Collinson, 198; and colonial landscapes,

Bartram, William (cont.)
  196–97, 214, 215; and ecological politics, 204–13; and Fothergill, 198, 200, 202; and historical ecology, 196, 201, 215–25, 227–28, 231, 232; letter to Barton, 204; letter to J. Bartram, 198; and Native Americans, 196–98, 203, 206, 213–14, 215–18, 219–23, 224, 228; and natural history, 199–202, 203, 206, 212, 213, 215; and pre-Columbian mounds, 219–21; and Solander, 198; and G. White, 204, 205, 206; and W. Wordsworth, 196, 197, 205, 214, 223, 231–32, 233, 268. Works: "Dignity of Human Nature," 204, 205, 207–8; "Genetic Text," 208, 210, 213, 326; "Observations on the Creek and Cherokee Indians," 221; "Preface to a Catalogue," 199; "Report to Dr. Fothergill," 196; *Sarracenia Flava or Pitcher*, 211 (fig.); "Some Hints & Observations," 197, 213, 214, 220; *Travels*, 22, 196–225, 227, 231–32

Bateman, James, 346–47n26
Bath, Saint Thomas, 94–95
Baudin, Nicolas, 156, 345n5
Bauer, Ferdinand, 157
Bayly, William, 191
Beatson, Alexander, 308
Beaumont, Sir George, 266
Beilby, Ralph, 261–62
Benjamin, Walter, 51, 52, 238, 291
Bennett, George, 136
Berkenhout, John, 82
Beston, Henry, 176
Bewick, Thomas, 174, 261–62
Bhabha, Homi, 139
birds, 150–51, 180–81, 206, 256; and Clare, 277–78, 283–84, 294; global transfer of, 76–77, 101; sociality of, 155, 166, 168–76, 182, 206–8; translation of, 139–52; violence toward, 128, 150, 257. *See also* White, Gilbert
Blackett, Sir Edward, 57
Black Hawk, 229
Blackstone, Sir William, 26, 42
Blake, William, 53; annotations to *The Excursion*, 233; *A Family of New South Wales*, 125; "The Marriage of Heaven and Hell," 10, 173, 278
Bligh, William A., 78, 98, 101–2, 103, 125, 128; and Cook, 117, 118–19; and global plant transfer, 95–101, 109–22; inscription of, 114, 115, 116; as naturalist and illustrator, 125, 142, 345n1; and Royal Botanic Garden at Sydney, 135; *A Voyage to the South Sea*, 98, 100, 110, 111, 114, 118, 191
Blumenbach, Johann Friedrich, 330
Bohrer, Martha, 155
Bond, Phineas, 186
Bonpland, Aimé, 23
Boswell, Thomas D., 88
Botanical Magazine, The, 63–64
botanic gardens, 36, 65, 69–79, 91, 94–95, 116, 199, 200; and Bligh, 109, 110–11, 120, 121, 122, 135; and global transfer of plants, 69–78, 109. *See also* Royal Botanic Gardens at Kew
botany, xiii, 62, 71, 75, 92, 102, 105–6, 123, 124, 154, 237, 241; and J. Bartram, 200; and Bryant, 230–31; and Clare, 275–76; and E. Darwin, 65, 70, 79; and Kincaid, 90, 91; and Park, 39; and Stork, 106; and W. Wordsworth, 234, 244. *See also* Bartram, William; natural history
Botany Bay, 36, 124–25, 128–30, 134
Boulton, Matthew, 54, 55
Bourdieu, Pierre, 7
Braudel, Fernand, 55
breadfruit, 78, 90, 94, 95, 96–100, 109, 116, 120, 121
Bridgewater Treatises, 178
Bristow, Abraham, xii
British Army Engineering Corps, 50
Brooke, Frances, 33, 34
Brown, Robert, 135, 156–57
Browne, Patrick, 44, 191
Bryant, William Cullen, 226–31
Buckland, William, 258
Buffon, Comte de, 257, 335–37
Bunbury, C. J. F., 319
Burke, Edmund, 81, 98
Burn, Peter, 146, 148
Bute, Lord, 74, 104
Byron, Lord, 53, 98, 99, 271

Caley, George, *Reflections*, 25
Campbell, Captain, 147, 148
Camper, Petrus, 330
Candolle, Augustin-Pyramus de, 349n3
capitalism, 39, 108, 239, 282, 283
Carey, Matthew, 185
Caribbean, 34, 43, 78, 87–95, 106, 121–22, 125
Carlyle, Thomas, 198
Carmarthen, Lord, 186

Carroll, Lewis, 158
Carter, Landon, 186
Catton, Charles, Jr., 140
Cavendish, Thomas, 310
Chambers, Douglas, 58, 276, 288, 291
Chambers, William, 73–74
Chesterfield, Lord, 73
Christian, Fletcher, 98
Churton, Ralph, 181
Clare, John, xiv, 5, 270–95; and ecology, 273, 275, 277, 279, 282, 285, 286, 291, 293, 294; and enclosure, 270, 272, 273, 279, 280, 281, 282–83, 290, 292; and exile, 270, 271, 280, 281, 287, 289, 290, 295; and ghosts, 6, 293, 294–95; and Goldsmith, 279–80; and Helpston, xiii, 271, 272, 273, 274, 275, 276, 277, 279, 280, 283, 286, 290, 291, 292, 293, 295; at High Beach Asylum, 270, 295; and homelessness, 270, 272, 280, 283, 284, 285; journal of, 281, 283; and Joyce, 271; and language, 275, 281, 282, 286, 287–88, 290, 291, 292, 294; letter to Elizabeth Kent, 276; letter to Mary Joyce, 270, 294; and local natures, 272, 273, 274, 279, 280, 290, 291; and mobility, 271–72, 274, 275, 279, 281, 283, 284, 287; and modernity, 270, 272, 277; and names, 276, 277, 283, 288, 292; and natural history, 275–77; at Northampton Asylum, 270, 290, 295; and Northborough, xiii, 286, 288; and place, 270, 274, 275, 280, 282, 287, 291, 292; and poetry and local names, 276–77, 285, 289–95; and rural people, 272, 273, 274–75, 277, 282, 283, 289, 293, 294; and Taylor, 275, 291; and time, 277–81; and traditional rights of tenancy, 282–86; and P. Turner, 271; and G. White, 154, 272; and W. Wordsworth, 277, 282. Works: *Child Harold,* 270, 271, 294, 295; "Decay A Ballad," xiii, 288, 293; "The Eternity of Nature," 277–78; First Natural History Letter, 291; "The Flitting," xiii, 286–88, 289, 293; "Helpston," 273, 279–81; "Helpston Green," 293–94; "I Am," 270; "The Lament of Swordy Well," 284–86; *The Midsummer Cushion,* 293; "The Mores," 273; "Natural History of Helpstone," 275; "Obscurity," 286; "Pastoral Poesy," 290; "The Progress of Ryhme," 287; "Remembrances," 283, 292–93, 294; "The Robins Nest," 283–84, 287; "The Round Oak," 290; "Shadows of Taste," 276–77, 287; *The Shepherd's Calendar,* 275;

"Sighing for Retirement," 278; "Songs Eternity," 278; "To Wordsworth," 282
Clark, Thomas, 83
Clarke, Thomas, 94
Clarkson, Catherine, 254
Coleridge, Samuel Taylor, 53–54, 60, 64, 69, 180, 196, 198–99, 208–9, 231, 232, 233–34; *Biographia Literaria,* 233; "Kubla Khan," 58, 211; *Marginalia,* 180, 231; *Notebooks,* 58, 64; *Shakespearean Criticism,* 69; *Table Talk,* 198–99; "This Lime-Tree Bower My Prison," 208–9, 231
Collinson, John, 167
Collinson, Peter, 30, 42–43, 57, 58, 62, 74, 198, 199, 200, 224, 241
colonialism, xii–xiii, 6–9, 36, 37, 102, 108, 122, 134, 180, 192, 194–95, 237; and botanic gardens, 71; and Botany Bay, 124; and Bryant, 230; and Coleman, 125; and ecology, 15, 19–20, 23–24, 28, 298, 306, 307–10, 338; and gardening, 64; global natural history of, 312; and Hooker, xi, xii; and indigenous languages and knowledge, 44–46; and Kincaid, 91; and Lyell, 303, 304–5, 310–11; and mobility, xiii, 20–25, 27, 28–33; and natural history, xiii, 8, 34, 44, 49, 126; and natural selection, 316, 325; and Nugent, 95; and Pennant, 164; and remaking of natures, 77; and Seven Years' War, 73; and success of European plants, 28–33; and translation, xiv, 18, 34, 43, 44–46, 49–50; and transplantation, 16–17, 19–20; and transport, 106; and traveling natures, 87–122; and West Indies, 88; and G. White, 161, 162, 167; and J. White, 138. *See also* Bartram, William; Darwin, Charles; empire and imperialism; Wordsworth, William
Columbus, Christopher, 89, 90, 121
commerce, 5, 8, 14, 20, 26–27, 55, 108, 110, 121–22, 194–95; and botany, 78, 105; and natural history, 26, 44, 48, 103, 106; and G. White, 160, 180; and W. Wordsworth, 6, 83, 265, 266. *See also* Darwin, Erasmus; natural history, and consumerism; natural history, and fashion
Conrad, Joseph, *Heart of Darkness,* 325
Constable, John, 268
Cook, James, 31, 78, 100, 111, 114, 115, 116–19, 136, 144, 153, 156, 157–58, 344n18; and cockroaches, 190–91, 195; *Journals,* 116, 118, 124, 191; scientific expeditions of, 156, 157–58; voyages of, 125, 153, 156, 157–58, 171; *Voyage to the Pacific Ocean,* 117,

Cook, James (*cont.*)
  118, 190, 191; *Voyage Towards the South Pole*, 171, 192
Cooper, Susan Fenimore, *Rural Hours*, 29–30
cosmopolitan natures, 24, 26, 56–62, 80–81, 99, 109, 125, 134, 192–93, 328, 329; and Banks, xiv, 154; and botanic gardens, 71, 76, 78; and H. Jones, 76, 77; and Kincaid, 91; and Lunar Society, 55; and P. B. Shelley, 81. *See also* Darwin, Erasmus
Cowper, William, *The Task*, 58, 343n19
Crabbe, George: *The Borough*, 166, 242; *The Parish Register*, 242–43
Creeks, 221, 222–23
Crewe, Emma, *Flora at Play with Cupid*, 60–61, 62
Crewe, Frances Anne (Greville), 60
Crewe, John, 60
Curtis, William, 62, 63–64, 65
Curwen, John Christian, 83, 84, 266
Cuthbert, Lewis, 92

Dadswell, Ted, 166
Dalrymple, Alexander, 125
Dalrymple, John, 104
Dampier, William, 15
Dana, Richard, 228, 230
Dancer, Thomas, 94
Darwin, Charles, xiv, 300, 305–6; and artificial selection, 302–3, 323; and W. Bartram, 197; and *Beagle*, xiv, 156, 296, 297, 298, 300, 325; biogeography of, 303, 312–26; at Cambridge, 297; and colonialism, xiv, 28–29, 30, 296–97, 298, 299–300, 306–7, 308, 309, 310–11, 312, 315, 316, 325, 327; diary of, 306; and dung beetles, 298–99; and ecology, 296, 298, 300, 305–11; and evolution, xiv, 296, 297, 312–26, 327, 332–33; and Haast, 31; and islands, 313, 315, 320, 321; letters to J. D. Hooker, 28–29, 313, 322, 323; letter to C. J. F. Bunbury, 319; letter to Caroline Darwin, 307; letter to W. Fox, 300; letter to A. Gray, 30; letter to Leonard Horner, 300; and Lyell, 297, 300, 305–7, 309, 310, 312, 313, 349–50n10; and Maldonado, 299; and Malthus, 296; and Manellin, 31–32; and migration, 297, 310, 311, 313, 314–15, 319, 322, 323, 324; and mobility, 27, 296–97, 300, 307, 312–26; and natural history, 296–98, 310–11; and natural selection, 303, 310, 315, 319, 321, 326; and M. Shelley, 332; and South America, 310, 311; and speciation, 310, 311, 312, 313, 316–17, 318, 319, 323, 324; and species, 297, 298, 299, 300, 302–3, 304, 306, 308, 310, 311, 312–26; and theory of modern natures, xiv, 27, 297, 311, 312, 325; and Tree of Life metaphor, 317–18. Works: *Autobiography*, 296, 297; *Beagle Diary*, 306, 308, 309; *Correspondence of Charles Darwin*, 31–32, 136; *Essay of 1844*, 317–18; *Journal of Researches*, 298, 299, 300, 306, 308, 309, 310, 311; *Natural Selection*, 314; *Notebooks*, 296, 313, 319, 320, 325; *On the Origin of Species*, 311, 313, 314, 315, 316, 318, 319, 320, 321, 322, 323, 324, 325–26, 327; "Ornithological Notes," 313; *The Power of Movement in Plants*, 321; *Zoology Notes*, 298
Darwin, Erasmus, 74, 135, 163, 332–34; *The Botanic Garden*, 53–54, 56, 60, 61–62, 64–70, 76, 78, 79, 343n11; and commerce, 54–55, 72, 78; and cosmopolitan natures, xiv, 54, 56, 76, 78, 79, 131–32; *The Economy of Vegetation*, 53, 65, 67–73; letter to Watt, 65; *The Loves of the Plants*, 53, 54, 60–61, 65, 66, 67, 79, 81, 343n12; and Mathias, 81–82; and natural history, xiv, 55, 65; *Phytologia*, 66, 332–33, 350n6; and plants, 65, 66, 69, 72–73, 79, 81; and Ryder, 343n10; and science, 54, 65, 67–68, 69, 70, 76, 78, 79; *The Temple of Nature*, 334; translation of Linnaeus's *Genera Plantarum*, 65; translation of Linnaeus's *Systema Vegetabilium*, 65; "Visit of Hope to Sydney Cove," 130, 131–33, 137
Darwish, Mahmoud, 270–71, 295
Davis, Samuel, 146, 147, 148
Dawes, William, 151
Defoe, Daniel, 20, 160
de Galaup, Jean François, comte de La Pérouse, 114–15, 116, 156, 347n27
Denham, John, "Cooper's Hill," 70
De Quincey, Thomas, *Confessions*, 18
Derham, William, 178–79
Descartes, René, 168, 170
Dickens, Charles, 38
Dickson, John, 342n16
Dilke, Charles Wentworth, 32
Dillard, Annie, 3
Driver, Abraham and Felix, 167
Dryden, John, 81
Duffin, Dr., 22

East, Hinton, 99, 101
East India Company, 87, 120
Eaton, Amos, *Manual*, 230–31
ecology, 2, 5, 16, 25, 27, 28, 78, 179–80, 297–98, 307–10, 318; and Josselyn, 19; and Leopold, 208; and Lyell, 300–307; and M. Shelley, 338, 339; and G. White, 154, 155, 178, 179. *See also* Bartram, William; Clare, John; colonialism; Darwin, Charles; historical ecology; Wordsworth, William
Edwards, Bryan, *History*, 99
Edwards, George, *Gleanings*, 347n5
Egmont, John James Perceval, second Earl, 224–25
Emerson, Ralph Waldo, 198
empire and imperialism, 9, 16, 24–25, 64, 99, 115, 125, 128, 283, 288, 312, 335–36; and Banks, 22, 103, 134, 161; and W. Bartram, 201, 221; and botanic gardens, 71, 78; and British identity, 106–8; and Bryant, 227, 229, 230; and Cowper, 343n19; and E. Darwin, 56, 69–70, 72, 79; and globalization, 103, 126, 135; and islands, 107–8; and H. Jones, 73, 74–75, 77; and Kew Gardens, 71; and Lyell, 304; and Makdisi, 6; and natural history, 8, 26, 35, 43, 47, 48, 100–109, 155–56; and Nugent, 95; and translation, 17, 34, 38, 43, 47; and G. White, 160; and W. Wordsworth, 230. *See also* colonialism
enclosure. *See* Clare, John
Enclosure Act of 1809, 285
Engels, Friedrich, 312
Evelyn, John, *Sylva*, 56–57
evolution. *See* Darwin, Charles
exchange, 6, 7, 20, 26, 28, 37, 42, 126, 132, 160, 272; and Banks, xiv, 22, 71, 78, 110, 111, 156; and Bligh, 111, 118, 120, 121; and Cook, 116–19; globalized system of, 137; and Kew Gardens, 71, 77; and G. White, 164, 165, 167. *See also* plants; transfer
extinctions, 25, 43, 187, 229, 256–58, 259, 260–61, 264, 285, 301–2, 312; and C. Darwin, 307, 309–10, 324; and Fleming, 256–58, 302; and human beings, 302–3; and Saint Helena, 308; and M. Shelley, 327, 340; and G. White, 180, 183; and J. White, 143. *See also* Lyell, Charles; Wordsworth, William

Faulkner, William, 233
Fenwick, Isabella, 260

Fernández de Oviedo, Gonzalo, *Natural History of the West Indies*, 34
First Nations, 19, 46, 215, 228, 231. *See also* Native Americans
Fisher, James, 160–61
Fisher, John, 265
Fleming, John, 260, 264, 302, 304, 336; "Remarks," 256–58
Flinders, Matthew, 111, 135, 136, 152, 156–57
Florida, 105–6, 198, 199, 200, 208, 210, 211–12, 213, 223, 224, 225
Forbes, Edward, 322–23
Forster, J. R., 163, 194, 344n18
Fort Frederica, Georgia, 218–19
Fortitude, 87–88
Fothergill, John, 198, 200–201, 202
Fox, Charles James, 249, 252
Fox, Will, 300
France, 8, 28, 59, 80, 100, 103, 121, 345n5
Franklin, Benjamin, 46, 68, 125, 156, 203
French Revolution, 181–82
Friel, Brian, *Translations*, 1, 50–51
Fuseli, Henry, *Flora*, 62, 63

Galapagos Islands, 296, 313
gardens and gardening, 1, 6, 8, 14, 23, 29, 90, 101, 118, 123, 241, 274, 324; and Bligh, 114, 119; and Bryant, 228, 230; and Clare, 275, 276, 277, 287, 288; and colonialism, 36, 38, 52, 106; and E. Darwin, 62, 64, 65; and George III, 72, 75, 78; and Lake District, 84, 265; landscape, 56–59, 74, 83, 240; and Nugent, 92, 93, 94; popularity of, 35, 36, 56–60, 62–64; and weavers, 240–43; and G. White, 163, 177, 184; and W. Wordsworth, 83, 84, 95, 231, 235, 236, 237, 238–39, 243, 244, 245, 246, 248, 266; and working class, 240–43
Geoffroy Saint-Hilaire, Isidore, 28
La Géographie, 345n5
geology, 67, 216, 234, 238, 251, 254, 256, 258, 300, 309, 311, 313
George I, 184
George III, 21, 58, 73, 74, 75, 78, 98, 111, 116, 118, 185
Georgia, 214, 218–19, 222, 228
Goethe, Johann Wolfgang von, 80
Goldsmith, Oliver, 279–80
Good, Peter, 136
Gosse, Edmond, *Gossip in a Library*, 163

Gosse, Thomas, 134, 142; *Founding*, 127–30, 129, 130; *Transplanting*, 96–97, 108, 128
Gough, Thomas, 254
Gould, Augustus A., 314–15
Grafton, Duke of, 187
Graham, Andrew, 45–46
Grahame, Kenneth, 158
Grainger, James, *The Sugar-Cane*, 189
Grainger, Margaret, 272
Gray, Asa, *Manual*, 30
Gray, Jane Loring, 30
Gray, Thomas, xi
Great Britain, 30–31, 82, 86, 103, 107, 135, 141, 160, 163; and identity, 20, 26, 75, 95, 202, 271–72; island nature of, 96, 106–8; and landscape gardening, 56–59
Griffin, Harry, 249–50

Haast, Julius von, 31, 32
Haeckel, Ernest, 318
Hamelin, Emmanuel, 15
Hamilton, Henry, 26–27
Hanbury, William, *Complete Body*, 241
Hartog, Dirck, 15
Hawkesworth, John, 98
Henderson, Joseph, 275
Herder, Johann Gottfried von, *Outlines*, 10
historical ecology, 18, 23–24, 234, 247–48, 249, 254, 262–68; and W. Bartram, 196, 201, 215–25, 227–28, 231, 232; and W. Wordsworth, 234–35, 247–49, 254–55, 262–68
history, xii–xiii, 9, 11, 12, 32, 79, 98, 233, 278, 297, 316; and W. Bartram, 203, 206, 215–25, 228, 232; and British identity, 20; and Bryant, 227; and Clare, 277, 291–92; natural and human, 254–66; and natural history, 14; and nature, 203, 232; and nature and mind, 255; nature as obstacle to, 10; spatial, 15–16; and translation, xiv, 42; and G. White, 342n11. *See also* Wordsworth, William
HMS *Bounty*, 98, 110–11, 112, 191
HMS *Providence*, 96, 103, 110–11, 113, 115, 120, 121, 122
HMS *Resolution*, 111, 116, 117, 190, 191, 192, 195, 344n18
Holland, 56, 57, 59, 60
Hooker, Joseph Dalton, xi–xiii, 38, 313, 322, 323; *Botany*, 28–29; "Distribution," xii, 29, 31; *Flora Antarctica*, xi; *Flora Novae-Zelandiae*, 136; "Insular Floras," 308–9, 310
Hooker, William, *The British Flora*, 82
Houghton, John, *Collection for Improvement*, 106
Howe, Michael, 123
Howe, Sir William, 185
Huahine, 118, 191, 192
Hudson's Bay Company, 45
Hume, David, 170
Hunt, Leigh, 81
Hunter, John, 125, 140–41
Hutton, John, 260
hybridity, 28, 76, 78–79, 95, 122, 135, 330, 333–34, 338

India, 64, 73
Indian laborers, 88
Indian Removal Bill, 228
indigenous peoples, 21, 34, 44–46, 48, 50–51, 89, 115, 171, 206, 327; and W. Bartram, 208, 213–14, 215; and C. Darwin, 310; and Lyell, 349–50n10. *See also* Aboriginal people; Native Americans
industry, 5, 6, 54, 55, 69, 160
invasive species, 23–25, 31–33; broad-leaved plantain, 19–20, 29, 135; cardoon, 306–7; cat, 24, 118, 134; cockroaches, 189–95, 346–47n26, 347n27–28; common housefly, 32–33; and Darwin, 305–11; European honey bee, 230; fennel, 306; Hessian fly, 185–88, 304–5; indigenous peoples' attitudes towards, 19–20, 32–33; mongoose, 43–44; prickly pear cactus, 134, 135; rabbits, 25, 101, 116, 134, 135; rats, 32, 43, 118, 183–84, 344n18; and M. Shelley, 328–29, 339; wax scale insect, 184–85; woolly aphid, 188. *See also* weeds
Ireland, 43, 74, 75, 77
islands, 96–97, 106–9, 160, 315, 320, 321, 323, 324

Jackson, Andrew, 228, 229
Jamaica, 78, 92–95, 96, 97, 99, 101, 105, 121, 122
James, Governor, 225
Java, 96, 100
Jefferson, Thomas, 26, 187, 227, 346n25
Johnson, Joseph, 65
Johnson, Richard, 134
Johnson, Samuel, 65, 242, 243, 287
Jones, Henry: *Kew Garden*, 53, 73–77; *The Royal Vision*, 77

Jones, Sir William, 79, 80, 220
Josselyn, John, *New-England's Rarities,* 19
Joyce, Mary, 270, 271

Kalm, Peter, *Travels,* 19
Keats, John, 72–73, 81, 123, 227–28
Keir, James, 54
Kennion, Edward, 140
Kew Gardens. *See* Royal Botanic Gardens at Kew
Kilburn, William, 62
Kincaid, Jamaica, 90–91, 100
King, James, *Voyage,* 117
King-Hele, Desmond, 54
Kipling, Rudyard, 9, 43
Kirby, W. and W. Spence, 188, 194, 299
Knight, Thomas Andrew, 188
Kyd, Robert, 109

Labillardière, Jacques-Julien, 116, 347n27
Lake District, 83–84, 230, 231, 247, 249–50, 253–56, 262–63, 264, 267, 268
landscape, xii, xiii, 2, 12, 14, 30, 58, 235, 246, 254, 263, 325; and Australia, 128, 130, 133, 138, 152; and W. Bartram, 196, 197, 198, 201–2, 203, 215–25, 228, 231; and Bryant, 226–31; Caribbean, 89–90, 95; and Clare, 272, 273–74, 280, 283, 292, 294; and colonialism, 18–19, 24, 327; and J. D. Hooker, xii; and Nugent, 92–94; and Ordnance Survey of Ireland, 50; and Rowley, 267; social dimensions of, 252; transport of, 23, 95. *See also* gardens and gardening; Wordsworth, William
language, 47, 53, 69, 80, 81, 92, 173, 177, 196, 204, 291, 328. *See also* Clare, John: and poetry and local names; White, Gilbert; Wordsworth, William
Latham, John, 144
Latour, Bruno, 11, 37, 38, 138, 345n7
Lavater, Johann Caspar, 330
Layman, William, "Hints," 88
Leckie, Gould Francis, 108
Leverian Museum, 139, 140
Lévi-Strauss, Claude, 51
Lightfoot, John, 104, 185
Ligon, Richard, 189
Lind, James, 22
Linnaeus, Carl, 44, 47, 57, 58, 62, 68, 102–3, 179, 180, 193, 194, 333; and Clare, 275, 291; and E. Darwin, 79; *Genera Plantarum,* 65; *Philosophica Botanica,* 40, 291; *Species Plantarum,* 40; *Systema Vegetabilium,* 65; taxonomic system of, 39–41; and G. White, 180, 184, 194
Linnean Society, 82, 140, 241, 342n16
Locke, John, 169, 170, 171
Long, Edward, 190, 195
Longfellow, Henry Wadsworth, 19–20
Lorrain, Claude, 274
Loudon, John C., 59, 64, 240–41, 244
Louis XVI, 181–82
Lovejoy, Arthur O., 12
Lowell, James Russell, 158, 162, 174
Lunar Society, 54–55, 62
Lyell, Charles, 22, 25, 258, 297, 349n3, 349–50n10; and C. Darwin, 297, 300, 305–7, 309, 310, 312, 313, 317–18, 349–50n10; and extinctions, 301, 302–3, 304, 305, 312, 349n3, 349–50n10
Lyly, John, 42

Macaulay, Thomas, 107
Magazine of Natural History, 35–36, 194
Malthus, Thomas, 296, 301, 312, 315, 319, 349n3
Manellin, W. B. D., 31–32
Maori, 32
Marx, Karl, 10, 20, 312
Matthews, William, 239
Melville, Herman, 3
Milne, A. A., 158
Milton, John, 173, 174, 182
Milton, Thomas, 140
Mitchell, Samuel, 186
Moira, Earl of, 223
More, Hannah, 60
Morgan, George, 187
Morton, Timothy, 2
Mouffet, Thomas, 194
Mount Royal, 224–25
Mueller, Baron von, 136
Murchison, Robert, 302

names, 16, 18, 50, 51, 83, 127, 140–41, 151, 171; and Clare, 276, 277, 283, 288, 292; and Linnaeus, 40, 291; and W. Wordsworth, 260. *See also* language
Napoleon I, 58, 98, 108
Napoleonic Wars, 108
nationalism, 107, 159, 160, 164, 193, 314

Native Americans, 24, 33–34, 202, 203, 208, 228–29, 230, 289, 290, 349–50n10. *See also* Bartram, William; indigenous peoples
natural history, 1, 6, 14, 104, 141, 156–57, 178; and Australia, 124, 126, 127; and colonialism, xiii, 8, 34, 44, 49; and commerce, 26, 44, 48, 103, 106; and consumerism, 4, 54–55, 64, 73, 255, 265–67; and fashion, 36, 55, 59–60, 65, 93–94; and Flinders, 136; and globalization, 25–33, 35, 39, 49, 102–6, 155–56, 185; and imperialism, 8, 26, 35, 43, 47, 48, 100–109, 155–56; and indigenous peoples, 44–46, 48, 49, 50–51, 138; and knowledge and action at distance, 27, 34, 35, 37, 38, 152, 275; and La Pérouse, 114–15; and mobility, 25–33, 38; and print, 35, 36, 39–40; and translation, 33–52, 137–52
"Nature": limited usefulness as nonpluralistic idea, 9–20
natures, xiv, 1, 8, 16, 48; colonial, xiv, 18, 20, 24, 28, 35–38, 77, 96, 127, 296–98; and contemporary ideas of nature, 3–5; cosmopolitan, xiv, 71, 76–78, 79–80, 91, 109, 125, 328; domesticated, 14, 23–24, 89–90, 116, 128, 176, 227, 257, 264, 305–6, 335; English, 158–61; globalization of, 17, 25–33; imperial, 69–70, 72, 75, 221; indigenous, xiii, 17, 32, 46, 49–51, 78, 90, 95, 102, 115, 119, 126, 128, 136, 196–97, 214, 325; local, 27, 40–41, 44, 47, 49, 78, 137, 159, 161, 273, 290–91, 327; lost, 51, 196, 222–23, 270, 289, 295; modern, 20–21, 23, 28, 51, 71, 121, 325, 327; traditional, xiv, 6, 18–19, 24, 54, 155–56, 273; traveling, 24–26, 78, 87, 94–95, 96, 108, 127, 132, 135, 325, 327, 340
Nelson, David, 117
networks, 104, 118, 137, 139, 154, 160, 199–201, 274, 331; information, 22, 24, 35–36, 164, 198; and Latour, 37–38; and local natures, 17; and natural history, 8, 342n11; transportation, 21, 71, 77, 107–9, 121, 132–33, 135; and G. White, 160
New England, 19, 30, 31
New Historicism, 236–37, 246
New South Wales, 125, 134, 135, 138, 139–51
Newton, Matthew, 84
New Zealand, 32, 117–18, 323
Nietzsche, Friedrich, 17
Nodder, Frederick Polydore, 140
Nomucka, 118

novelty, 55, 56–67, 69, 81
Nugent, Maria Skinner, 92–95
nursery businesses, 56, 62, 83, 241

Ogeeche nation, 222
Okey, William, 146, 147–48
Omai, 191, 192
Ordnance Survey of Ireland, 50
Osbeck, Pehr, 163
Ovid, 66

Paley, William, 178, 297
Pannwitz, Rudolf, 52
Park, Mungo, 39, 103, 342n16
Penn, Thomas, 58
Pennant, Thomas, 45–46, 104, 153, 163–64, 168, 174, 183
Petre, Robert James Petre, eighth Baron, 57
Philip II, 241
Phillip, Arthur, 128, 130–33, 134–35
plantations, 88, 93, 99, 101, 106, 108
plants, xi, xiii, 116, 128, 129–30, 305–6, 312; and W. Bartram, 199, 201, 205, 210, 217, 218, 219, 222, 223, 224, 225; and Bligh, 95, 96, 98, 100, 101, 103, 109, 110–11, 114, 115–16, 117, 119–21, 122; and Clare, 275, 276, 277–78, 286, 287, 292; exchange of, 20, 22, 23, 26, 35, 59, 71, 77, 80, 96, 105, 109, 110, 111, 118, 121, 134–35, 156, 199; exotic, 59, 62–64, 65, 69, 72, 74, 81, 82–86, 95, 102–3, 266; migration of, 23, 41, 322; mobility of, 72, 79, 300, 321–26; native/indigenous, 18, 28, 29, 45, 59, 71, 81, 84, 109, 121, 126, 135, 136, 171, 308, 327; naturalized, 30, 59, 71, 75, 77, 78, 82, 83, 103, 307, 323; transfer of, 22, 23–25, 26, 56–60, 71, 74, 75, 76, 96, 99, 100–101, 104, 105, 111, 115, 121, 135, 283; transplantation of, 27, 76, 97, 104–5; transportation of, 37, 103, 134–35, 266; and Walker, 104–5. *See also* Darwin, Erasmus; species
Playfair, John, 300
Plenty Coups, 289, 290
Poivre, Peter, 100
Polwhele, Richard, 60
Pope, Alexander, 62, 81; "Essay on Criticism," 60; *Essay on Man*, 178; *The Rape of the Lock*, 73; "Windsor-Forest," 132–33, 343n14
Porteus, Henry, 120
postcolonialism, 9, 18, 192, 247, 326
Potanos, 222

Poussin, Nicolas, 274
Pulteney, Richard, 104

race(s), 31, 106, 314, 329–32, 339, 340
Rawdon Hall, 223
Ray, John, *The Wisdom of God*, 178
reproduction, 79, 328, 331, 332–39. *See also* sex
Richardson, John, 46
Roberts, Charles G. D., 172
Robinson, Henry Crabb, 72, 154
Rolle, Denys, 215, 216, 218
Romanticism, 1–3, 37, 126, 202, 203, 226
Rosas, Juan Manuel de, 310
Ross, James Clark, xiii
Rousseau, Jean-Jacques, 5, 62
Roxburgh, Sir William, 308
Royal Botanic Gardens at Kew, 57, 69–78, 94, 103, 134, 137, 200; and Anderson, 121; and Australia, 124; and Bligh, 109, 122; as clearinghouse and administrative center, 77; and Schleiermacher, 80
Royal Society, 103, 168, 172
rural people, 38, 279–80, 282. *See also* Clare, John; Wordsworth, William
Rushdie, Salman, 52, 161
Ruskin, John, 5, 167
Russell, Emily W. B., 347n11
Ryder, Dudley, 343n10

Said, Edward, 290
Saint Helena, 96, 120, 122, 307–10
Saint Vincent, 96, 97, 120–21, 122
Salisbury, William, *Hints*, 188
Schleiermacher, Friedrich, 79–80
science, 8, 47, 99, 100, 109, 124, 157, 312; and W. Bartram, 196, 201; British hegemony in, 141; and colonialism, 17, 37, 44; and C. Darwin, 142, 296; and E. Darwin, 67; and H. Jones, 76; and La Pérouse, 115; and Latour, 37; and Pacific, 126; and Park, 39; and M. Shelley, 328; and G. White, 154, 158; and J. White, 141, 142–43; and W. Wordsworth, 85. *See also* Darwin, Erasmus
Scopoli, Giovanni, 168, 345n11
Second Treaty of Augusta (1773), 214
Sedgwick, Adam, 254
Selborne, 153, 155, 159, 160, 189–95
Seminoles, 213–14
Seton, Ernest Thompson, 172

Seven Years' War, 73, 107
Seward, Anna, 65, 117–18
sex, 56, 79, 328, 331, 332–33, 337. *See also* reproduction
Shakespeare, William, 42, 88–89, 106–7, 132, 133
Shaw, Sir George, 140, 141, 143
Sheffield, William, 157
Shelley, Mary Wollstonecraft: *Frankenstein*, xiv, 327–40; *History of a Six Weeks' Tour*, 81; *The Last Man*, 340; and modernity, 327, 330–31, 334; and natural philosophy, 328, 329–30; and species, xiv, 327, 328–29, 331, 332, 337–38, 339–40
Shelley, Percy Bysshe, 3, 72, 80–81, 86, 336
Short, Thomas, 31
Sibly, Ebenezer, 194, 347n28
Silliman, Benjamin, 306
Sitting Bull, 290
slavery, 2, 43, 87, 88, 90, 99–100, 106, 122, 215, 277, 310
Sloane, Sir Hans, 36
Smellie, William, 104
Smith, Adam, 107, 170
Smith, Charlotte, 2, 36, 154
Smith, Christopher, 109, 114, 116, 119–20
Smith, John, 189
Smith, Sir James Edward, 82, 140, 141, 241
Society for Promoting Natural History, 139, 140
Solander, Daniel, 58, 124, 198
Sonnerat, Pierre, *Voyage à la Nouvelle Guinée*, 44, 45
South America, 121, 135, 305–7
Southey, Robert, 54, 67, 152, 229
Sowerby, James, 62
speciation. *See* Darwin, Charles
species, 276, 298, 299, 306, 308; competition among, 301, 303, 310, 311, 312, 314–16, 317, 318–19, 321, 323, 324, 325–26, 329, 339, 349–50n10; cooperation among, 301, 326; distribution of, 180, 297, 304, 312–26; habits *vs.* forms of, 320–21; and human selection of, 305; and Lyell, 301, 312; migration and settlement of, 319, 321; mobility of, 27, 179–80, 312–26; and M. Shelley, xiv, 327, 328–29, 331, 332, 337–38, 339–40
Spitalfields, Norwich, 239, 240, 243
Stace, Clive, 342n6
Stedman, John Gabriel, 189–90
Steiner, George, 47
Stevens, William Bagshaw, 343n11

Stone, Sarah, 139–40, 144–46; *Great Brown King's Fisher,* 145; *Port Jackson Thrush,* 147; *Yellow Eared Fly Catcher,* 149
Stork, William, 28, 105–6, 223
Stukeley, William, 220
Swainson, William, 46
Sydney, Australia, 21, 124, 132
Sydney Cove, 131, 133
Sydney Domain, 135

Tahiti, 96, 98, 100, 108–9, 118–20, 122
Tasmania, 33, 96, 111, 114, 115–17, 120, 122, 123, 310
Taylor, John, 275, 291
Taylor & Hessey, 275
Temple, John, 187
Tennyson, Alfred, Lord, 287
Thelwall, John, 53, 239–40
Thomas, Nicholas, 126
Thoreau, Henry David, 5
Townshend, Peter, 25
transfer, 9, 23–25, 26, 40, 108, 126, 134, 135, 137, 283; and Banks, 100, 101, 105, 135, 185; and Bligh, 111, 115; and Clare, 283; and Gosse, 128; and M. Shelley, 331; as term, 42. *See also* exchange; plants
translation, xiv, 42, 58, 71, 97, 103, 114, 126, 127, 274; and Clare, 288, 291; collaborative, 139, 141, 144, 148; and colonialism, xiv, 18, 34, 43, 44–46, 49–50; and culture, 44, 46–47, 172; definition of, 41; and imperialism, 17, 34, 38, 43, 47; and natural history, 33–52, 137–52; possession through, 48, 49, 126; and Schleiermacher, 79–80; and M. Shelley, 331; and G. White, 172, 174, 176; and J. White, 139, 141, 144, 148
transplantation, 16–17, 19–20, 41, 88, 95, 96, 98, 102, 111, 115, 119, 121
travel narratives, 36, 48–49, 51, 166
Trinidad, 87–88
Turner, Joseph Mallord William, 268
Turner, Patty, 271

uniformitarianism, 256, 302, 303
United States. *See* Bartram, William; Native Americans

Vaughan, Benjamin, 187
Vaux, James Hardy, *Memoirs,* 21
Venuti, Lawrence, 47

Victoria, Queen of England, 262
Virgil, 182
Voyage of Governor Phillip to Botany Bay, 124

Walcott, John, *Flora Britannica Indigena,* 163
Walker, George, *The Vagabond,* 82
Walker, John, *Essay . . . of the Translation,* 104–5, 344n12
Wallace, Alfred Russell, 31, 305
Wallen, Matthew, 87, 94, 101
Wallis, Samuel, 108
Walpole, Horace, 53, 74
Walsh, B. D., 194
Watling, Thomas, 125
Watson, Richard, 83
Watt, James, 54, 65, 68–69
weavers, 240–43
Wedgwood, Josiah, 54, 55, 60, 68, 73, 130, 133
weeds, 23–25, 29–31, 89–90, 134–36; and Clare, xiii, 284, 287–88; and C. Darwin, 29–30, 32, 306–7, 323–24; and J. D. Hooker, xi–xiii; and W. Wordsworth, 230–31, 235, 237, 244–46
Weil, Simone, 272
Wells, H. G., 33
Welsh, Stanley L., 109
West, Thomas, 255
Westall, William, 152, 157; *Chasm Island, Native Cave Painting,* 151
West Indies, 87, 88–95, 99, 101, 102, 106, 121
Weston, Richard, 59, 60
White, Benjamin, 163
White, Gilbert, xiv, 60, 153–85, 272, 297, 299, 345n5; and Banks, 153, 154, 155–56, 161; and Barrington, 168–69, 170; and W. Bartram, 196, 204, 205, 206; and bird migration, 158, 161, 162, 164, 166, 176, 180–81, 184, 189, 195, 345–46n12; and birds, 166, 168, 170–78, 180–81, 183, 189, 346n19; and Churton, 181; and cockroaches, 189–95; and English nature, 159–62; "Flora Selborniensis," 163; *Journals,* 179, 181–82, 183, 194–95; and language, 168–78, 182; and natural history, 154–59, 163–65, 166, 168, 345n6; *Natural History and Antiquities of Selborne,* 153, 157; *The Natural History of Selborne,* 105, 155, 157–85, 189–95, 275, 297, 299, 342n11, 345n6, 345n11; and B. White, 163; and J. White, 165, 168, 169, 181
White, John, 125, 165, 168, 169, 184; *Journal,* 124, 137–38, 139–51; and natural history, 124, 137–39

White, John (Gilbert White's brother), 165, 168, 169, 181
White, Mrs. John, 195
Wilde, Oscar, 66
wilderness, 14, 26, 95, 217, 222, 255, 274
Wiles, James, 109, 114, 116, 119–20
Wilkinson, Joseph, 253, 254
Willdenow, D. C., 23
Williams, Raymond, 12
Wilson, Thomas, 139–40, 143, 144
Winckelmann, Johann Joachim, 330
Withering, William, 19, 62, 82, 84, 163
Wollaston, T. V., 323
Woolf, Virginia, 162, 346n15
Wordsworth, Dorothy, 83, 154, 239, 254, 259, 343n21; *Grasmere Journal*, 84, 265; *Journals*, 251, 265
Wordsworth, William, xi, 5, 60, 83–84, 90, 91, 343n21; and W. Bartram, 196, 197, 205, 214, 231–32, 268; and colonialism, 230, 238, 249, 255, 256, 262–63; and deforestation, 262–63, 265–66; and ecology, 233, 234, 235, 236, 247–48, 249, 254, 259, 262–68; and extinctions, 256, 259, 260–61, 282; and history, 232, 234, 236–37, 243, 246, 247, 254–66; and landscape, 84, 95, 152, 231, 233–35, 237, 245–46, 248, 249, 250, 251, 252, 253, 254, 255, 259, 260, 263, 264, 265, 266, 267, 268; and language, 54, 83, 85, 236, 238, 266, 267, 282; and natural history, 85, 234–35, 254; and past, 6, 232, 234, 236, 238, 243–47, 256, 268; and ruins, 243, 244–45, 246, 247, 251, 259; and rural people, 238–39, 248; and time, 243, 251, 268–69; and West, 255; and Westal, 152; and G. White, 154, 158, 159, 160, 162. Works: Alfoxden Notebook, 232; "The Brothers," 248–49, 250; *The Excursion*, 230, 232, 233, 244; "Fenwick note to The Thorn," 235; *A Guide through the District of the Lakes*, 95, 247–48, 253–56, 262–68, 282; "Hart-Leap Well," xiv, 258, 259–62; "Home at Grasmere," 233; "Intimations Ode," 162; letter to C. J. Fox, 249, 252; letter to Isabella Fenwick, 260; letter to J. Pering, 247; letter to Sir George Beaumont, 266; letter to William Matthews, 239; "Lines Written in Early Spring," 205; "Love lies bleeding," 83; *Lyrical Ballads*, 230, 232; "Matron's Tale," 249; "Michael," 229–30, 231, 232, 247–54; "A Night on Salisbury Plain," 246; "Nutting," 258; preface to *Lyrical Ballads*, 54, 248; *The Prelude*, 1, 7, 54, 83, 214, 226, 236, 248, 249, 251, 258, 263, 266; "The Ruined Cottage," 11, 230, 231–33, 235–39, 240, 243–47, 248, 250, 253, 348n6; "A slumber did my spirit seal," 11; "The Thorn," 223, 235; "Tintern Abbey," 235, 248, 296; "To the Cuckoo," 269; "To the Same Flower," 85; "To the Small Celandine," 84–85; "The world is too much with us," 6

Yamasees, 222–23
Yellow Bear, 289
Yonge, Sir George, 109
Young, Arthur, 188